Cloud and Fog Optimization-based Solutions for Sustainable Developments

Cloud and Fog Optimization-based Solutions for Sustainable Developments discusses the integration of fog computing and the Internet of Things to provide scalable, secure, and cost-effective digital infrastructures for smart services in diverse domains:

- Highlights resource management solutions for the Internet of Things devices in fog computing architectures.
- Discusses waste management using cloud and fog computing for sustainable development, and optimization of the Internet of Things in fog computing for fault tolerance.
- Covers smart surveillance and monitoring using cloud and fog computing, and energy-efficient smart healthcare.
- Explains energy-efficient frameworks for cloud-fog environments for sustainable development, and smart grid infrastructure using cloud and fog computing.
- Presents the management of metropolitan mobility for public transport and smart vehicles with cloud and fog computing.

The text is primarily written for senior undergraduates, graduate students, and academic researchers in the fields of electrical engineering, electronics and communications engineering, computer science and engineering, and information technology.

Cloud and Fog Optimization-based Solutions for Sustainable Developments

Edited by
Shilpi Harnal
Rajeev Tiwari
Lalit Garg
Ashish Mathur

CRC Press
Taylor & Francis Group
Boca Raton London New York

CRC Press is an imprint of the
Taylor & Francis Group, an **informa** business

Designed cover image: Shutterstock

MATLAB® and Simulink® are trademarks of The MathWorks, Inc. and are used with permission. The MathWorks does not warrant the accuracy of the text or exercises in this book. This book's use or discussion of MATLAB® or Simulink® software or related products does not constitute endorsement or sponsorship by The MathWorks of a particular pedagogical approach or particular use of the MATLAB® and Simulink® software.

First edition published 2025
by CRC Press
2385 NW Executive Center Drive, Suite 320, Boca Raton FL 33431

and by CRC Press
4 Park Square, Milton Park, Abingdon, Oxon, OX14 4RN

CRC Press is an imprint of Taylor & Francis Group, LLC

© 2025 selection and editorial matter, Shilpi Harnal, Rajeev Tiwari, Lalit Garg, and Ashish Mathur; individual chapters, the contributors

ISBN: 978-1-032-79909-4 (hbk)
ISBN: 978-1-032-79910-0 (pbk)
ISBN: 978-1-003-49443-0 (ebk)

DOI: 10.1201/9781003494430

Typeset in Sabon
by SPi Technologies India Pvt Ltd (Straive)

Contents

About the editors

Shilpi Harnal is an Associate Professor in the Department of Computer Science & Engineering at Chandigarh University, Mohali, Punjab, India. She specializes in cloud computing and security. She has worked on fog computing, underwater wireless sensor networks (UWSN), and artificial intelligence. She has around 14 years of teaching experience and has published 20 patents and 35+ research papers in various international, national, and peer-reviewed journals, books, and conferences.

Rajeev Tiwari is a Professor and Dean in the School of Computer Science and Engineering at IILM University Greater Noida, India. He is a senior IEEE member. He has more than 16 years of research and teaching experience. His broad areas of research include cloud computing, MANET, VANET, fog computing, QoS in such network environments, cache invalidation techniques, the Internet of Things (IoT), big data analytics, and machine learning. He has 70+ international publications in SCI, and Scopus-indexed journals and conferences.

Lalit Garg is an Associate Professor in the Department of Computer Information Systems, at the University of Malta, Malta, and an Honorary Lecturer at the University of Liverpool, UK. He has been a researcher at Nanyang Technological University, Singapore, and Ulster University, UK. He has published 150+ high impact publications in refereed journals/conferences/books, 17 edited books, and 22 patents. His research interests are business intelligence, machine learning, data science, deep learning, cloud computing, mobile computing, the Internet of Things (IoT), and information systems.

Ashish Mathur is a Professor in the Center for Inter-disciplinary Research and Innovation (CIDRI), at UPES Dehradun, India. He worked as a research associate at the Nanotechnology and Integrated Bio-Engineering Center (NIBEC), at the University of Ulster, UK. His noteworthy contributions in the field of research are reflected in his published 116 research papers in high impact journals including Biosensors and Bioelectronics, Scientific Reports, IEEE Sensors, Applied Surface Science, and Sensor. His research focuses on developing novel sensors using nanotechnology for societal benefits, which include women's safety, drug abuse, and environmental issues.

Contributors

Akash
Chitkara University Institute of
  Engineering and Technology
Chitkara University
India

Rajni Aron
NMIMS University
India

Jatin Arora
Chitkara University Institute of
  Engineering and Technology
Chitkara University
Punjab, India

Mayank Arora
Seth Jai Parkash Mukand Lal
  Institute of Engineering and
  Technology
Radaur, Haryana, India

David Asirvatham
Taylor's University
Malaysia

Vidhu Baggan
Chitkara University School of
  Engineering and Technology
Chitkara University
Punjab, India

Sana Bharti
Chitkara University Institute of
  Engineering and Technology
Chitkara University
Punjab, India

Shantanu Bindewari
SCSE, IILM University
Greater Noida, Uttar Pradesh, India

Priyanka Chawla
NIT Warangal
Telangana, India

Xiaochun Cheng
Department of Computer Science
Swansea University
Bay Campus, Fabian Way, Swansea,
  Wales, UK

Premkumar Chithaluru
Department of Information
  Technology
Mahatma Gandhi Institute of
  Technology
Hyderabad, India

Monika Choudhary
Seth Jai Parkash Mukand Lal
  Institute of Engineering and
  Technology
Radaur, Haryana, India

Vinay Gautam
Chitkara University Institute of
  Engineering and Technology
Chitkara University
Punjab, India

Rupali Gill
Chitkara University Institute of
  Engineering and Technology
Chitkara University
Punjab, India

Gaurav Goel
Chandigarh Engineering
  College-CGC,
Landran, Punjab, India
and
School of Computer Science, UPES
Dehradun, Uttarakhand, India

Aditi Goyal
Chitkara University Institute of
  Engineering and Technology
Chitkara University
Punjab, India

Nitin Goyal
Central University of Haryana
Mahendergarh, Haryana, India

Sonia Goyal
Punjabi University
Punjab, India

Shilpi Harnal
Department of Computer Science &
  Engineering
Chandigarh University
Mohali, Punjab, India

Sonam Jain
Chitkara University Institute of
  Engineering and Technology
Chitkara University
Punjab, India

Sonam Kamboj
CSE Deptt, JMIETI
Radaur, Haryana, India

Kanika
Chitkara University Institute of
  Engineering and Technology
Chitkara University
Punjab, India

Gagandeep Kaur
Chitkara University Institute of
  Engineering and Technology
Chitkara University
Punjab, India

Gagandeep Kaur
Punjabi University
Patiala, Punjab, India

Gaganpreet Kaur
Chitkara University Institute of
  Engineering and Technology
Chitkara University
Punjab, India

Harpreet Kaur
Chitkara University Institute of
  Engineering and Technology
Chitkara University
Punjab, India

Harpreet Kaur
Punjabi University
Punjab, India

Mandeep Kaur
Chitkara University Institute of
  Engineering and Technology
Chitkara University
Punjab, India

A V Senthil Kumar
Hindustan College of Arts and
  Science
Hindusthan Gardens,
  Nava India
Coimbatore, Tamil Nadu, India

Mohit Kumar
Department of IT
Dr. B.R. Ambedkar National
  Institute of Technology
Jalandhar, Punjab, India

Rajender Kumar
Computer Science and
  Engineering
Punjabi University
Patiala, India

Tajinder Kumar
CSE Department, JMIETI
Radaur, Haryana, India

Shalini Kumari
Chitkara University Institute of
  Engineering and Technology
Chitkara University
Punjab, India

Sachin Lalar
Department of Computer Science
  and Applications
Kurukshetra University
Kurukshetra, India

Swati Malik
Chitkara University Institute of
  Engineering and Technology
Chitkara University
Punjab, India

Swati Malik
Jindal Global Law School
OP Jindal Global University
Sonipat, Haryana, India

Raja Kumar Murugesan
Taylor's University
Subang Jaya, Selangor, Malaysia

Pallati Narsimhulu
Chaitanya Bharathi Institute of
  Technology
Hyderabad, Telangana, India

Htet Ne Oo
Chitkara University Institute of
  Engineering and Technology
Chitkara University
Punjab, India

Shikha Saxena
SoHST, UPES, DDN
Uttarakhand

Anand Singh
SCSE, IILM University
Greater Noida, Uttar Pradesh,
  India

N Sudhakar Yadav
Department of Information
  Technology
Chaitanya Bharati Institute of
  Technology
Hyderabad, Telangana, India

Righa Tandon
Chitkara University Institute of
    Engineering and Technology
Chitkara University
Punjab, India

Rajeev Tiwari
School of Computer Science and
    Engineering
IILM University
Greater Noida, India

Shamik Tiwari
School of Computer Science
UPES
Dehradun, Uttarakhand, India

Shuchi Upadhyay
SOHST, UPES
Dehradun, Uttarakhand, India

Ajay Verma
Jaypee University of Information
    Technology,
Solan, Himachal Pradesh, India

Heena Wadhwa
Chitkara University Institute of
    Engineering and Technology
Chitkara University
Punjab, India

Chapter 1

Resource management solutions for IoT devices in fog computing architecture

Harpreet Kaur

Chitkara University Institute of Engineering and Technology,
Chitkara University, Rajpura, India

Shilpi Harnal

Chandigarh University, Mohali, Punjab, India

Vidhu Baggan

Chitkara University School of Engineering and Technology,
Chitkara University, Rajpura, India

Rajeev Tiwari

IILM University, Greater Noida, India

1.1 INTRODUCTION

With the ability to access it from anywhere, cloud computing offers limitless processing and storage capacity. The profits that accompany cloud computing demand resources to mitigate the delay that may occur while data processing in real-time [1]. To serve the end customers with speed and efficiency, the load on the Cloud has been distributed among the framework of Fog as well as edge computing. Further to meet the demands of cognitive operations in IoT-connected devices, fog computing is a new dispersed computing model that has most recently attracted the interest of business and educational institutions [2]. As, IoT-based smart applications generate vast information and data from mobile devices, the traditional cloud environment faces difficulties with response time, bandwidth, and security in areas like vehicular smart transport, smart grids, healthcare homes connected with smart appliances, and other delay-sensitive applications. In 2012, Cisco unveiled a cutting-edge distributed computing system called "fog computing" to address the aforementioned problems. Fog computing acts as a conduit between the cloud and end devices to meet the processing demands of delay-sensitive applications. Furthermore, fog computing needs to work independently to keep running even when there are sporadic connections

DOI: 10.1201/9781003494430-1

to the cloud, but it also needs to properly integrate with the cloud once full resource connectivity has been restored. Geographically dispersed computing resources, known as fog devices, are used in fog computing nearer to the network edge. IoT devices can use decentralized fog nodes for offloading, which drastically reduces delay and prevents the cloud from becoming a bottleneck. Similar to how computing chores are offloaded on the cloud layer, end devices can send computing work to neighboring nodes. Bringing specific cloud computing capabilities closer to the end user at the network's edge is the primary objective of fog computing to decrease latency, improve location awareness, and facilitate mobility, among other benefits that are currently lacking in cloud computing. For instance, a user using a handheld device would need to request the stream from the cloud to view the most recent CCTV footage from a locally installed IoT security camera because the camera lacks storage. This might take some time, but with fog computing, it can be avoided by using a nearby fog node for much faster video streaming. Fog computing is diverse, necessitating a hybrid management strategy that may offer both centralized and distributed options. As a result, effective resource management must be implemented to ensure that applications and services utilize cloud capabilities as intended [3]. To finish all of the tasks assigned by end users, the resource management categories such as resource allocation, job offloading, load balancing, application placement, resource scheduling, and resource provisioning are therefore implemented.

1.1.1 Fog computing architecture

The distributed approach known as fog computing enlarges the cloud platform to the network's edge. At the network's edge, fog computing provides the processing of data, memory, surveillance, and service capabilities. The proper three-tier architecture of fog is shown in Figure 1.1.

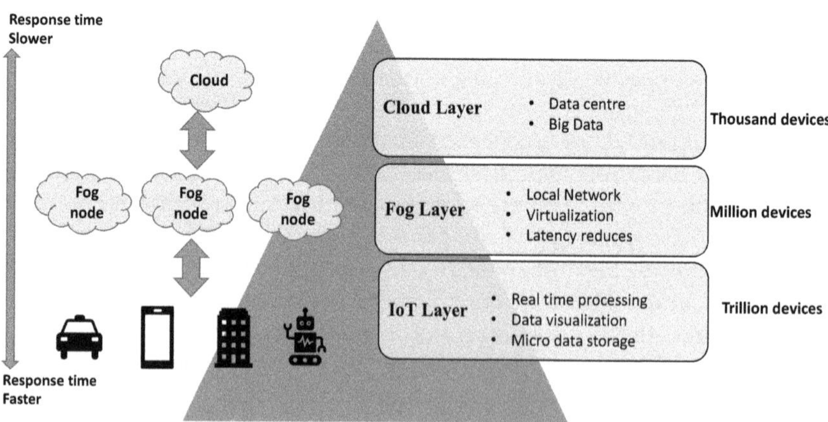

Figure 1.1 Architecture of Fog computing.

This architecture is composed of three layers namely the Cloud layer, Fog layer, and IoT layer. IoT layer-carrying end devices are present in huge numbers but have a small capacity in terms of memory and processing of data. This layer, represented by the terminal layer, includes sensors, actuators, objects, mobile smartphones, and iPads, along with desktop and portable computers, smart meters, airplanes, and intelligent transportation systems. They offer a variety of computing capabilities and can be thought of as human-operated resources. These components work together to create a communications infrastructure, and information from this network is sent to the cloud's data center via the fog layer. Fog nodes are millions of devices, that have a large capacity, in terms of memory and processing power and are present at the intermediate layer between Cloud Layer devices and terminal devices. A fog processing device is any kind of device with the capacity to process, record, and join (connect) a network. Thus several gadgets, including cell phones, can be categorized as both IoT and fog gadgets. Between network edges and clouds, a layer of processing, gateway, and networked (routers and switches) equipment is called the fog layer. A variety of services, including computing, storage, and network services, are provided to users via networked fog resources. Fog devices shape a decentralized system that manages information transferred by these end devices and delivers services for a particular collection of IoT devices at a particular place. The cloud offers high memory and processing capacities. The cloud layer contains actual data center nodes. Each node satisfies user requests for resources by providing a processor, storage, i.e., main memory, and network bandwidth. By using control techniques, the balancing of the load has been done to manage and schedule the cloud resources. Wide Area Networks (WANs) are connected to clouds, which offer financial advantages, dynamic services, data-demanding analysis for clients, Quality of Service (QoS) provided, and a high level of intrusion detection. However, because of present WAN connectivity, clouds experience increased latency and low bandwidth availability. Furthermore, because cloud computing is centralized, context-aware computing for end devices cannot be assisted by it. The cloud layer is present at a huge distance from end devices while the fog layer is near it; hence, the fog layer is easily approachable to end devices with less delay as compared to the cloud layer with high latency. Fog computing can successfully supply delay-critical applications with all the required processing resources by combining these three layers. The step-by-step discussion of the computing process is as follows:

- To automate IoT devices, signals are linked from those devices to an automation controller, which then runs a control system program on those devices.
- Data is wired by a control system program through a protocol gateway.
- Data is converted into a protocol, like HTTP, so that web-based services may readily understand it.

- The information is gathered by a fog layer for further processing.
- It filters the information and stores it for later use. The fog architecture provides a brief description of all three layers, and the computing process is also described.

1.1.2 Features of fog computing

The National Institute of Standards and Technology (NIST) summarizes the key elements of fog computing (as shown in Figure 1.2), are as follows:

- Less Delay: In comparison to cloud computing, fog computing nodes are located closer to the end users and can provide a quicker analysis and response [4].
- Geographic Distribution: Fog computing provides geographical deployment and management to the services and applications running on it.
- Heterogeneity: It enables the gathering and processing of data gathered from many sources using a variety of network connection techniques.
- Federation and Interoperability: Services, applications, and tasks must be federalized across domains and resources must be able to communicate with one another.
- Real-Time Interactions: Rather than only batch processing, real-time interaction is a feature of fog computing services, applications, and tasks.
- Scalability: It supports resource flexibility by enabling quick detection of fluctuations in workload response times as well as detecting any kind of change in the device and network's current situation.

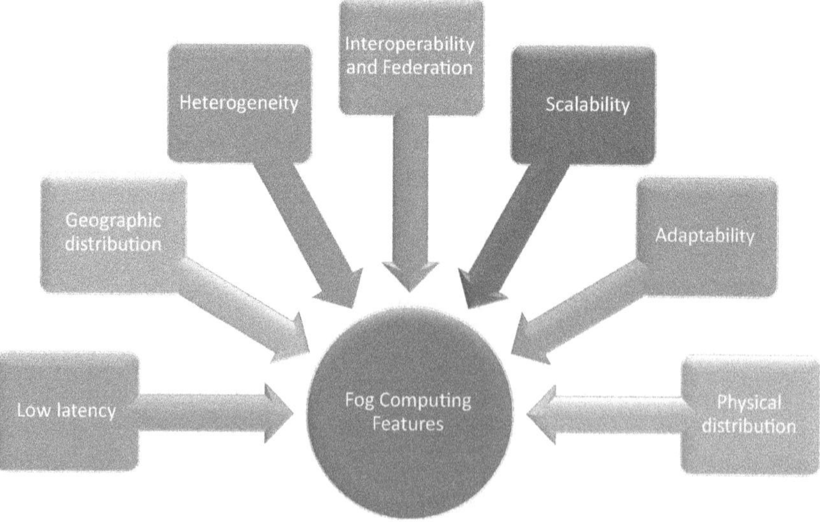

Figure 1.2 Features of Fog computing.

- Adaptability: The fog serves as a reservoir and dispersed computational resources for use with large-scale network sensors.
- Physical distribution: In contrast to the centralized cloud, fog offers decentralized programs and apps that can be deployed anywhere [5].

Fog computing becomes easily accessible to clients and provides more key features as compared to the cloud.

1.1.3 Objective

Despite so many features, fog computing still has many challenges related to the placement of applications, scheduling of resources, offloading tasks, balancing of load, allocation, and provisioning of resources [6–8]. Additionally, more research has to be done in these disciplines. The following are the goals of this study:

- To identify the placement of the application in the fog module.
- To investigate the best way to distribute various jobs for resource scheduling on fog nodes.
- To examine task offloading for efficient utilization of resources such as memory, CPU, etc.
- To analyze the load balancing technique, and distribution of incoming workload among the available fog nodes.
- To analyze effective resource allocation to a group of dispersed fog devices in a particular area, competitors' end-device applications are assigned to these nodes.

To examine resource provisioning from the available fog nodes to manage the submitted end devices load to decrease the hefty amount.

The organization of this paper inculcates a discussion on the introduction of fog computing and its architecture in Section 1. In Section 2, the literature review is provided. In a subsequent section, an introduction to resource management in fog computing is given, and a discussion of categories of resource management, furthermore presentation and identification of future work that can be carried out in these directions. The last section presents a conclusion.

1.2 LITERATURE REVIEW

The literature review of all resource management categories by different authors, the resource management categories, the techniques, and the evaluation tools used with pros and cons are presented in Table 1.1.

Table 1.1 Literature review based on resource management categories

Authors	Category	Technique	Evaluation tool	Pros	Cons
[2]	Application placement	Genetic Algorithm	Simulation (iFogSim)	i) Presenting a solution to the issue of service placement ii) Considering the diversity of applications and resources	i) Without taking into account resource costs The suggested framework's reliance on a particular framework.
[9]	Application placement	Heuristic-based	Prototype	i) Simple computational architecture ii) Considering the shifting conditions of the network topology	i) Limited scalability ii) The latency hasn't been assessed
[3]	Application placement	Heuristic-based	Simulation (iFogSim)	i) Fit for latency-aware Internet of Things applications ii) Shortening the deployment period	i) No actual case study has examined the specific proposal. ii) Mobility and users' specific settings have not been taken into account
[10]	Application placement	Monkey algorithm + genetic algo	Simulation (NA)	i) Simple computational architecture ii) Low price	i) Latency, energy use, and diversity have not been taken into account Minimal scalability
[11]	Resource scheduling	NSGA-II	Simulation (MATLAB®)	i) Fast execution ii) Excellent scalability	Hefty price

[12]	Task offloading	Matching theory	Simulation (NA)	i) Shortening the duration ii) Improving precision	Energy utilization latency has not been assessed.
[13]	Task offloading	Heuristic-based	Simulation (QualNettool)	Lowering the latency, lag, and energy usage	Outdoor settings are not suitable for the suggested strategy.
[14]	Task offloading	Queuing theory + IPM-based algorithm	Simulation (NA)	Lowering the cost, inconvenience, and energy usage	High computational complexity
[15]	Task offloading	Queuing theory + Approximate based	Simulation-(NA)	Enabling online traffic on systems for the Automotive Internet of Things (IoT)	Energy consumption has not evaluated
[16]	Resource allocation	Priced Timed Petri nets	Simulation (NA)	i) Dynamic formal method ii) Fast execution	i) Ignoring assessments of fairness and accuracy ii) Spare time for analysis
[17]	Resource allocation	Game theory	Simulation (MATLAB)	i) Minimal latency Excellent efficiency i) Failing to consider time	i) Failing to consider costs
[18]	Resource allocation	Heuristic-based	Simulation (CloudSim)	i) Quick response ii) Reducing the number of VMs	i) Limited adaptability High price
[19]	Resource provisioning	Queuing theory	Simulation (Java ModelingTools (JMT))	Using a mathematical queuing model	Distribution of the request arrival rate's dependence
[20]	Resource provisioning	Fuzzy theory	Simulation (Apache JMeter)	i) Low price ii) high precision iii) minimal overhead	i) Energy usage has not been assessed ii) live mobility has not been taken into account

(*Continued*)

Table 1.1 (Continued) Literature review based on resource management categories

Authors	Category	Technique	Evaluation tool	Pros	Cons
[21]	Resource provisioning	Linear programming	Simulation (NA)	Supporting IoT applications for mass monitoring	Data analytics tools that protect privacy have not been taken into consideration
[3]	Resource provision ing	Heuristic-based	Simulation (Apache JMeter	i) Lowering service delay ii) Making it simple to implement iii) Having little overhead	Application migration across nodes is not taken into consideration.
[22]	Load balancing	Graph partitioning theory	Simulation (NA)	ii) A low level of computational complexity ii) Decrease service lag	Energy use has not been assessed
[23]	Load balancing	Parallel Othello Group	Simulation (Extended Hadoop)	i) Lower migration overhead ii) Simple to implement	Without taking into account the graph re-partitioning properties
[24]	Load balancing	Heuristic- based	Simulation (Java)	i) High Durability ii) Memory-efficient	i) Energy usage has not been assessed ii) Bottleneck

1. Resource Management: The state-of-the-art mechanisms have been studied to identify the list of research gaps in this domain [1]. Various techniques have been evaluated and compared based on their performance, benefits, and drawbacks.

2. Application Placement: Install the applications in various fog nodes while considering resource heterogeneity and the QoS requirement of the particular application is a complex process, authors have suggested a genetic algorithm for addressing the same, and the findings show that the strategy is efficient in utilizing fog resources and IoT application running time [25]. In addition, a simple service placement mechanism has been suggested for community network microcloud infrastructures [9]. According to the evaluation's findings, the proposed algorithm's complexity is polylogarithmic and can deliver adequate bandwidth performance and reaction time. To carry on, to address the varied service delivery delay requirements for various IoT applications, the authors have suggested an approach for latency-sensitive application modules that may be deployed systematically over distributed fog nodes. Two algorithms make up their proposed method: the first algorithm deals with application module deployment, while the second algorithm is concerned with application module relaying to the idler components' idle resources. Additionally, the issue of implementing intelligent computing systems to improve the efficiency of computation in a logistics hub has been researched by the writers [10]. To address placement issues in combined Fog-to-Cloud situations, based on the concept of service atomization, which divides composite Internet of Things services into smaller, atomic services, this author has investigated suitable solutions [26].

3. Resource Scheduling: Resource management is the basic technology of fog computing. It indicates the process of organizing the fog nodes among the different user requests according to certain rules and regulations of resource usage under a specified fog environment. To decrease latency and increase job execution stability, proposed two-level resource scheduling, which is used on upgraded non-dominated sorting genetic algorithm-II [11]. To achieve similar goals, split the resources using fuzzy clustering and particle swarm optimization techniques [27]. They also offer a more effective fuzzy clustering-based resource scheduling strategy.

4. Task offloading: To compute workloads that require acute execution in the fog computing environment, the IoT clients unload the data either partially or completely to the local fog computing nodes, to accomplish this process, task offloading is required. Hence, to handle the fair matching between customers and service providers, proposes a special architecture for networking and an offloading method across several tiers of the foggy architecture. This framework uses matching

algorithms [12]. The offloading technique proposed and put into practice by [13], is for indoor mobile cloud networks. Cooperative code offloading is the name of the proposed offloading technique in a femtolet-based fog network. The femtolet does the requested computation if none of the mobile devices can do it. The use of a femtolet gives mobile devices access to cloud services and voice calling capabilities.

5. Load Balancing: In resource management, load distribution among different resources is also a challenging process. To equally distribute the workload among numerous resources, has been accomplished by load balancing, to maximize the throughput, low latency, and to manage traffic efficiently. To address the problems that occur in cloud computing, software-defined networking (SDN) and fog computing have been merged [22]. Fog computing enables geographic monitoring and functionality while extending processing and storage to the network's edge, which potentially reduces latency. SDN, meanwhile, gives the network adaptable centralized control and global knowledge. In the fog computing architecture, examine and transform physical nodes at different stages into VM nodes using cloud atomization technology [23]. This paper uses the graph partitioning theory to construct a fog computing load balancing technique that relies on dynamic graph partitioning. A brand new, MEC-compliant design is proposed by [24] for a Scalable and Dynamic Load Balancer (SDLB), the primary load balancing method used by SDLB is excellent minimal hashing.

6. Resource Allocation: In the resource management process, after balancing the load, further resource allocation is an important task to accomplish by improving the network delay and managing the latency. In this resource allocation strategy, a priced timed Petri nets (PTPNs) model is put forth by [16], which enables the users to independently choose the right resources from a pool of pre-allocated resources by carefully considering the cost and time needed to complete a task. For all FNs (fog nodes), DSOs, and DSSs, [17] provides a collaborative optimization methodology to achieve the optimal distributed resource allocation strategies. The authors of the framework first create a Stackelberg game to analyze the price problem for the DSOs and the problem of allocating resources for the DSSs. The architecture for delegating IoT devices and assigning resources that rely on collaboration between fog and cloud computing is proposed by [18], to balance workload and offer decision rules of a linearized decision tree.

7. Resource Provisioning: To ensure application performance, resource provisioning involves selecting, installing, and maintaining software (such as load balancers and database server management systems) and physical servers (such as processors, memory, and network). MIST, a fog-based data analytics approach with a cost-effective resource provisioning optimization approach is introduced by [21], by allowing

prospective quick response at three different levels (e.g., edge, fog, and cloud level). The MIST scheme's powerful computational power and intelligence will assist future smart cities and can be applied to IoT crowd-sensing apps. [3], offers a "Fog Computation" architecture to relocate computation closer to the edge of the network, which is closer to the user devices, to reduce transmission delay, information flows to the cloud, and mode of interaction between user devices and the cloud. The comparison of resource management categories is shown in Table 1.1.

Further, resource management is discussed in brief with the processing of the categories.

1.3 RESOURCE MANAGEMENT

The cloud servers are mostly not able to deal with the bulk of resources from end devices, while fog machines have the extra power to keep and interpret information from end nodes. The management of resources in a foggy environment is a tedious job. The management of resources to fulfill the needs of end devices becomes a time-consuming process, that can be improved with more creative processes and approaches. So, due to the dynamic, heterogeneous, and uncertain behavior of the fog computing environment, a resource management technique is required to implement fog computing in actuality. The resource management approaches are classified into six main categories (as shown in Figure 1.3) given below:

- Application Placement
- Resource Scheduling
- Task Offloading
- Load Balancing
- Resource Allocation
- Resource Provisioning

The location and method of application placement directly affect how effectively hardware and networks are used. For instance, a decentralized fog architecture with an inconsistent deployed application approach for a large amount of workload may cause network congestion. The first category that has to be managed in fog computing is the placement and development of application schemes, which may have an impact on how to place the applications on the various resources in the fog environment called the application placement technique. In resource scheduling, the tasks are assigned by the scheduler to accomplish the job [4]. For resource management in a foggy environment, an important role is played by the scheduler. When the end-user asks the fog server to provide services, it schedules their requests and assigns the resources. Further, the requests are scheduled by the

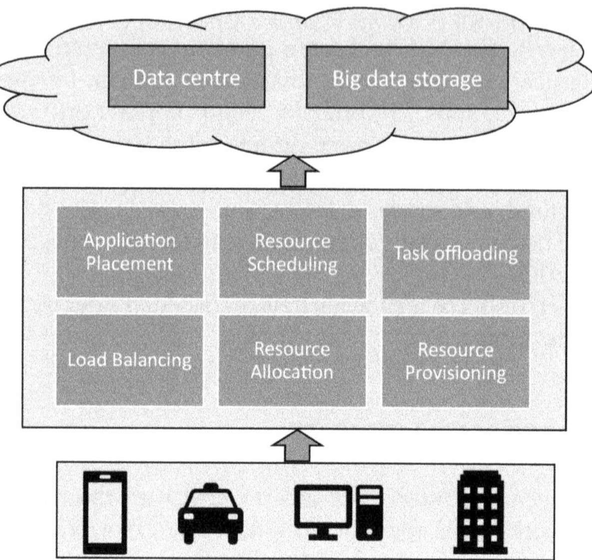

Figure 1.3 Categories of Fog computing.

server to relay them to the resource allocator, so that client gets their desired resources as per demand [28]. Schedulers assist in assigning tasks and processing them quickly in real-time applications for fog. The clients request the services from resource applications, resource scheduling will determine whether all client requests will be executed [4]. IoT requests can therefore be fulfilled using near, far, or collaborative scheduling depending on how the application is deployed. To deal with computationally demanding task processing in a fog computing environment, the IoT users use the task offloading categories to offload, the information/data either completely or partially to the local fog computing node [26]. If a single fog node can handle, analyze, and deliver workloads from end users, it would be the optimal solution. Another approach came with load balancing, in the provisioning and de-provisioning instances of applications and optimizing resource usage [29]. Load balancing enables the fog system to distribute workload among resources equally, aiming to continue providing services even if the service component fails. To improve the performance of applications and network utilization in fog computing, a suitable load-balancing technique is required because data centers procure differences between hosts and exhibit unique aspects of traffic. As a mechanism, load balancing divides the burden among several resources to avoid any overloading or underloading of those resources. Hardware or software can be used to implement load balancing, which spreads the workload across several resources. Network management, latency reduction, and bandwidth maximization are a few of the objectives of load balancing. In resource allocation, by properly utilizing the resource skills and by taking care of the requirement specification, the resources are allocated to lessen the delay management and match network bandwidth. Resource provisioning refers to the distribution of compute, network, and storage capacity, which is required for the instantiation and deployment of services and applications that clients and devices request through the Internet [21]. In other words, to optimize the fog resource management in terms of energy, reaction time, cost, etc. The term "resource provisioning" describes the scalability of the available resources. Fog computing has been developed, to solve the inherent difficulties of allocating computing resources for IoT services in Smart City installations. Given that present sensors and gateways lack sufficient processing power, storage space, and memory, also services can be offered farther away from sensors, which would increase connection latency. IoT is not a good fit for centralized solutions since the high bandwidth requirements make transmitting all of the collected information to the cloud impractical. Local data processing and analytics operations are provided by fog computing, which drastically lowers the volume of data that must be sent to the cloud. The six resource management categories in fog computing have also been covered in upcoming paragraphs.

1.3.1 Application placement

The best deployment of applications across several tiers of fog computing nodes is known as application placement. In contrast, the optimum method to put applications is to map their modules to the resources that exist to enhance the Quality of Experience (QoE) and QoS matrices. QoE and QoS matrices mean to map resources to end user requests to provide the best experience and services [30]. The type of resources, placement techniques, orientations, mapping approaches, placement control systems, and placement parameters are just a few of the parts that make up the placement of applications. Applications are distributed among nodes with varying amounts of resources. These resources can be divided into three categories: raw equipment made of metal, storage boxes such as containers, and compute nodes, also known as virtual machines (VMs). The workload sent by end users varies according to requests, which act as a main factor in the placement algorithms [7, 9, 25]. Therefore, stationary and non-stationary placement methods, also known as static and dynamic, can be used. In the stationary method, also known as static, applications are executed continuously after being deployed just once utilizing the application placement algorithms [8, 31]. In non-stationary placement methods, also called dynamic, apps are either added, removed, or completely transferred due to mobility needs. The architecture model of nodes in fog computing is scattered, which is one of the crucial factors of this computing. As a consequence, it is impossible to ignore considering node orientations when deciding where to position an application because they may have an impact on the QoS. Based on goal-oriented metrics like priority of services, optimization, and multi-objective trade-off, other mapping strategies can also be implemented. The placement of an application requested by end users based on the minimization or maximization of a specific objective measure is known as optimization, where priority is the placement of an application (requested by IoT users) on a specific node [32]. Let F be a collection of fog nodes situated in fog regions and A be the service provider application to the end users with a set of QoS requirements denoted by S.

The projections from each end-user service of application placement manager A to certain fog nodes F that are placed in fog region should meet all QoS requirements S, and maximizing the objectives to measure their performance is the answer to the placement problem for the fog services. The one fog node can invite and host single or multiple end-user services, such as S1 to S8 service data being mapped on single fog node F1, and an end-user service can be installed upon single or multiple fog nodes, such as S1, which can be deployed on any of the fog devices (as shown in Figure 1.4).

The basic goal of application placement is to either minimize or maximize the QoE and QoS [33]. Figure 1.4 describes the example of the application placement process in fog computing. The application placement approaches

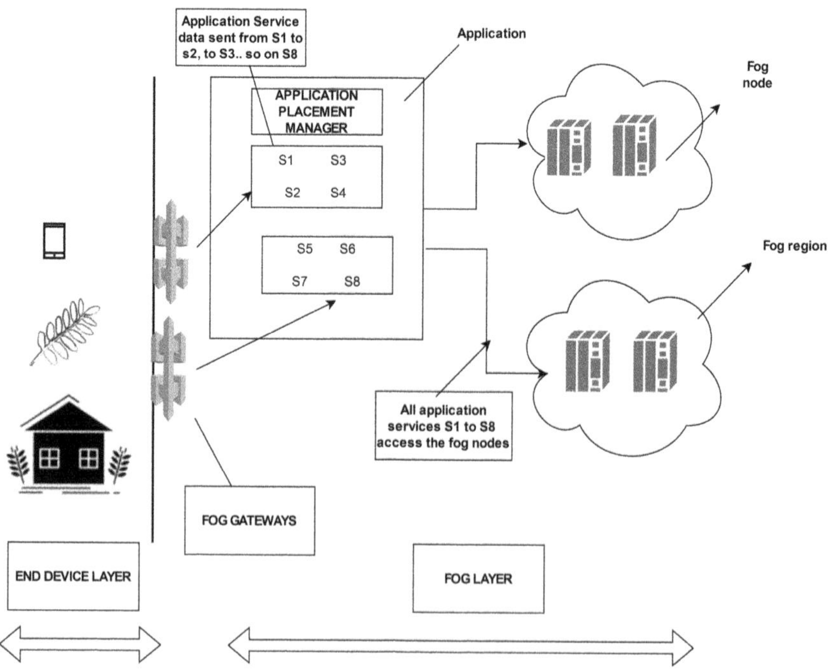

Figure 1.4 Application placement process in fog computing.

based on broker management are divided into three subcategories, including centralized, decentralized, and hierarchical [19, 20].

1.3.1.1 Centralized approach

The first type of application placement is centralized, and it operates as follows:

i. To make a universal optimization conclusion, the centralized strategy gathers data from all the auxiliary devices (such as fog nodes, customers, cloud servers, and IoT devices).
ii. The application placement manager collects knowledge regarding all fog nodes and end-user applications, and all choices must be taken by a centrally managed core. The centralized application placement strategies cannot ensure effective optimal solutions related to implementation overhead and low latency.
iii. A global centralized manager incurs any kind of error at any point, which increases its susceptibility to errors.
iv. The problems occur in the case of a centralized manager if the number of end devices and fog node devices increases because data has been sent to this manager from different devices, it becomes a cumbersome task to handle all the data.

The scalability for system implementation in the form of bandwidth and processing the capacity of a centralized approach is limited, as a centralized manager is in charge of processing and managing choices.

1.3.1.2 Decentralized approach

The decentralized approach of application placement works as follows:

i. The area having few numbers of components and carrying data of smaller regions is often suitable for the decentralized category.
ii. The numerous optimizations in the local area may increase the scalability of decentralized placement strategies.
iii. In the decentralized category, the scalability does not rely on only one managerial core, as numerous managerial cores are added in the decentralized category.
iv. Decentralized systems are less effective than centralized ones in this regard due to the communication costs across managerial cores.
v. The knowledge has been transferred by all IoT devices to the managerial core of the centralized category, causing traffic overhead, and overall task completion time in this category clearly shows how the decentralized category does better work as compared to the centralized one.

1.3.1.3 Hierarchical approach

The advantages of both centralized and decentralized techniques are provided by the hierarchical solutions, which create many semi-global, many local, and both managers who collaborate. The processing of resource scheduling is described with an example in the subsequent paragraph.

1.3.2 Resource scheduling

There are two categories used in scheduling those are resource scheduling and task scheduling to solve the challenges that arise in scheduling. Resource scheduling aims to meet scheduling objectives such as optimal resource usage while locating the finest resources for client needs. The set of readily available resources is distributed among the jobs throughout the task-scheduling process, based on the QoS requirements of the tasks. While task scheduling involves allocating a group of tasks to available resources while taking into account their QoS needs. In general, task scheduling gives clients a satisfying experience, while resource scheduling helps service providers achieve their objectives. Static, dynamic, and hybrid scheduling algorithms can all be included under this category [4, 34]. The static algorithms necessitate previous knowledge (the number of tasks and resource availability) for scheduling, which is rarely a practical method. As a result, dynamic or hybrid techniques are more suited and can react to the scheduling of arriving tasks more correctly and instantly. The execution of a collection of tasks T = T1, T2 to till Tn with various QoS requirements (e.g., low latency, time limit, cost) on a collection of fog devices F = F1, F2 to till Fm with various attributes (e.g., computation power, memory utilization, bandwidth utilization, etc.) must be planned; a scheduling solution looks for the best solution in a large solution space [11, 27].

In Figure 1.5, tasks are first sent from end devices through gateways toward the resource scheduler, where all tasks are queued according to priority, then in fog region T1, T2 . . . Tn tasks are assigned to fog nodes F1, F2. . . Fm. Doing this makes it possible to meet the QoS commitments made to an IoT device user while lowering the task's execution time. In addition, in fog computing environment, the methodologies of task offloading are described in the subsequent paragraph.

1.3.3 Task offloading

In a computing context, offloading is the process of shifting resource-demanding tasks from devices with limited resources to those with enough resources. Because mobile devices frequently have scarce resources, they must frequently delegate some of their activities to the fog or cloud in order to boost efficiency and save power consumption. Another entity is required to carry out functions on behalf of the IoT devices and deliver the results to

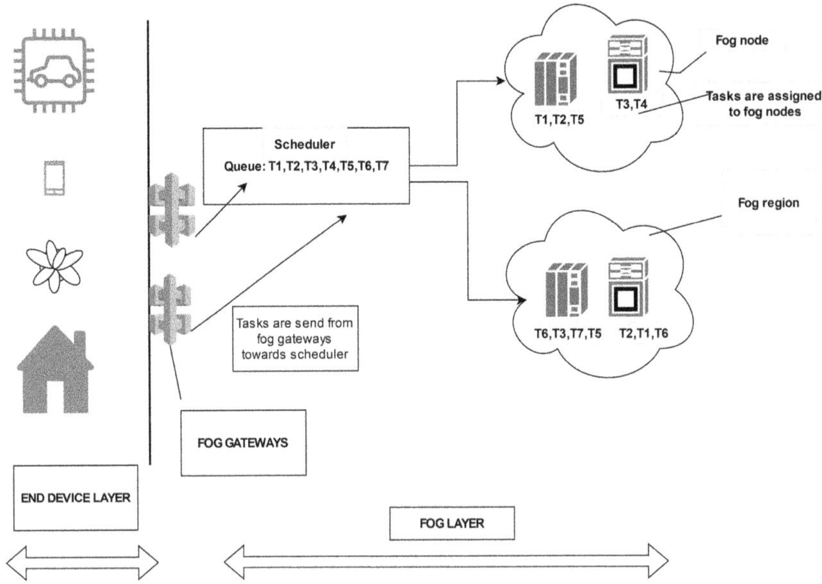

Figure 1.5 Resource scheduling process in fog computing.

them. Task offloading is the name of this procedure [12, 15]. For instance, vertical offloading refers to device-to-fog or fog-to-cloud, while horizontal offloading can be system-to-system or one fog node to another node [2]. In particular, for computationally heavy operations on mobile devices, the task offloading technique can address resource constraints like CPU power, battery life, and client device storage space [35]. Moving gadgets or vehicles must transfer some of their operations to the fog or cloud to optimize performance and prolong battery life because they frequently have limited resources. The job offloading methods can be categorized into two primary groups: single and multiple, depending on the quantity of offloading targets. The single-type offloading approaches allow computing of tasks to be transferred to only a single fog node for sequence-wise processing in contrast to the ways to multiple-type offloading, which comprises multiple fog nodes processing for parallelization to meet the QoS demands (e.g., less delay in responding) [10, 12]. Single-type offloading techniques are utilized for IoT devices wherein execution is consecutive, i.e., here transmission between serial components is rapid. However, IoT systems with parallel processing modules and repeated computation are better suited for multiple-type offloading techniques. If a single-type offloading technique is insufficient to meet the timing constraints of the client application, more fog nodes or cloud services (i.e., multiple-type offloading) are made accessible to implement the IoT application [13, 14].

As described in Figure 1.6, offloading has taken place from gateways toward VMs, and further, it goes toward the fog nodes situated inside the

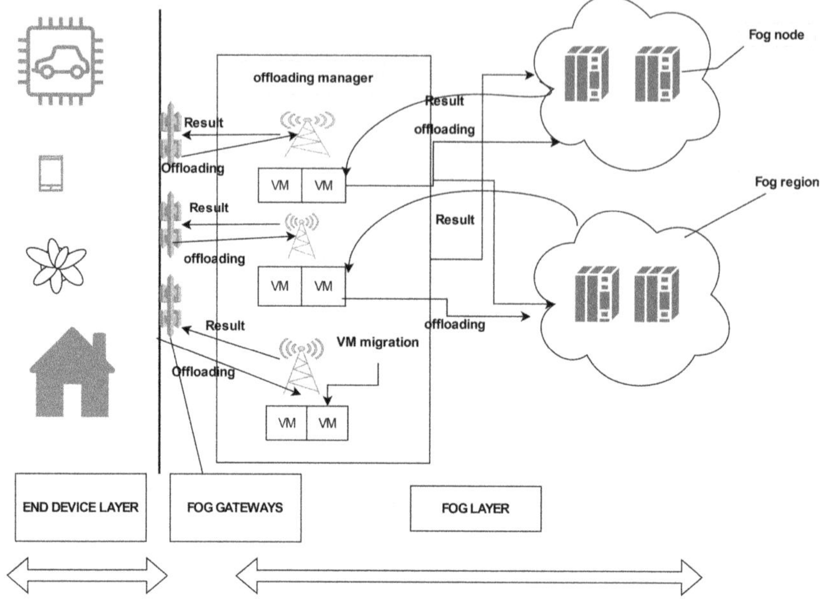

Figure 1.6 Task offloading processing in fog computing.

fog region. Then the result is passed back toward VMs, and it gets relayed to the gateways so that end users get the desired result [36]. The subsequent paragraph has been designated as the discussion of load-balancing techniques in fog computing.

1.3.4 Load balancing

By provisioning and de-provisioning instances of applications and optimizing resource usage, load balancing enables the fog system to distribute workload among resources equitably (as shown in Figure 1.7), intending to continue providing services even if the service component fails.

Applications' efficiency and network utilization in fog computing can be improved by an adequate load balancing system because data centers display distinctive elements of traffic and procure disparities across hosts [23, 24]. As a mechanism, load balancing divides the burden among several resources to avoid any overloading or less loading of those resources [37–39]. Hardware or software can be used to implement load balancing, which spreads the workload across several resources [5, 6, 39]. With the right load balancing, reaction time and energy usage can be decreased while throughput can be increased [22]. The three main types of load-balancing architecture are centralized, decentralized, and semi-decentralized [40]. In a centralized system, a middle node serves as the network's core load-balancing node and keeps track of all the resources that are accessible. In contrast, the decentralized method

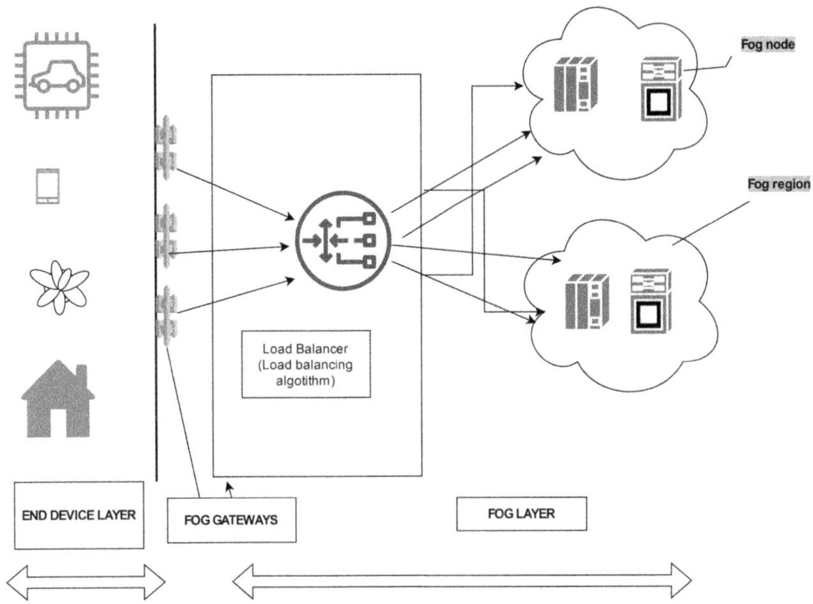

Figure 1.7 Load balancing process in fog computing.

uses several load-balancing nodes [41]. This kind of architecture is typically used in networks with many clusters. In a semi-decentralized architecture, information is communicated to and from local load balancers by a global load balancer, while each node has its own local load balancer. However, because fog computing is heterogeneous, the effective load-balancing mechanism is also based on the load-balancing approaches (such as static approach, dynamic approach, and hybrid approaches), algorithms, and parameters considered to attain particular performance objectives (i.e., reducing reaction times, asset use, energy usage, cost, or increasing performance). Regardless of the storage and processing power of the available fog nodes, the workload is spread evenly across them in a static strategy. When using dynamic load balancing, the knowledge base of the load balancer is frequently upgraded with the currently available resources. Although this technique aids in creating a quick response time and high throughput, it uses more bandwidth than the static strategy due to its excessive information sharing. While hybrid load balancing, which offers minimal bandwidth consumption, incorporates the characteristics of both static and dynamic techniques. The upcoming section contains a description of resource distribution in fog computing systems.

1.3.5 Resource allocation

In the cloud and fog computing environments, there are distinct issues of resource allocation. It is necessary to efficiently identify a group of

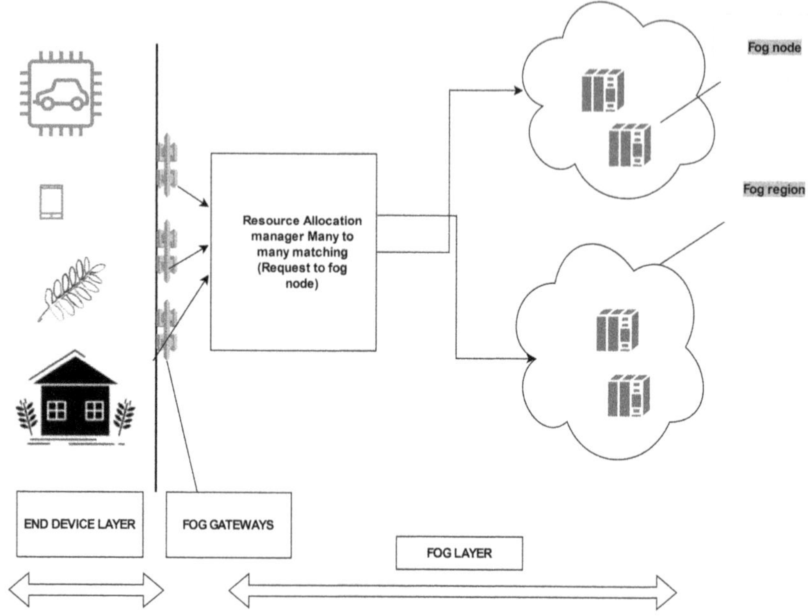

Figure 1.8 Resource allocation process in fog computing environment.

geographically dispersed diverse fog nodes while taking service prioritization and equity into account for rival IoT services with different QoS requirements [42, 43]. In cloud computing, the only entities in the cloud network are clients and cloud servers, and there is a marketplace for servers where they are competing for customers. Since there are numerous scattered elements, including the public cloud, clients, and fog nodes in the fog network, the allocation of resources becomes more complicated. As a result, to resolve the problem of resource allocation in fog computing networks the fog nodes, and Internet of things users are paired for cloud servers, while fog nodes and cloud servers are paired for Internet of things users (as shown in Figure 1.8).

In other words, while coupling the layers, the resource allocation is characterized as a double-matching resource allocation problem in the optimization methods. The double-matching resource allocation problem is thought to be the best method for properly allocating fog nodes to carry out IoT services with various QoS criteria. The allocation of resources, which is resource provisioning in fog computing, is briefly discussed in the section that follows.

1.3.6 Resource provisioning

The process of selecting, deploying, and managing hardware resources, such as CPU, storage, and networks, as well as software (such as load balancers and database server management systems), is known as resource provisioning. The user sends requests for hardware and software res that the provisioning of the resources has taken place [44, 45]. Resource provisioning is divided into three categories namely, user-centric, dynamic, and static models. In the user-centric model, based on the user demands, resources have been provisioned. Using such a paradigm leads to overpriced and underused service adaptations. In dynamic models, auto-scaling dynamically satisfies the resource demand for the fluctuating workload. Due to exact projections of resource requests, this could result in over- or under-provisioning of resources. As an outcome, service-level agreements are either expensive or voluntary. Continuous monitoring thereby guarantees that resources will be available when needed. The resources are assigned steadily for a specific amount of time in the static model as well [46]. However, if this approach is not used appropriately, resources are wasted. Dynamic provisioning, also known as auto-scaling, can assist in scaling up and down to provide the right resources for changing demand.

As described in Figure 1.9, there are different computing fog nodes containing resources that have different efficiencies, such as some nodes provide good computation and storage properties, while others carry better memory and bandwidth, so according to the user QoS requirements, that particular resource is allocated to fulfill the IoT service requirements.

By implementing multiple sorts of policies, including reactive, proactive, and hybrid ones, dynamic provisioning can be accomplished. Real-time resource provisioning in response to user demands based on the condition

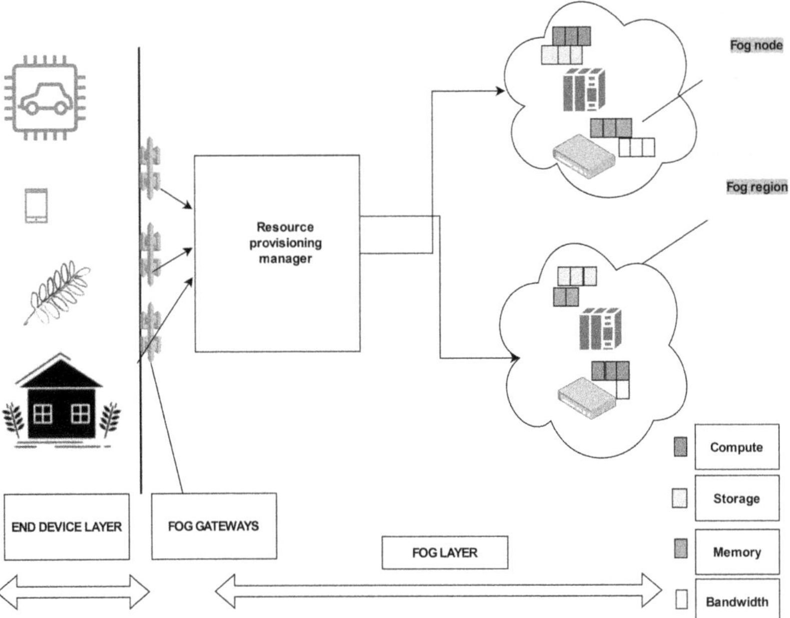

Figure 1.9 Resource provisioning in fog computing.

of the system is known as reactive. While the proactive approach anticipates future resource demands from users. However, the hybrid policies use both reactive and proactive measures to scale in and out to release extra resources and to add more resources. Predictive methods are mostly dependent on proactive policies. Therefore, ML-based methods like neural networks (NN), reinforcement learning (RL), deep learning (DL), and others can play a crucial role in enhancing resource provisioning accuracy by predicting the necessary resources for future user demand.

1.4 CHALLENGES

The aforementioned sections make clear that considerable effort has been made to manage computational resources in the fog computing environment efficiently [47]. The multitude of issues ought to be handled immediately in a variety of ways. The following is a description of the main problems with resource management in fog computing environments:

- To assess the application placement technique in fog computing, the energy and scalability factors (e.g., dealing with heavy workload under pressure, cost, etc.) have not been considered [48, 49, 26].
- One of the key contemporary concerns in resource management for fog computing that has received scant attention in research is self-adaptive resource scheduling.
- To figure out the exact number of resources needed to run the compute-intensive mobile applications, this depends upon the present workload of the mobile applications.
- The safety and confidentiality of mobile information transferred from the resource-constrained mobile device to the resource-rich servers is another difficult task offloading challenge.
- In the load balancing approach, strengthening the energy consumption is the main issue to be taken into account.
- It is difficult to evaluate how well resource provisioning systems perform in a fog computing environment because there isn't an actual workload being recorded by IoT devices. This is a significant difficulty in resource provisioning.
- A single fog node computes and storage capabilities are inadequate for handling all the processes and data by the wait time. The fog node either identifies the local supporting fog node or uploads the job to a far-off cloud for additional computing capabilities. Both situations nevertheless present difficulties since, a) the capacity of the assisting fog node isn't always sufficient, and b) offloading to the cloud still puts a strain on the uploading connection and slows down job processing. It consequently becomes challenging to decide where to unload and how much incomplete job data to offload under the latency assurance established by the end customers.

1.5 CONCLUSION AND FUTURE WORK

In the framework of the fog computing environment, we have discussed resource management solutions. It provides brief information regarding fog computing architecture and its features. Fog computing technology aids in offering an adequate method for time-sensitive IoT-based devices. As opposed to the processing at a cloud server, it allows for substantial data processing near the end devices, which decreases power consumption, latency, network traffic, etc. Fog nodes have restricted computing power, so it's important to create efficient resource management plans that consider several things, including application placement, provisioning of computing resources, load balancing, scheduling of computing resources, and task offloading, all while maintaining the required level of service in a fog computing environment. As the outcome of the analysis, resource management categories make delay-sensitive end devices to get the desired result without any kind of holdup.

REFERENCES

1. Ghobaei-Arani, M., Souri, A., & Rahmanian, A. A. (2020). Resource management approaches in fog computing: a comprehensive review. *Journal of Grid Computing*, **18**(1), 1–42.
2. Bali, M. S., Gupta, K., Rani, S. & Ratna, R. An energy-efficient partial data offloading-based priority rate controller technique in edge-based IoT network to improve QoS. *Wireless Commu- nications and Mobile Computing* 2022 (2022).
3. Wang, N., Varghese, B., Matthaiou, M. & Nikolopoulos, D. S. ENORM: A framework for edge node resource management. *IEEE transactions on services computing* **13**, 1086–1099 (2017).
4. Malik, S. & Gupta, K. Resource scheduling in fog: Taxonomy and related aspects. *Journal of Com- putational and Theoretical Nanoscience* **16**, 4313–4319 (2019).
5. Harnal, S., Sharma, G. & Mishra, R. D. in *Mobile Radio Communications and 5G Networks: Pro- ceedings of Second MRCN 2021*, 331–344 (Springer, 2022).
6. Harnal, S., Sharma, G., Seth, N. & Mishra, R. D. Load balancing in fog computing using qos. *Energy Conservation Solutions for Fog-Edge Computing Paradigms*, 147–172 (2022).
7. Mahmud, R., Ramamohanarao, K. & Buyya, R. Latency-aware application module management for fog computing environments. *ACM Transactions on Internet Technology (TOIT)* **19**, 1–21 (2018).
8. Souza, V. B. et al. Towards a proper service placement in combined Fog-to-Cloud (F2C) architec- tures. *Future Generation Computer Systems* **87**, 1–15 (2018).
9. Selimi, M. et al. A lightweight service placement approach for community network micro-clouds. *Journal of Grid Computing* **17**, 169–189 (2019).

10. Lin, C.-C. & Yang, J.-W. Cost-efficient deployment of fog computing systems at logistics centers in industry 4.0. *IEEE Transactions on Industrial Informatics* **14**, 4603–4611 (2018).

11. Sun, Y., Lin, F. & Xu, H. Multi-objective optimization of resource scheduling in fog computing using an improved NSGA-II. *Wireless Personal Communications* **102**, 1369–1385 (2018).

12. Tran, D. H. et al. OaaS: offload as a service in fog networks. *Computing* **99**, 1081–1104 (2017).

13. Mukherjee, A., Deb, P., De, D. & Buyya, R. C2OF2N: a low power cooperative code offloading method for femtolet-based fog network. *The Journal of Supercomputing* **74**, 2412–2448 (2018).

14. Liu, L., Chang, Z., Guo, X., Mao, S. & Ristaniemi, T. Multiobjective optimization for computation offloading in fog computing. *IEEE Internet of Things Journal* **5**, 283–294 (2017).

15. Wang, X., Ning, Z. & Wang, L. Offloading in Internet of vehicles: A fog-enabled real-time traffic management system. *IEEE Transactions on Industrial Informatics* **14**, 4568–4578 (2018).

16. Ni, L., Zhang, J., Jiang, C., Yan, C. & Yu, K. Resource allocation strategy in fog computing based on priced timed petri nets. *IEEE Internet of Things Journal* **4**, 1216–1228 (2017).

17. Zhang, H. et al. Computing resource allocation in three-tier IoT fog networks: A joint optimization approach combining Stackelberg game and matching. *IEEE Internet of Things Journal* **4**, 1204–1215 (2017).

18. Alsaffar, A. A., Pham, H. P., Hong, C.-S., Huh, E.-N. & Aazam, M. An architecture of IoT service delegation and resource allocation based on collaboration between fog and cloud computing. *Mobile Information Systems* 2016 (2016).

19. El Kafhali, S. & Salah, K. Efficient and dynamic scaling of fog nodes for IoT devices. *The Journal of Supercomputing* **73**, 5261–5284 (2017).

20. Tseng, F.-H. et al. A lightweight autoscaling mechanism for fog computing in industrial applications. *IEEE Transactions on Industrial Informatics* **14**, 4529–4537 (2018).

21. Arkian, H. R., Diyanat, A. & Pourkhalili, A. MIST: Fog-based data analytics scheme with cost- efficient resource provisioning for IoT crowdsensing applications. *Journal of Network and Computer Applications* **82**, 152–165 (2017).

22. He, X., Ren, Z., Shi, C. & Fang, J. A novel load balancing strategy of software-defined cloud/fog networking in the Internet of Vehicles. *China Communications* **13**, 140–149 (2016).

23. Ningning, S., Chao, G., Xingshuo, A. & Qiang, Z. Fog computing dynamic load balancing mechanism based on graph repartitioning. *China Communications* **13**, 156–164 (2016).

24. Yu, Y., Li, X. & Qian, C. *SDLB: A scalable and dynamic software load balancer for fog and mobile edge computing* in *Proceedings of the workshop on mobile edge communications* (2017), 55–60.

25. Skarlat, O., Nardelli, M., Schulte, S., Borkowski, M. & Leitner, P. Optimized IoT service placement in the fog. *Service Oriented Computing and Applications* **11**, 427–443 (2017).

26. Mukherjee, M. et al. Task data offloading and resource allocation in fog computing with multi-task delay guarantee. *IEEE Access* **7**, 152911–152918 (2019).

27. Li, G., Liu, Y., Wu, J., Lin, D. & Zhao, S. Methods of resource scheduling based on optimized fuzzy clustering in fog computing. *Sensors* **19**, 2122 (2019).

28. Ahmed, K. D., Zeebaree, S. R., et al. Resource allocation in fog computing: A review. *International Journal of Science and Business* **5**, 54–63 (2021).

29. Kashani, M. H., Ahmadzadeh, A. & Mahdipour, E. Load balancing mechanisms in fog computing: A systematic review. *arXiv preprint arXiv:2011.14706* (2020).

30. Mahmud, R., Srirama, S. N., Ramamohanarao, K. & Buyya, R. Quality of Experience (QoE)- aware placement of applications in Fog computing environments. *Journal of Parallel and Distributed Computing* **132**, 190–203 (2019).

31. Smolka, S. & Mann, Z. Á. Evaluation of fog application placement algorithms: a survey. *Computing* **104**, 1397–1423 (2022).

32. Souza, V. B. et al. Towards a proper service placement in combined Fog-to-Cloud (F2C) architec- tures. *Future Generation Computer Systems* **87**, 1–15 (2018).

33. Verma, P., Tiwari, R., Hong, W.-C., Upadhyay, S. & Yeh, Y.-H. FETCH: a deep learning-based fog computing and IoT Integrated environment for healthcare monitoring and diagnosis. *IEEE Access* **10**, 12548–12563 (2022).

34. Arisdakessian, S., Wahab, O. A., Mourad, A., Otrok, H. & Kara, N. FoGMatch: an intelligent multi- criteria IoT-Fog scheduling approach using game theory. *IEEE/ACM Transactions on Networking* **28**, 1779–1789 (2020).

35. Hamdi, A. M. A., Hussain, F. K. & Hussain, O. K. Task offloading in vehicular fog computing: State-of-the-art and open issues. *Future Generation Computer Systems* (2022).

36. Harnal, S. & Chauhan, R. Efficient and flexible role-based access control (EFRBAC) mechanism for cloud. *EAI Endorsed Transactions on Scalable Information Systems* **7**, e1–e1 (2020).

37. Kaur, M. & Aron, R. Focalb: Fog computing architecture of load balancing for scientific workflow applications. *Journal of Grid Computing* **19**, 40 (2021).

38. Kaur, M. & Aron, R. *Equal distribution based load balancing technique for fog-based cloud computing* in *International Conference on Artificial Intelligence: Advances and Applications 2019: Proceedings of ICAIAA 2019* (2020), 189–198.

39. Kaur, M. & Aron, R. An energy-efficient load balancing approach for scientific workflows in fog computing. *Wireless Personal Communications* **125**, 3549–3573 (2022).

40. Kaur, M. & Aron, R. Focalb: Fog computing architecture of load balancing for scientific workflow applications. *Journal of Grid Computing* **19**, 40 (2021).

41. Singh, S. P., Kumar, R., Sharma, A., Abawajy, J. H. & Kaur, R. Energy efficient load balancing hybrid priority assigned laxity algorithm in fog computing. *Cluster Computing* **25**, 3325–3342 (2022).

42. Naha, R. K., Garg, S., Chan, A. & Battula, S. K. Deadline-based dynamic resource allocation and provisioning algorithms in fog-cloud environment. *Future Generation Computer Systems* **104**, 131–141 (2020).

43. Wadhwa, H. & Aron, R. TRAM: Technique for resource allocation and management in fog computing environment. *The Journal of Supercomputing* **78**, 667–690 (2022).

44. Ghobaei-Arani, M., Khorsand, R. & Ramezanpour, M. An autonomous resource provisioning frame- work for massively multiplayer online games in cloud environment. *Journal of Network and Computer Applications* **142**, 76–97 (2019).

45. Durga, S., Daniel, E., Onesimu, J. A., Sei, Y., et al. Resource provisioning techniques in multi- access edge computing environments: Outlook, expression, and beyond. *Mobile Information Sys- tems* 2022 (2022).
46. Pg. Ali Kumar, D. S. N. K. *et al.* Green demand aware fog computing: A prediction-based dynamic resource provisioning approach. *Electronics* **11**, 608 (2022).
47. Xiong, Z., Kang, J., Niyato, D., Wang, P. & Poor, H. V. Cloud/edge computing service management in blockchain networks: Multi-leader multi-follower game-based ADMM for pricing. *IEEE Transactions on Services Computing* **13**, 356–367 (2019).
48. Tiwari, R., Mittal, M. & Goyal, L. M. *Energy Conservation Solutions for Fog-Edge Computing Paradigms* (Springer, 2022).
49. Tiwari, R., Mittal, M., Garg, S. & Kumar, S. Energy-aware resource scheduling in FoG environment for IoT-based applications. *Energy conservation solutions for fog-edge computing paradigms*, 1–19 (2022).

Chapter 2

Applications and challenges for sustainable development with cloud/fog/edge computing

Gagandeep Kaur

Chitkara University Institute of Engineering and Technology,
Chitkara University, Chandigarh, India

Shilpi Harnal

Chandigarh University, Mohali, India

Aditi Goyal and Akash

Chitkara University Institute of Engineering and Technology,
Chitkara University, Chandigarh, India

Rajeev Tiwari

SCSE, IILM University, Greater Noida, India

Xiaochun Cheng

Swansea University Bay Campus, Swansea, UK

2.1 SUSTAINABLE

2.1.1 Introduction

During the past ten years, there has been a significant transformation in the field of computing paradigms. Among them, the most widely recognized and established one is cloud computing, which emerged as a solution for the demand of utilizing computing as a service [1–3]. Making it easier to develop new Internet services. With the advent of the IoT revolution, researchers have become increasingly interested in decentralized paradigms. Consequently, edge computing arose to offer the capability of cloud computing at the network edge, tackling problems of bandwidth, latency, and connectivity that cloud computing alone cannot resolve. Within the realm of edge computing, fog computing emerged as the most advanced manifestation of edge computing principles. Fog computing is a framework that disperses resources both horizontally and vertically along the Cloud-to-Things spectrum. This implies that it is not merely an extension of the Cloud, but rather a novel entity that

DOI: 10.1201/9781003494430-2

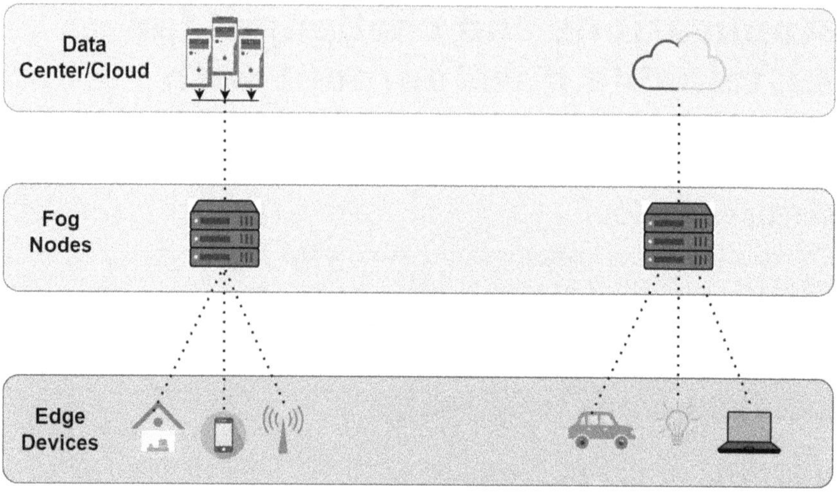

Figure 2.1 Hierarchy of cloud, fog, and edge computing.

collaborates with the Cloud and IoT to provide support and data. The hierarchy of cloud, fog, and edge is shown in Figure 2.1.

2.1.2 Cloud computing

In 2011 [4], the National Institute for Standards and Technology (NIST) defined cloud computing as a framework that allows for widespread, hassle-free, on-demand network access to a shared collection of adaptable computing resources. These resources comprise networks, servers, storage, applications, and services, which can be promptly configured and de-provisioned with minimal management effort or involvement from the service provider.

2.1.3 Types of cloud computing

Cloud computing is classified into four distinct categories, namely public cloud, private cloud, community cloud, and hybrid cloud. Each of these models delivers a unique set of services to users, contingent on factors such as customization capabilities, sharing of cloud services, security requirements, and location of the hosted services. The deployment of cloud computing is influenced by several factors, and its classification is determined by the extent of services that are provided to users [5, 6].

1. Public Cloud
   A public cloud is an online service accessible to the general public and offered by third-party providers who share these services with other organizations. Most providers have their own policies, values, pricing,

and billing structures. According to Apostu et al. [5], the services provided are manageable and consistently available, although users are considered untrustworthy. Research has shown that security and privacy are prominent concerns in public cloud computing [7, 8]. Examples of public cloud services include email and photo storage. Some of the largest public cloud providers are Amazon Web Services (AWS), Google Cloud, IBM Cloud, Microsoft Azure, and Alibaba Cloud.

2. Private Cloud

   A private cloud service is intended solely for the use of an organization. Compared to public cloud, private cloud provides enhanced security since only trusted users or third parties are granted access. An instance of deploying a private cloud in an organization is to share customer data with other branches [9]. This type of cloud service is less susceptible to risks, offers superior security, is more energy-efficient, dependable, cost-effective, and less complex [10, 11]. Researchers have confirmed these claims by putting organizations in precarious situations. However, creating a private cloud necessitates more investment in equipment and software.

3. Community Cloud

   Community clouds are positioned between public and private clouds, and they are difficult to differentiate from private clouds. Resources are shared by organizations with similar objectives and requirements. The services can be managed either by third-party providers or the organizations themselves. According to Goyal [9], community clouds are less expensive to set up than private clouds since the costs are shared among organizations. Additionally, they provide enhanced security and privacy. One limitation of community clouds is that data storage must be shared among users. An example of a community cloud service is the educational cloud, which can be shared by universities globally for research purposes.

4. Hybrid Cloud

   A hybrid cloud is a combination of two or more clouds that can be community, private, or public. The service is provided by using both physical hardware and server instances. Hybrid cloud services offer advantages over both public and private clouds. This type of cloud service allows organizations to easily address data security concerns by storing sensitive data in private storage. Typically, it is used for backup purposes [12]. An organization can host critical applications on a private cloud and use a public cloud for less security-concerned applications [13].

2.1.4 Features of cloud computing

Cloud computing has had a noteworthy impact on the contemporary world. According to Sahandi et al. [14], this innovative technology helps organizations to maintain competitiveness in their respective fields.

Cloud computing offers numerous advantages to businesses, including improved capabilities that are not available with traditional IT solutions.

2.1.4.1 Cost Efficiency

According to Shandi et al. [14], organizations adopt cloud computing mainly to reduce costs. With cloud service providers providing "in-house" provision of these services, organizations reap the benefits of reduced costs. Cloud computing offers organizations a pay-as-you-go model where they only pay for the services they use. According to research, approximately 45.5% of organizations adopt cloud computing with the goal of reducing costs.

The deployment of resources in cloud computing is rapid, particularly for SaaS, as the provider handles everything. Providers manage, patch, and upgrade services, freeing enterprises from IT-related issues. Providers also offer technical assistance, which reduces the workload of IT staff. As a result, enterprises can focus on their core business functions, while cloud providers take care of the IT infrastructure.

2.1.4.2 Automatic software/hardware upgrades

According to Xue et al. [15], implementing an IT solution in a business can lead to financial difficulties due to the high cost of hardware and software purchases and maintenance, which can ultimately have negative consequences. Cloud computing, however, offers a solution to these issues by enabling businesses to convert their capital costs into operating costs. This not only reduces expenses but also enhances relationships, keeps pace with technological advancements, increases profits, and provides customers with standardized, affordable services. The availability of low-cost subscriptions for supply chain management and customer relationship management software encourages more businesses to use these applications, which are instantly accessible to employees. Rapid access without the need for capital investments enables a quicker time to market, and decreasing prices make the applications more accessible, increasing opportunities and encouraging more organizations, particularly in countries that are lagging behind in the IT revolution, to start their business.

2.1.4.3 Scalability

Cloud computing offers the advantage of allowing customers to adjust their resource allocation based on changing business needs. This is made possible by the user-friendly interfaces of most cloud computing services, which can be extended to expand the computing infrastructure. In contrast, traditional IT systems do not support scalability, which is inconvenient for businesses. These systems cannot handle fluctuating resource requirements and lead to wasted resources and dissatisfied customers. Cloud computing's scalability allows for rapid resource availability and customer satisfaction. Capacity planning becomes unnecessary because of the rapid availability of resources in the cloud. Cloud computing has several advantages, especially for smaller businesses, as it allows for the addition of resources as needed. Additionally, due to its processing power, cloud computing enables the analysis of vast amounts of data in a matter of minutes. This has piqued the interest of business analysts, who use cloud computing to study the market and predict consumer behavior and purchasing patterns.

2.1.4.4 Flexibility

Most businesses opt for cloud computing due to the flexibility it provides in their operations, as stated by Goyal [9]. Employees are empowered to have more flexibility both in and outside of work since the majority of cloud computing services offer easy access to information from any device with an Internet connection. The industry offers a wide range of services tailored to meet the varying needs of enterprises. With cloud computing, employees can work together on files and documents by sharing them online, which allows every member to access the latest version of the document. This benefit is particularly significant for employees who work remotely or are frequently on the go, as they can access their data from any location. As a result, business owners can work from multiple locations, freeing up their time to concentrate on other business ventures. Multiple employees can access the same resources simultaneously, thanks to the virtual storage of information and documents online.

2.1.4.5 Agility

To remain competitive in the current business environment, companies must be able to adapt quickly to changing customer needs. According to Xue et al. [15], cloud computing is an effective way to achieve this goal. Because cloud computing is always accessible via the internet, it can be utilized for quick development, allowing companies to provide services in a timely manner.

Cloud computing enables agility by providing three distinct low-level services:

- System Infrastructure – maintenance of machines and spare parts
- Backup Policy – management of backups
- Single Application – management of software (upgrades and application support)

Cloud computing increases agility, or the ability to respond quickly to changes in the business environment. In contrast to a physical server, which can take days to acquire and provision, a cloud server can be set up and ready to use in a matter of minutes.

2.1.5 Applications of cloud computing

2.1.5.1 Cloud computing in education

Cloud computing has enabled the development of a contemporary network of teaching resource libraries that can benefit many individuals seeking to learn. It is necessary to plan this online learning resource library carefully to ensure sustainable and healthy development, as it is a long-term project. The main objective is to construct teaching resources that facilitate teaching and research. This can be achieved through a comprehensive planning process, delegation of responsibilities, and emphasis on utilizing existing resources optimally, while applying innovative mechanisms. The classification, organization, and integration of the latest advancements with cloud computing services can provide ideas and services for IT architecture, innovation, scientific quality, and distribution of appropriate and efficient teaching resources. The cloud computing environment teaching resource library exhibits stability and fast response time, along with standardized data storage, clear categories, and easy user editing and usage, all of which promote sustainability. Furthermore, the resources are secure [16–18] in the cloud, and there is no need to store or backup any information or data. The library also serves as a resource-based platform with a user-friendly interface, making it easily accessible to users of all levels. The ultimate goal of this library is to achieve maximum openness and sharing of educational resources.

2.1.5.2 Business applications

In typical business settings, there are four primary classes of applications: transactional systems, collaborative tools, multimedia applications, and data mining. Transactional systems have been a cornerstone of various businesses for many years, such as web shop applications and stock exchange transaction management. Outsourcing transaction processing allows companies to focus on their core business and benefit from cost reductions through economies of scale. Machine-to-machine (M2M) applications, such as those used in stock exchange systems, require increasingly short interaction times, which can push the limits of optical networks and computation speeds.

Efficient collaboration across different geographic locations is made possible by virtual meeting systems and collaborative frameworks. Virtual meeting systems provide more than just video conferencing capabilities, allowing for the processing of streams from multiple sources and rendering a single image of all participants or their avatars in a virtual space. Additional features include automatic point-of-interest detection, gesture and mood detection, and client-side functionality. Collaborative frameworks enable shared working on objects like text documents and presentations or design drafts, with dedicated remote rendering facilities available for generating different perspectives of the shared object as required. Depending on the type of collaboration, individual operations on the shared object may require significant computational resources.

2.1.5.3 Multimedia

The multimedia sector relies heavily on decentralized infrastructure for both storing and retrieving high-quality multimedia content (e.g., recorded raw material for further processing, querying multimedia archives to identify sequences that meet specific thematic search criteria) and processing this material (e.g.). These systems are obviously very demanding in terms of providing robust storage capacity and networking.

2.1.5.4 Data mining

As electronic data becomes available in nearly every application domain, the need to understand the underlying structure of this data and identify trends in large amounts of data (e.g., to profile expensive infrastructure usage, understand customer behavior, make meaningful recommendations, build search indexes for large amounts of data, etc.)

2.1.6 Challenges of cloud computing

2.1.6.1 Privacy and security

Cloud services offer a convenient way for users to access and share their personal data across different platforms. Identity management (IDM) mechanisms are employed to authenticate both users and services using various credentials and characteristics. However, interoperability issues arising from the use of different identity tokens and negotiation protocols are major concerns with IDM in the cloud environment. Currently, password-based authentication is the most common method used, but it has limitations and significant risks. To protect private and sensitive information related to users and processes, an effective IDM system is necessary [19]. Authentication mechanisms are used to verify the identity of cloud resource users, but many cloud providers still rely on weak password-based authentication. This could be improved by using trusted authentication mechanisms based on public keys, X.509 certificates, or the Lightweight Directory Access Protocol (LDAP).

Auditing mechanisms provide information about activities on cloud resources, identifying users accessing them, time of access, and actions taken. According to the Cloud Auditing Data Federation (CADF) Working Group of the Distributed Management Task Force, cloud providers are required to provide specific audit events, logs, and reporting information for each user. They should also provide standard mechanisms and information for users to self-assess their application security. The hardware, software, and network infrastructure used by the provider to run user applications should also be audited [20].

2.1.6.2 Interoperability

The adoption of cloud computing is hindered by various challenges, with interoperability being a major concern according to research. The risk of being locked into a particular vendor is one of the issues discussed, attributed to the lack of standard interfaces and open APIs, as well as open standards for VM formats and service deployment interfaces. These challenges cause problems with integrating services from different cloud providers and internal legacy systems. When organizations want to switch to another cloud-based solution, they often face the challenge of having to rewrite applications that were written for a particular cloud using specific application stacks. This process can be lengthy, and the complexity involved in data movement, security setup, and networking adds to the inflexibility when transitioning between cloud solutions.

2.1.6.3 Load balancing

A load balancer is used for load balancing, which involves redirecting incoming requests to different servers based on certain parameters such as current load and availability. The load balancer uses various scheduling algorithms to determine which server should handle the request and forward it to the selected server. In order to ensure that the selected server is capable of handling the request, the load balancer obtains information about its health and current workload. Load balancing solutions can be categorized into software-based and hardware-based load balancers. Hardware-based load balancers are specialized boxes that contain an application-specific integrated circuit (ASIC) customized for a specific use [21].

Load balancing is a critical concern in cloud computing environments, particularly in relation to storage usage and download performance. The primary goal is to establish an algorithm for effectively allocating tasks to cloud nodes while considering existing constraints such as high communication latency and heterogeneity. The challenges and issues associated with load balancing in cloud computing environments typically fall into four main categories: the spatial distribution of cloud nodes, data replication, performance, and points of failure [22].

2.2 EDGE COMPUTING

In their publication [23], Shi et al. introduced the concept of edge computing as a novel mode of network computing, which involves executing tasks at the network's edge. According to their description, the data sent downwards in edge computing represents cloud services, while the data sent upwards represents the Internet of Things (IoT). Additionally, the "edge" of edge computing refers to the computing and network resources situated between the data source and the path of the cloud computing center [24].

Satyanarayanan, a professor at Carnegie Mellon University, provided his own definition of edge computing as a new computing model that deploys computing and storage resources, such as cloudlets, micro data centers, or fog [25] nodes, at the network's edge closer to mobile devices or sensors.

2.2.1 Application of edge computing

2.2.1.1 Cognitive assistance

Cloudlets facilitate the execution of tasks with minimal response time, which makes them suitable for powering real-time cognitive assistance applications that can be used on wearable devices like smart glasses. These applications are supported by a server-side infrastructure that comprises multiple virtual machines (VMs). Each VM is responsible for providing distinct subservices, such as face recognition and optical character recognition (OCR), which together form the major components of the framework. The serverside infrastructure also includes control and user guidance mechanisms that collectively contribute to the framework's overall functionality.

2.2.1.2 Body Area Networks (BANs)

According to Bakir et al. [26], the primary objective of a Body Area Network (BAN) is to ensure the reliable and prompt monitoring of the data collected by its nodes. Since the nodes in a BAN generate a significant amount of data, it necessitates robust computational resources and a vast storage area. Furthermore, the generated data is often critical to the individual's wellbeing and requires immediate monitoring. In such cases, Cloudlets appear to be an ideal solution for analyzing the collected sensor data with minimal response time and storing it for subsequent analysis.

2.2.1.3 Military and hostile environments

Sensors deployed in hostile environments, such as military fields, are susceptible to attacks. To mitigate such risks, intrusion detection systems can be used, and the data collected by the sensors can be analyzed by fog servers to detect malicious nodes. The benefits of using Cloudlets in such environments extend beyond providing low-latency services through fog computing. For instance, Cloudlets can help reduce the vulnerability to DoS (Denial of Service) attacks in military operations by being located within a single wireless hop, thereby enhancing the security of the overall system.

2.2.1.4 Language and speech processing

Applications that facilitate multilingual processing, wherein voice input is translated into different languages, require a persistent internet connection and access to significant computational resources. Cloudlets can provide the necessary resources, thereby reducing the load on mobile devices with limited capacity. Additionally, applications based on a Cloudlet infrastructure can eliminate the need for a continuous internet connection and mitigate WAN latency. An example of such an application is collecting speech data from patients with Parkinson's Disease (PD) using an Android smartwatch, which is then transferred to a fog server for processing [26]. The fog server processes the collected speech data and transfers it to cloud servers for further analysis and storage.

2.2.1.5 Smart grid

Fog computing is a promising technology that can facilitate the implementation of smart grid concepts. Specifically, energy load balancing applications can be executed on edge servers that consider the availability and demand for alternative energy resources. Given the multi-tier nature of smart grid applications, both fog and cloud architectures are needed to support their functionalities. The Cloud architecture provides global coverage, storing data for long periods for business intelligence analysis. On the other hand, fog collectors process the generated data or send a portion of it to higher tiers for visualization or real-time reporting. Besides fog computing, MEC servers can also serve similar functionalities for smart grid applications.

2.2.2 Challenges of edge computing

We have described five potential applications of edge computing in the last section. To realize the vision of edge computing, we argue that the systems and network community need to work together. In this section, we will further summarize these challenges in detail [27].

2.2.2.1 Programmability

Cloud computing enables users to develop and deploy code to the cloud, while the cloud provider determines where the data processing occurs. This provides users with the benefit of transparent infrastructure, as they typically have minimal knowledge about how their application will function. Typically, programs are written in a specific programming language and compiled for a particular platform, given that they are exclusively executed within the cloud. Conversely, edge computing involves the offloading of computation from the cloud, leading to a greater diversity of hardware and software platforms at the edge nodes. This disparity can result in differing transit times among nodes, leading to significant hurdles for developers attempting to create applications suitable for edge computing environments.

2.2.2.2 Naming

Edge computing is founded on the idea of having a massive number of devices, where multiple applications run on edge nodes, each with its own method of service delivery. In programming, addressing, thing identification, and data communication, naming schemes play a significant role in edge computing, as in any computing system. Nonetheless, creating an efficient and standardized naming scheme for edge computing paradigms is still an unresolved challenge. Practitioners dealing with edge computing need to be knowledgeable about various communication and networking protocols to connect with different devices in their systems. The naming scheme for edge computing must consider factors such as the mobility of devices, dynamic network topologies, security and privacy protection, and scalability for a large number of undependable things.

Although traditional naming mechanisms such as DNS and Uniform Resource Identifier are appropriate for most current networks, they lack the flexibility required for the highly mobile and resource-limited devices found at the edge. Moreover, for resource-constrained devices located at the edge of the network, an IP-based naming scheme may be too complex and burdensome to support.

2.2.2.3 Data abstraction

The use of edgeOS allows for the deployment of various applications that can provide services or consume data by utilizing radio location indicators from the service management layer. While data abstraction has been extensively discussed and explored in wireless sensor networks and cloud computing paradigms, it presents a challenge in the context of edge computing. In the IoT ecosystem, multiple data generators exist, and the smart home environment serves as a relevant example. In a smart home, a large number of devices report data to edgeOS, including those used for daily operations. However, typically only periodically collected data is transmitted to the gateway at the edge of the network. For instance, a thermometer may record temperature every minute, but such data may only be relevant to a user a few times a day. Similarly, a home security camera may record and send video data to the gateway, but it is often stored in a database for a limited time and is not used further.

2.2.2.4 Privacy and security

Ensuring privacy and data security are critical services at the network edge. When IoT devices are deployed in homes, usage data collected from them can reveal sensitive information, such as whether the house is empty. The challenge is to provide such services without compromising privacy. One approach is to remove personal information from the data before processing it, such as masking all faces in a video. Keeping computing resources at the edge, such as in the home, can be an effective way to protect privacy and data security. However, several challenges still need to be addressed to ensure data security and privacy at the network edge.

2.2.2.5 Optimization metrics

Edge computing comprises several layers, each with varying computing power, and therefore distributing workload becomes a significant concern. Developers must determine how to allocate tasks across different layers, deciding on the number of tasks each layer should handle. Various allocation strategies exist to balance the workload, such as evenly distributing the workload across layers or maximizing the amount of work completed in each layer. In some cases, extreme approaches are used, such as running the workload entirely on the endpoints or entirely in the cloud.

2.2.3 Fog computing

The first formal definition of fog computing in 2012 by Bonomi et al. [28] states: "Fog computing is a highly virtualized platform that provides compute, storage and networking services between end devices and traditional cloud computing data centres, typically, but not exclusively located at the edge of the network" [29].

In [16, 30, 31] fog computing is defined as "a model to complement the cloud for decentralizing the concentration of computing resources (for example, servers, storage, applications and services) in data centers towards users for improving the quality of service and their experience." Machine learning [32] and fog computing are two technologies that are increasingly being used together in various applications. Fog computing can be used to support machine learning [33, 34] by providing a local computing infrastructure that can process and analyze data in real-time, without having to send it to a remote cloud server for processing. This can significantly reduce the latency and bandwidth requirements for machine-learning applications [35–39], especially in situations where the data is generated at the edge of the network, such as in IoT devices or sensors.

2.2.4 Application of fog computing

For tasks where bandwidth, latency, and reliability are constrained, fog computing may be a viable solution. In general, fog computing can be applied to a wide range of applications [40].

2.2.4.1 Smart cities

Smart agriculture, as a component of smart cities, utilizes fog computing to monitor plant growth and climate conditions using intelligent sensor nodes [40]. In this context, the incorporation of context information and the ability to provide location awareness and real-time response are essential. Additionally, fog computing can be employed in smart transportation and waste management in fog-enabled smart cities, where data collection and aggregation models are crucial. Other potential areas where fog computing can be widely adopted in smart cities include water management, greenhouse gas control, and retail automation.

Fog computing has significant potential in smart urban surveillance, a crucial and ever-changing application of smart cities. A fog-based system presented by Kaur and Aron in [40] can monitor real-time vehicle [41] movements and multi-target tracking through a single algorithm. A future plan proposes creating a uniform system architecture to interconnect disparate network edge devices. The text suggests that specific service requests within SIoT can be effortlessly addressed due to social relationships between end users, where cyber counterparts can reside in the cloud. A four-level SIoT cloud-based architecture is recommended, where the virtualization layer can be implemented in both traditional cloud and fog architectures to ensure trustworthiness among edge resources.

2.2.4.2 Smart energy management

Cloud computing is a centralized computing approach that offers infrastructure, platforms, software, and sensor networks to customers. However, as reliability and performance are critical for energy management platforms, increasing the number of devices in the system may not always be suitable for some delay-sensitive devices with timing requirements. In contrast, fog computing can pre-process data while meeting low-latency requirements.

An energy management system implemented on a fog computing platform can be applied to different domains of operation, such as homes or microgrids. Its objectives may include monitoring and metering power consumption of each device (e.g., home power consumption) and managing energy consumption efficiently by controlling devices (e.g., intelligent lighting, Electric Vehicle (EV) charger, Heating, Ventilation, and Air Conditioning (HVAC) management). Since an energy management platform is a system of systems, the hardware, software, and communication architectures should be properly defined and integrated to design the platform for these systems.

2.2.4.3 Healthcare

Zao et al. [40] proposed a brain–computer interaction system based on fog computing as a healthcare application. The system pre-processes Electroencephalogram (EEG) data on smartphones to reduce service latency.

Low latency and fast processing of context-aware data are crucial for emergency alerts in health and environment. Cloud computing platforms are useful in emergency applications to store and process large amounts of data with flexible and scalable virtual servers and networks. However, fog computing, with its low latency, context-awareness, and localized nature, is expected to support emergency and healthcare-related services. Aazam and Huh introduced a smartphone-based service, the Emergency Help Alert Mobile Cloud (EHAMC) system, where fog computing acts as an offloading layer and sends location and other emergency details to the concerned department, taking advantage of fog computing's inherent nature.

2.2.4.4 Intelligent transportation system

Fog computing has significant potential in the Intelligent Transportation System (ITS) domain, particularly in Vehicular Ad-hoc Networks (VANETs). Cloud computing has previously been used in vehicular networks to enhance safety, traffic efficiency, and public convenience. Cloud-based applications execute on cloud infrastructure, with data collected from the vehicular network.

A new architecture called Fog-based Vehicular Crowdsensing (FVCS) has been proposed, which integrates mobile crowdsensing with fog computing. This architecture is suitable for several applications, including parking navigation, road surface monitoring, and traffic collision monitoring. The security and privacy of data collected from the surroundings are protected through privacy-preserving measures. The Vehicular Fog Computing (VFC) architecture is also introduced, which utilizes the near-use edge devices to optimize the use of vehicular computation and communication resources. This architecture is different from the fog computing architecture presented by Kaur and Aron in [40] where vehicular and mobile devices belong to the TNs. The VFC architecture is more efficient in supporting critical situations compared to Vehicular Cloud Computing (VCC), as it relies on near-location instead of sending data to a remote server. Estimating the mobility model of the vehicles can help better utilize resources. Future directions include addressing security and incentive issues for the EUs and vehicular devices.

2.2.4.5 Augmented and virtual reality

In the domain of Augmented Reality (AR) applications, even a small delay can severely impact the user experience, making it critical to minimize latency. As a result, solutions based on fog computing hold immense potential in this field, as they enable computing resources to be deployed closer to the edge, reducing the distance data needs to travel and minimizing latency. Similarly, fog computing can also benefit connected Virtual Reality (VR) or VR-based games. In a recent study, Zao et al. proposed an AR-based brain–computer interaction game that utilized the fog and cloud infrastructure. The fog was responsible for real-time analysis tasks such as signal processing and classification of the brain state, while the cloud infrastructure was used for model classification updates. However, the study only focused on the fog infrastructure and neglected many important aspects related to platform and application development.

2.2.5 Challenges of fog computing

2.2.5.1 Location privacy preservation

Local real-time services, data management, and content distribution are all enhanced by fog computing's location awareness and geographic distribution. Various location-based services and functions have been made more attractive by fog computing. Users' locations are evidently exposed in fog computing because of localization's feature. For example, a wearable devices upload collected data to fog nodes, which in turn deliver the data or summaries of it to the cloud. During these processes, both fog nodes and clouds can learn the location of wearable devices worn by users. Even if reliable anonymization techniques are used to effectively prevent Nebula Nodes and Clouds from identifying you, positioning techniques will still reveal your location if your device connects to multiple Nebula Nodes will be B. Three-point positioning. If the exact location information is leaked, the anonymous user can be re-identified at four points in time and space with a probability of 90%, so the risk of being identified is high.

Therefore, the anonymous user's global location privacy is reduced to local location privacy due to the inclusion of fog nodes. As a result, the curious cloud can constantly learn the user's harsh local knowledge, depending on the geographic location of the Nebula Nodes, making it difficult to protect the user's location data [42].

2.2.5.2 Security and privacy

Distributed in locations and not near monitored and protected locations. This makes it vulnerable to many types of attacks. Fog computing is also vulnerable to many security attacks because it is built on top of traditional network components rather than modern components. Fog computing is subject to the following types of attacks: Data theft and eavesdropping, man-in-the-middle attacks; intrusions with fog devices acting as gateways; malicious attacks; Fog Node data is unfair and fake; malicious nodes; and DoS. There are many services and requests in the fog. This makes it difficult for fog to handle many services at the same time. As a result, the network becomes too busy to serve end users.

Fog computing also has privacy issues. This is due to the lack of resources to assist in encrypting or decrypting data. There is also a problem with data management, where you need to ensure and validate that your nodes offer and support the same services to your users [43].

2.2.5.3 *Decentralized and scalable secure infrastructure*

Fog computing is a dynamic and scalable paradigm where fog nodes and IoT devices can join and leave the architecture as needed. However, building a secure [31, 44, 45] infrastructure in such a distributed framework is challenging due to the lack of a centralized server. The absence of a trusted leader to regulate network operations makes it difficult to establish a trustworthy network with reliable service. Additionally, traditional security mechanisms are not very effective in the distributed and dynamic fog computing framework.

One of the major challenges is user authentication and access delegation, as each fog node must authenticate a user's identity and grant access to services. This can be inefficient if each fog node stores a copy of authentication data, leading to reduced storage efficiency and high communication overhead. Therefore, developing an efficient approach for fast authentication and delegation in fog computing is difficult.

Another significant issue is ensuring the accuracy of computational results when distributed computation is performed across multiple fog nodes. Not all fog nodes can be trusted, making it challenging to guarantee the accuracy of computational results. These issues can be addressed separately, but ensuring compatibility among these solutions can be another major concern.

2.2.5.4 *Privacy exposure in data combination*

In IoT applications, devices generate data of varying sensitivities, some of which may be inherently sensitive, such as data from a heart rate sensor. However, even seemingly benign data can cause serious security and privacy issues when combined with other data. The use of fog computing in IoT exacerbates this problem [46], as it enables collaboration between fog nodes that can aggregate and process data from a large number of devices. For example, some mobile social apps can record social interactions and detect coughing or sneezing through voice recognition, which can be combined with health data to improve infection analysis. However, health and social data are privacy-sensitive and not in the same trusted domain, posing a trade-off between the functional benefits of combining data and privacy concerns. To address this trade-off, techniques such as differential privacy and fully homomorphic encryption can be employed during data aggregation to prevent privacy risks [47–49].

2.2.5.5 *Detection of fog nodes and IoT devices*

The architecture of fog computing is susceptible to cyber-attacks, which puts fog nodes and IoT devices at a high risk of being compromised. Malicious fog nodes and IoT devices can deceive users by pretending to be legitimate. DoS attacks, wireless jamming, and man-in-the-middle attacks are examples of cyber-attacks that can target fog nodes. Stojmenovic et al. have demonstrated the feasibility of man-in-the-middle attacks in fog computing. Even if the fog nodes and IoT devices are not compromised, they can switch to malicious nodes due to their own motives [50, 51]. These malicious nodes can manipulate user data and pose a major threat to the security and privacy of user data. Detecting such malicious nodes in fog computing is difficult due to diverse trust models and the dynamic and distributed environment. Robust anomaly detection systems can be used to detect such nodes, but they can only detect external attacks with a certain probability. It is challenging to detect fog nodes and untrusted or broken IoT devices using these systems [52]. Thus, developing intrusion detection and protection systems specifically designed for fog computing is crucial to enhancing the security and privacy of user data.

2.3 CONCLUSION

In conclusion, cloud computing, edge computing, and fog computing are three distinct computing paradigms that have emerged in response to the growing demand for distributed computing and the IoT. Cloud computing is a centralized computing paradigm that provides on-demand access to a shared pool of computing resources over the internet, while edge computing brings computation and data storage closer to the edge of the network where IoT devices generate and consume data. Fog computing is a hybrid computing paradigm that combines the benefits of cloud computing and edge computing. This analysis delved into the intricacies of each computing paradigm, providing a comprehensive understanding of their unique characteristics and differences. The study shed light on the complex interplay between these paradigms and the potential synergies that can be harnessed through their integration to unlock new opportunities and overcome challenges in the rapidly evolving landscape of distributed computing. The insights from this analysis can help stakeholders make informed decisions when choosing a computing paradigm for their specific use cases.

REFERENCES

1. De Donno, M., Tange, K., & Dragoni, N. (2019). Foundations and evolution of modern computing paradigms: Cloud, iot, edge, and fog. *IEEE Access*, 7, 150936–150948.
2. Dillon, T., Wu, C., & Chang, E. (2010, April). Cloud computing: issues and challenges. In *2010 24th IEEE international conference on advanced information networking and applications* (pp. 27–33). IEEE.
3. Kaur, M., & Aron, R. (2020, February). Equal distribution based load balancing technique for fog-based cloud computing. In *International Conference on Artificial Intelligence: Advances and Applications 2019: Proceedings of ICAIAA 2019* (pp. 189–198). Singapore: Springer Singapore.

4. Tandon, R., & Gupta, P. K. (2019). Optimizing smart parking system by using fog computing. In *Advances in Computing and Data Sciences: Third International Conference, ICACDS 2019, Ghaziabad, India*, April 12–13, 2019, Revised Selected Papers, Part II 3 (pp. 724–737). Springer Singapore.

5. Apostu, A., Puican, F., Ularu, G., Suciu, G., & Todoran, G. (2013). Study on advantages and disadvantages of Cloud Computing–the advantages of Telemetry Applications in the Cloud. *Recent advances in applied computer science and digital services*, 2103.

6. Harnal, S., & Chauhan, R. K. (2016, January). Multimedia support from cloud computing: A review. In *2016 International Conference on Microelectronics, Computing and Communications (MicroCom)* (pp. 1–6). IEEE.

7. Harnal, S., Sharma, G., Malik, S., Kaur, G., Simaiya, S., Khurana, S., & Bagga, D. (2023). Current and Future Trends of Cloud-based Solutions for Healthcare. *Image Based Computing for Food and Health Analytics: Requirements, Challenges, Solutions and Practices: IBCFHA*, 115–136.

8. Youssef, A. E. (2012). Exploring cloud computing services and applications. *Journal of Emerging Trends in Computing and Information Sciences*, 3(6), 838–847.

9. Goyal, S. (2014). Public vs private vs hybrid vs community-cloud computing: a critical review. *International Journal of Computer Network and Information Security*, 6(3), 20–29.

10. Ghanbari, H., Simmons, B., Litoiu, M., & Iszlai, G. (2012). Feedback-based optimization of a private cloud. *Future Generation Computer Systems*, 28(1), 104–111.

11. Zhang, Q., Cheng, L., & Boutaba, R. (2010). Cloud computing: state-of-the-art and research challenges. *Journal of Internet Services and Applications*, 1(1), 7–18.

12. Tarannum, N., & Ahmed, N., 2013. Efficient and reliable hybrid cloud architecture for big database. *International Journal on Cloud Computing Services and Architecture*, 3(6), pp. 17–29.

13. Ali, H. (2015). Cloud computing security: An investigation into the security issues and challenges associated with cloud computing, for both data storage and virtual applications. *International Research Journal of Electronics and Computer Engineering*, 1(2), 15.

14. Sahandi, R., Alkhalil, A., & Opara-Martins, J. (2013). Cloud computing from SMEs perspective: a survey based investigation. *Journal of Information Technology Management*, 24(1), 1–12.

15. Xue, C. T. S., & Xin, F. T. W. (2016). Benefits and challenges of the adoption of cloud computing in business. *International Journal on Cloud Computing: Services and Architecture*, 6(6), 01–15.

16. Kaur, M., & Aron, R. (2022). A novel load balancing technique for smart application in a fog computing environment. *International Journal of Grid and High Performance Computing (IJGHPC)*, 14(1), 1–19.

17. Kaur, V., & Singh, A. (2013). Review of various algorithms used in hybrid cryptography. International *Journal of Computer Science and Network*, 2(6), 157–173.

18. Kaur, G., Choudhary, P., Sahore, L., Gupta, S., & Kaur, V. (2023). Healthcare: In the Era of Blockchain. In *AI and Blockchain in Healthcare* (pp. 45–55). Singapore: Springer Nature Singapore.

19. Takabi, H., Joshi, J. B., & Ahn, G. J. (2010). Security and privacy challenges in cloud computing environments. *IEEE Security & Privacy*, 8(6), 24–31.

20. Tiwari, R., & Kumar, N. (2016). An adaptive cache invalidation technique for wireless environments. *Telecommunication Systems*, 62, 149–165.

21. Harnal, S., & Chauhan, R. K. (2019). Hybrid cryptography based E2EE for integrity & confidentiality in multimedia cloud computing. *International Journal of Innovative Technology and Exploring Engineering (IJITEE), Scopus*, 8(10), 918–924.

22. Goel, G. and Tiwari, R., (2023). Resource scheduling techniques for optimal quality of service in fog computing environment: a review. *Wireless Personal Communications*, 131(1), 141–164

23. Cao, K., Liu, Y., Meng, G., & Sun, Q. (2020). An overview on edge computing research. *IEEE Access*, 8, 85714–85728.

24. Roman, R., J. Lopez, and M. Manbo. (2018). Mobile edge computing, fog et al.: A survey and analysis of security threats and challenges. *Future Gener. Comput. Syst.*, 78, 680–698.

25. Kaur, M., & Aron, R. (2021). Focalb: Fog computing architecture of load balancing for scientific workflow applications. *Journal of Grid Computing*, 19(4), 40.

26. Baktir, A. C., Ozgovde, A., & Ersoy, C. (2017). How can edge computing benefit from software-defined networking: A survey, use cases, and future directions. *IEEE Communications Surveys & Tutorials*, 19(4), 2359–2391.

27. Shi, W., Cao, J., Zhang, Q., Li, Y., & Xu, L. (2016). Edge computing: Vision and challenges. *IEEE Internet of Things Journal*, 3(5), 637–646.

28. Kaur, M., & Aron, R. (2022). An energy-efficient load balancing approach for fog environment using scientific workflow applications. In *Distributed Computing and Optimization Techniques: Select Proceedings of ICDCOT 2021* (pp. 165–174). Singapore: Springer Nature Singapore.

29. Stojmenovic, S. Wen, X. Huang, and H. Luan. (2016). An overview of fog computing and its security issues. *Concurrency Comput. Pract. Exp.*, 28(10), 2991–3005.

30. Varghese, B., Wang, N., Nikolopoulos, D. S., & Buyya, R. (2020). Feasibility of fog computing. In *Handbook of Integration of Cloud Computing, Cyber Physical Systems and Internet of Things* (pp. 127–146). Springer, Cham.

31. Harnal, S., & Chauhan, R. K. (2020). Towards secure, flexible and efficient role based hospital's cloud management system: case study. *EAI Endorsed Transactions on Pervasive Health and Technology*, 6(23), 1–13.

32. Kaur, G., Kaur, V., Sharma, Y., & Bansal, V. (2022, December). Analyzing various machine learning algorithms with SMOTE and ADASYN for image classification having imbalanced data. In *2022 IEEE International Conference on Current Development in Engineering and Technology (CCET)* (pp. 1–7). IEEE.

33. Kaur, V., Kaur, G., Dhiman, G., Bindal, R., & Mishra, M. K. (2020). Adaptability of machine learning in cryptography. *Solid State Technology*, 63(4), 2874–2880.

34. Kaur, G., Kaur, H., & Goyal, S. (2022). Correlation analysis between different parameters to predict cement logistics. *Innovations in Systems and Software Engineering*, 1–11.

35. Verma, A., Shah, C. K., Kaur, V., Shah, S., & Kumar, P. (2022, September). Cancer detection and analysis using machine learning. In *2022 Second International Conference on Computer Science, Engineering and Applications (ICCSEA)* (pp. 1–5). IEEE.

36. Reddy, K., Reddy, M. A., Kaur, V., & Kaur, G. (2022). Career guidance system using ensemble learning. Akhila and Kaur, Veerpal and Kaur, Gagandeep, Career Guidance System Using Ensemble Learning (July 14, 2022).

37. Kaur, V., & Singh, A. (2021). Role of machine learning in communication networks. In *Intelligent Communication and Automation Systems* (pp. 1–10). CRC Press.

38. Kaur, G., Goyal, S., & Kaur, H. (2021, December). Brief review of various machine learning algorithms. In *Proceedings of the International Conference on Innovative Computing & Communication (ICICC)*.

39. Kaur, V., & Kaur, R. (2022). An elucidation for machine learning algorithms used in healthcare. In *Machine Learning for Edge Computing* (pp. 25–36). CRC Press.

40. Kaur, M., & Aron, R. (2022, February). Fog clustering-based architecture for load balancing in scientific workflows. In *Proceedings of International Conference on Computational Intelligence and Data Engineering: ICCIDE 2021* (pp. 213–221). Singapore: Springer Nature Singapore.

41. Tandon, R., & Gupta, P. K. (2023). A hybrid security scheme for inter-vehicle communication in content centric vehicular networks. *Wireless Personal Communications*, 129(2), 1083–1096.

42. Harnal, S., Sharma, G., & Mishra, R. D. (2022). QoS-Based Load Balancing in Fog Computing. In *Mobile Radio Communications and 5G Networks: Proceedings of Second MRCN 2021* (pp. 331–344). Singapore: Springer Nature Singapore.

43. Bani Yassein, M., Alzoubi, O., Rawasheh, S., Shatnawi, F., & Hmeidi, I. (2020). Features, challenges and issues of fog computing: A comprehensive review. *WSEAS Trans. Comput*, 19, 86–97.

44. Tandon, R., & Gupta, P. K. (2021). Security and privacy challenges in healthcare using Internet of Things. In *IoT-Based Data Analytics for the Healthcare Industry* (pp. 149–165). Academic Press.

45. Tandon, R., & Gupta, P. K. (2021). SV2VCS: a secure vehicle-to-vehicle communication scheme based on lightweight authentication and concurrent data collection trees. *Journal of Ambient Intelligence and Humanized Computing*, 1–17.

46. Narayanan and V. Shmatikov. (2010). Myths and fallacies of 'personally identifiable information. *Commun. ACM*, 53(6), 24–26.

47. Wadhwa, H., & Aron, R. (2018, December). Fog computing with the integration of internet of things: Architecture, applications and future directions. In *2018 IEEE Intl Conf on Parallel & Distributed Processing with Applications, Ubiquitous Computing & Communications, Big Data & Cloud Computing, Social Computing & Networking, Sustainable Computing & Communications (ISPA/IUCC/BDCloud/SocialCom/SustainCom)* (pp. 987–994). IEEE.

48. Wadhwa, H., & Aron, R. (2023). Optimized task scheduling and preemption for distributed resource management in fog-assisted IoT environment. *The Journal of Supercomputing*, 79(2), 2212–2250.

49. Kaur, M., Aron, R., Wadhwa, H., & Oo, H. N. (2023, April). CNN-based smart waste management system in fog computing environment. In *2023 IEEE 12th International Conference on Communication Systems and Network Technologies (CSNT)* (pp. 774–779). IEEE.

50. Wadhwa, H., Kaur, M., Kaur, A., & Tiwana, P. S. (2024). Analysis of IoT devices data using bayesian learning on fog computing. In *Artificial Intelligence, Blockchain, Computing and Security Volume 2* (pp. 194–199). CRC Press.
51. Kaur, G., & Kaur, H. (2017, July). Prediction of the cause of accident and accident prone location on roads using data mining techniques. In *2017 8th International Conference on Computing, Communication and Networking Technologies (ICCCNT)* (pp. 1–7). IEEE.
52. Harnal, S., Sharma, G., Malik, S., Kaur, G., Simaiya, S., Khurana, S., & Bagga, D. (2023). Current and future trends of cloud-based solutions for healthcare. *Image Based Computing for Food and Health Analytics: Requirements, Challenges, Solutions and Practices: IBCFHA*, 115–136.

Chapter 3

Management of metropolitan mobility for public transport and smart vehicles with fog computing

Righa Tandon
Chitkara University Institute of Engineering and Technology,
Chitkara University, Chandigarh, India

Ajay Verma
Jaypee University of Information Technology, Solan, India

Mandeep Kaur and Heena Wadhwa
Chitkara University Institute of Engineering and Technology,
Chitkara University, Chandigarh, India

David Asirvatham
Taylor's University, Malaysia

3.1 INTRODUCTION

In the rapidly urbanising world of today, managing the complex web of metropolitan mobility has become one of the most pressing issues in urban planning and development. The demand for effective transport infrastructure has expanded dramatically as cities' populations and economic activity increase. However, traditional methods of managing transport are frequently pushed beyond their breaking point, leading to clogged roads, protracted commutes, and environmental issues. Cities all over the world are reverting to cutting-edge solutions as a result, utilising the interplay between technology and strategic urban planning. This chapter, "Management of Metropolitan Mobility for Public Transport and Smart Vehicles with Fog Computing," explores the complex interactions among urban mobility, public transport systems, smart vehicles, and the potentially transformative effects of fog computing.

Metropolitan areas serve as hubs of social interaction, cultural exchange, and thriving economies in the dynamic environment of contemporary urban living. The complexity of the problems these thriving hubs provide, such as traffic congestion, lengthy travel times, and environmental deterioration, frequently outweigh their vibrancy and potential. A paradigm shift that reimagines the fundamental basis of mobility within urban environments is being

DOI: 10.1201/9781003494430-3

adopted by forward-thinking cities in an effort to overcome these challenges. Public transport networks and smart vehicles are at the centre of this revolution; they are essential elements that hold the key to achieving efficient, sustainable, and harmonious mobility experiences for citizens. Alongside these developments, fog computing has evolved as a ground-breaking architecture that enables metropolitan centres to efficiently handle and manage real-time data at the periphery of their extensive networks. Fog computing, which departs from the conventional cloud computing model, makes use of a distributed architecture to enable data processing to occur closer to the source. This closeness to data generation is particularly important in the dynamic world of metropolitan mobility, where split-second decisions can have significant effects on traffic flow, passenger safety, and the overall effectiveness of transportation systems. More than just a technical development, the fusion of public transportation, intelligent transportation systems (ITS), and fog computing represents a paradigm shift that has the potential to reinvent the core functions of urban transportation systems. Cities are able to give passengers personalised, up-to-date information, optimise routes to reduce delays, and quickly adapt to the changing conditions that are unique to urban transportation by fusing these elements smoothly. Through this combination, a mobility ecology is produced that is naturally intelligent, adaptable, and responsive to the changing needs of a contemporary city and its inhabitants. This chapter will take us on a journey through the sophisticated architecture of fog computing systems, through case studies and implementations in the real world, and into the optimisation of urban mobility. Ultimately, our voyage will add to the ongoing conversation about this topic. We hope to provide insights into the current status and potential future of smart, interconnected, and resilient urban transportation systems by casting light on the nexus of public transportation, smart automobiles, and fog computing. We hope that this investigation will add to the continuing conversation about rethinking urban mobility, turning it from a problem into a chance for cities to thrive in an era of perpetual change [1–5].

The efficient management of metropolitan mobility has emerged as a major puzzle for cities worldwide in an era marked by rising urbanisation and increasing connectedness. The demand for efficient transport networks is greater than ever because of the growth in urban populations and the concentration of economic activity. Traditional methods of mobility management are reaching their limits, as evidenced by the snarl of traffic, the lengthening of commutes, and environmental issues. Cities are searching for creative answers to these problems that use the interplay between technology and well-thought-out urban design. This chapter, "Management of Metropolitan Mobility for Public Transport and Smart Vehicles with Fog Computing," launches a thorough investigation of the complex interactions between metropolitan mobility, public transport networks, smart vehicles, and the ground-breaking potential of fog computing [6]. The overview of fog computing smart devices is shown below in Figure 3.1.

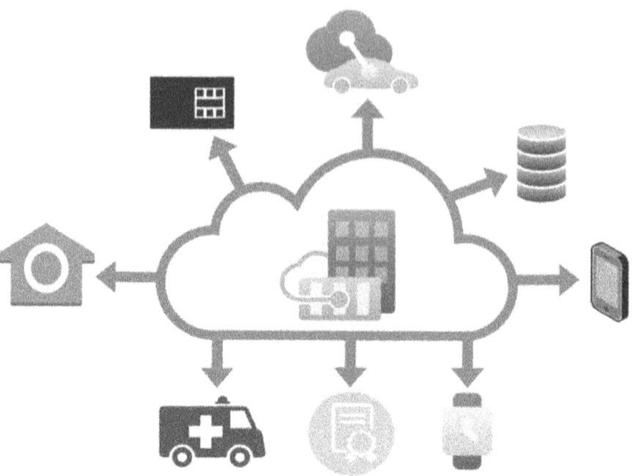

Figure 3.1 Overview of fog computing smart devices.

Alongside these developments, fog computing has developed into a ground-breaking architecture that enables cities to quickly process and handle real-time data at the edge of their extensive networks. By adopting a distributed architecture, fog computing departs from the traditional cloud computing approach and allows data to be processed closer to where it originated. This geographic closeness to data collection is especially important in the context of urban mobility, where quick judgements affect traffic patterns, passenger safety, and the overall effectiveness of transportation systems.

However, the combination of public transportation, smart cars, and fog computing symbolises more than just technological advancement; it denotes a paradigm shift that has the ability to completely rethink the structure of urban transportation systems. Cities may provide commuters with personalised, current information by smoothly integrating various components, optimising routes to reduce congestion, and swiftly reacting to the dynamic aspects that define urban mobility. As a result of this fusion, a mobility ecology is created that is intrinsically intelligent, adaptable, and sensitive to the changing needs of a modern city and its inhabitants. Our journey through the chapters of this in-depth investigation will take us deep into the complex design of fog computing systems. Examining actual cases of effective implementation can help us draw conclusions from instructive case studies. Our goal is to meaningfully contribute to the discussion that is currently taking place on how to improve urban mobility. Our goal is to shed light on how public transportation, smart cars, and fog computing interact in order to offer insightful information on the status quo and potential course of intelligent, networked, and resilient urban transportation networks. This voyage captures both the advancement of technology and the possibility for cities to flourish in the face of perpetual change, ultimately transforming the difficulty of metropolitan transportation into a basis for urban excellence [7].

3.2 PUBLIC TRANSPORT SYSTEMS

The demand for effective and sustainable transport networks has never been greater as cities continue to grow and urbanisation progresses. The cornerstone of urban mobility management in this environment is public transportation, which includes buses, trains, trams, and subways. The importance of these transit options is explored in depth in this section, which also shows how they solve issues like accessibility, sustainability, and traffic congestion. Buses are frequently regarded as the workhorses of urban transportation because they are adaptable and accessible for carrying big crowds of people over a variety of routes. They serve as a crucial link for commuters who depend on their daily journeys by navigating both major thoroughfares and smaller streets thanks to their agility. Buses have the ability to considerably reduce traffic congestion by allowing people to congregate in one vehicle, relieving the pressure on road networks and lowering the number of individual automobiles competing for scarce road space. Trains provide a high-capacity, rapid-transit solution that is suitable for medium- to long-distance travel in parallel. In order to avoid traffic congestion and allow passengers to move around the city efficiently, commuter rail systems and subways offer a dedicated right of way. Since trains are naturally more effective at transporting huge crowds of people than other modes of transportation, there are fewer cars on the road, which directly lessens traffic congestion [8].

Additionally, trams are a hybrid mode of transportation because they frequently run on specialised tracks while blending in perfectly with urban settings. In addition to offering ecologically-friendly transportation, these systems encourage urban redevelopment and growth along their lines, promoting mixed-use construction and limiting urban sprawl. By lowering individual vehicle emissions as a group, these public transport options promote sustainability, an important first step in addressing issues with urban air quality and the environment. Public transport has the potential to be a more environmentally friendly option with improvements in electrification and fuel efficiency, reducing the negative environmental effects of using private vehicles and promoting cleaner urban air. A key component of equitable urban mobility is the accessibility provided by public transit systems. They offer services to a wide spectrum of clients, including people who might not be able to access private vehicles because of financial or physical limitations. Public transit provides a more equal urban fabric and improves social inclusion as a result. The role of public transport networks is becoming more and more important as cities grow and struggle with escalating mobility issues. They can reduce traffic congestion, encourage sustainability, and improve accessibility, making them crucial parts of a holistic metropolitan transportation strategy [9].

Urban planners and politicians may handle the changing requirements of contemporary cities while working towards a more effective, sustainable, and accessible urban future by appreciating and developing the potential of buses, trains, trams, and subways. The importance of public transport networks has taken centre stage in the conversation about urban development in an era characterised by rapid urbanisation and the mobility issues that go along with it. Public transportation systems, which include buses, trains, trams, and subways, are playing a crucial part in resolving the growing transportation problems facing cities as they grow in population. This section aims to delve deeper into the public transportation systems' growing significance by explaining how they not only help to reduce traffic congestion but also promote sustainability and accessibility, positioning them as crucial elements of an all-encompassing metropolitan transportation strategy. Urban population growth puts more strain on already-strapped transport systems, leading to the bottleneck of heavy traffic that blights many large cities. This is where public transport systems may truly transform society. These systems can significantly lessen the number of individual vehicles clogging the streets by providing an effective and collective method of moving people. Buses, trains, trams, and subways can all carry a lot of people at once thanks to their combined capacity, which lessens the load on the transportation system and clears traffic on the streets. Reduced travel times, lessened commuter stress, and a more seamless urban mobility experience are all results of this congestion relief [10–12].

But the function of public transport networks goes well beyond just reducing congestion. These systems are now in the spotlight as crucial tools in the fight against climate change and air pollution thanks to the requirements of sustainability and environmental stewardship. Due to the fact that fewer private automobiles are on the road, public transit naturally minimises harmful emissions and leaves a smaller carbon imprint. The incorporation of effective public transport options becomes more important for a greener and more sustainable future as cities attempt to meet environmental goals and improve the quality of urban life. Public transport systems' potential to change urban accessibility and inclusivity is arguably one of their most transformational features. People without access to private vehicles, whether because of financial limitations or physical restrictions, have a lifeline provided by the interconnection of different modes. Public transportation facilitates social inclusion and fosters a more egalitarian urban fabric by offering a dependable and complete transportation system, removing barriers that insufficient transportation options would otherwise place in the way of access to education, employment, and leisure activities for people from all walks of life. Urban planners and politicians are responsible for navigating these intricacies as cities develop and deal with the dynamic difficulties of modern urbanisation. A comprehensive and integrated approach to metropolitan mobility that leverages the untapped potential of public transport networks is required due to the changing environment. Cities may not only address the current problems of congestion by recognising and utilising the possibilities of buses, trains, trams, and subways, but they can also foster a more productive, sustainable, and accessible urban future.

By doing this, they create urban settings that are lively, connected, and receptive to the changing demands of their population. The need for transport systems that are effective and environmentally friendly has gone beyond vital in an era marked by relentless urban development and the complexity of modern living [13–15].

The role of public transit systems has grown to be unmatched within this complex metropolitan structure. The basic foundation of how cities operate is supported by these systems, which are made up of a mosaic of buses, trains, trams, and subways. They have become leaders in metropolitan mobility management. This part launches an investigation into the multifaceted relevance of these many means of transport, shedding light on how crucial a role they play in tackling the urgent issues of traffic congestion, sustainability, and accessibility that cities face. The complicated network of streets can become a convoluted web of traffic congestion as urban centres grow in size and population, stifling movement and degrading the quality of life for city people. Here, public transit systems' significant effects may be seen in action. These systems, which are made to move a lot of people effectively, serve as pressure valves that can lessen the burden on the road systems. Because of their combined capacity, buses, trains, trams, and subways help reduce the number of private automobiles on the road, eradicating traffic jams and regaining movement flexibility. Cities may reduce the negative consequences of congestion by adopting these integrated forms of transport, creating avenues for their citizens to move more smoothly and quickly. Although they are the forerunners of sustainability in urban settings, public transport networks serve a purpose that goes beyond simply relieving traffic congestion. The ecological impact of mobility looms big as worries about climate change and environmental deterioration gain traction. The intrinsic capacity of public transport to carry a large number of passengers on a single trip acts as a counterbalance to the carbon emissions produced by individual motor vehicle travel. Together, buses, trains, trams, and subways can lessen the environmental effect of urban transportation by cutting greenhouse gas emissions and encouraging more energy-efficient forms of transportation. This ecologically friendly feature of public transport contributes to the larger idea of urban sustainability by enabling cities to use less energy and create healthier living conditions for their citizens [16–19].

The ability of public transit systems to facilitate accessibility and inclusivity, however, is where their greatest revolutionary power rests. These systems span social and economic boundaries, extending their reach to people without own vehicles or with physical limitations. Public transportation improves social fairness by offering a dependable and extensive network, creating opportunities for all facets of the population to pursue education, employment, and civic engagement. As cities try to build increasingly diverse neighbourhoods, the vast reach of public transport systems becomes a symbol of urban cohesion, promoting a setting where everyone can benefit from the benefits the city offers. The foundation upon which urban mobility management is based is public transport networks, which include an extensive mosaic of modes. These technologies usher in a transformative era in urban transportation by addressing the complicated issues of traffic congestion, sustainability, and accessibility. Cities can reduce traffic jams, promote sustainability, and pave the way for efficient, ecologically conscious, and openly accessible transportation in the future by embracing these modes of transportation. Public transport systems' capacity to foster accessibility and inclusivity, however, is where they truly exert their transformational potential. The reach of these networks extends to individuals without own vehicles or who have physical limitations by bridging socioeconomic and geographic barriers. Public transportation improves social fairness by offering a dependable and extensive network, opening up doors to civic engagement, jobs, and education for all facets of the population. The vast reach of public transport networks becomes a symbol of urban cohesion as cities try to build more diverse neighbourhoods, promoting a climate where everyone can benefit from the benefits the city offers [20].

3.3 SMART VEHICLES AND INTELLIGENT TRANSPORTATION SYSTEMS (ITS)

The introduction of smart vehicles and ITS has ushered in a new paradigm of mobility in an era marked by unrelenting technical advancement and the constantly changing needs of urban transportation. These ground-breaking systems, which feature cutting-edge communication technology, sophisticated automation, and advanced sensors, have the potential to fundamentally alter how cities manage user experiences, safety, and transportation efficiency. This section explores the complex role that smart vehicles, driven by ITS, play in determining the future of urban mobility. Smart vehicles— vehicles that are not only means of transportation but also intricate nodes within a connected urban ecosystem—are at the vanguard of this shift. These vehicles can gather real-time information about their surroundings, including road conditions and pedestrian activity, thanks to a variety of sensors they are equipped with. The ability of vehicles to react quickly to shifting traffic patterns, weather conditions, and potential hazards is made possible by this plethora of information, which serves as the cornerstone of intelligent decision-making. Smart vehicles can readjust routes, optimise speeds, and reduce congestion thanks to this data-driven approach, innovative algorithms, and machine learning, greatly boosting the efficiency of transportation. Additionally, in the urban environment, smart automobiles serve as safety beacons. These cars can comprehend their surroundings with a level of precision never before possible thanks to the integration of cutting-edge technology like computer vision, radar, and LiDAR (light detection and ranging). With improved collision avoidance, pedestrian identification, and adaptive cruise control as a result of this increased awareness, the entire system of urban transportation becomes safer. Additionally, the ability for vehicle-to-vehicle (V2V) and vehicle-to-infrastructure (V2I) communication permits intelligent cars to exchange crucial information in real-time, enabling coordinated reactions to potential risks and the smooth flow of traffic. Beyond effectiveness and safety, smart vehicles support a superior user experience appropriate for the digital era [21–23].

The seamless, information-rich ride is made possible for passengers by onboard multimedia systems, connectivity to mobile devices, and real-time navigation updates. Beyond specific cars, this improved user experience interacts with the larger world of urban mobility. Integrated systems that are personalised to individual preferences and needs are available to commuters. These platforms offer real-time traffic reports, suggest the best routes, and enable multimodal mobility. The digital infrastructure supporting this paradigm change is ITS, which serve as the foundation for smart vehicles. ITS makes it possible for information to be seamlessly exchanged between vehicles, infrastructure, and central control centres using a network of sensors, communication nodes, and data analytics. A comprehensive transportation management ecosystem is created as a result of cities' increased ability to choreograph traffic flow, improve signal timings, and handle emergency response. Urban transportation is being brought into a new era of efficiency, safety, and improved user experiences via smart cars and ITS. These cutting-edge technologies are transforming urban mobility into a responsive and proactive system, decorated with sensors and propelled by data analytics. By adopting these advances, cities can build transit networks that enhance productivity, boost safety, and meet the many demands of their residents, setting the groundwork for a smarter and more connected urban future [24–27].

With the integration of smart automobiles and ITS, the landscape of urban transportation is experiencing a significant upheaval. Urban mobility is entering an era characterised by previously unheard-of efficiency, elevated safety standards, and noticeably better user experiences thanks to the confluence of cutting-edge technologies. This section explores the complex interactions between these developments, demonstrating how they have changed the way that people travel in cities. A significant development in the transportation industry is the introduction of smart automobiles, which are outfitted with a variety of sensors, communication tools, and high levels of automation. In a dynamic urban environment, these vehicles serve as complex data-gathering and decision-making hubs, not just as means of transportation.

ITS, a sophisticated digital framework that significantly alters the entire mobility management environment, are at the centre of the revolutionary revolution in urban transportation. Smart vehicles are made possible by ITS. This paradigm shift's solid infrastructure is supported by ITS, which orchestrates a symphony of sensors, communication nodes, and data analytics to enable the smooth interchange of information. Through this interconnected system, ITS makes it possible for communication between vehicles, infrastructure, and central control points, giving cities previously unheard-of control over traffic flow, signal timing, and emergency response. A comprehensive transportation management ecosystem that is flexible, adaptable, and sensitive to the requirements of contemporary urban environments is the result of this complex interplay. At its foundation, ITS relies on a complex network of sensors that are strategically positioned within both smart vehicles and urban infrastructure. As the system's eyes and ears, these sensors are constantly gathering information on the flow of traffic, the state of the roads, the forecast for the day, and current events. By combining this abundance of data, ITS not only creates a comprehensive view of the urban mobility scene, but also gives cities the ability to make data-driven decisions that improve efficiency and safety. In the ITS ecosystem, communication nodes serve as the hubs that enable dynamic information exchange. These nodes provide real-time data exchange between cars and infrastructure by utilising wireless communication technologies like as cellular networks and dedicated short-range communication (DSRC). As a result, smart vehicles can collaborate to adapt to shifting traffic circumstances, communicate with one another, and foresee future crashes. Additionally, communication nodes work as channels for transmitting vital data to central control centres, letting authorities keep an eye on traffic, spot irregularities, and act quickly when necessary.

The actual strength of ITS is found in its sophisticated data analytics capabilities. ITS is able to handle and analyse information in real-time by utilising the massive amounts of data that are produced by sensors and communication nodes. Algorithms for machine learning analyse patterns, forecast traffic jams, and foresee possible bottlenecks. Cities are now able to dynamically optimise signal timings thanks to this predictive capability, ensuring that traffic flows are controlled to reduce congestion and maximise efficiency. These factors working together allow ITS to direct an integrated transportation management ecosystem. Cities have the ability to dynamically modify the timing of traffic signals based on current circumstances, lowering wait times and promoting more efficient traffic flow. As ITS helps emergency services in times of need, emergency response coordination becomes more dynamic. The development of autonomous vehicles and the realisation of a future in which vehicles seamlessly communicate and cooperate to navigate challenging urban environments are both made possible by the interconnectedness between infrastructure and vehicles that ITS fosters. The core of the seismic shift in urban transportation is ITS. ITS develops a complex ecosystem that enables cities to better manage mobility through the convergence of sensors, communication nodes, and data analytics. A holistic management paradigm that supports efficiency, safety, and adaptability in urban mobility is formed by the orchestration of traffic flow, improvement of signal timings, and coordination of emergency responses. The role of ITS is becoming more and more important in creating smarter, more responsive,

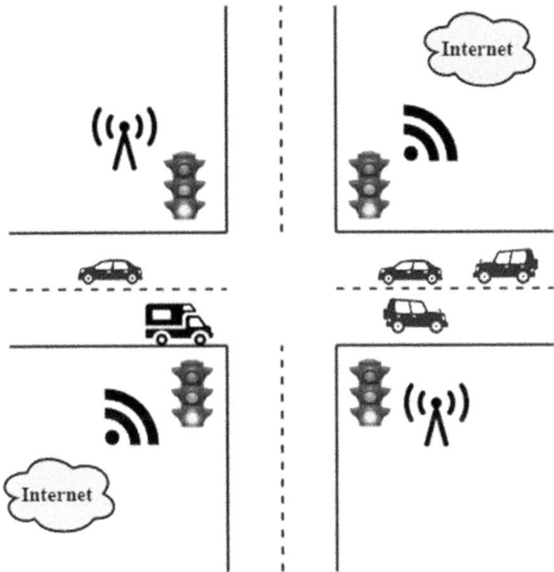

Figure 3.2 Intelligent transportation systems.

and more connected urban futures as cities continue to struggle with the problems posed by expanding populations and changing transportation needs [28]. The general ITS pictorial representation is shown below in Figure 3.2.

Smart cars gather information about their environment in real-time, including things like traffic patterns, road conditions, and potential hazards. This information forms the basis for intelligent decision-making, allowing these vehicles to adjust their itineraries, maximise their speeds, and eventually lessen traffic congestion. Smart automobiles help create a more effective transport system by adapting routes and reducing bottlenecks by dynamically adjusting to real-world situations. However, the change goes well beyond productivity. The integration of cutting-edge technologies within smart cars promotes a setting of increased safety. These vehicles develop a precision understanding of their environment that is unmatched thanks to the integration of radar, LiDAR, and computer vision technologies. Through the coordination of adaptive cruise control, pedestrian identification, and collision avoidance are made possible by this increased awareness. Additionally, incorporating vehicle-to-vehicle (V2V) and vehicle-to-infrastructure (V2I) communication allows for seamless information transmission, enabling coordinated responses to potential dangers and easing the smooth flow of traffic. Parallel to this, ITS act as the foundation for this shift. A smooth information flow is made possible by the ITS, a digital infrastructure that connects cars, sensors, and command centres. Due to this interaction, cities are able to regulate emergency response times, optimise traffic signals, and prevent future congestion. When smart cars and ITS are combined, a symbiotic connection is created in which the vehicles supply real-time data to the system, which in turn orchestrates traffic management for the best results [29].

With the help of data analytics and a plethora of sensors, the upshot of this technological convergence is a transportation system that is both responsive and proactive. Cities that adopt these innovations are better positioned to build transit systems that directly benefit the people who live there. Travel times are shorter and congestion is lessened, which increases productivity. A higher level of safety is achieved by sophisticated danger and collision avoidance systems. User-centric technology is seamlessly integrated to provide individualised and effective transportation experiences while meeting the needs of a diverse population of residents. Urban transportation is entering a new era marked by tremendous improvements in effectiveness, security, and user experiences thanks to the collaboration of smart cars and ITS. The foundation of a smarter, more connected urban future is this transition, which is driven by data-driven decision-making and technologically-enabled insights. Cities are paving the way for transport networks that go beyond the constraints of the past, addressing the issues of the present, and establishing the groundwork for a dynamic and intelligent urban landscape by embracing these advancements [30–32].

3.4 FOG COMPUTING ARCHITECTURE FOR METROPOLITAN MOBILITY

A crucial intersection of cutting-edge technology and urban planning can be seen in the architecture of a fog computing system designed for managing metropolitan mobility. An agile and effective environment for real-time data processing, analysis, and decision-making is produced by this architecture, which is based on the idea of dispersing computer resources closer to the data source. The details of this architecture are explored in this part, which also outlines the essential elements of a fog computing system intended to revolutionise the control of urban mobility.

- Edge Devices: The function of edge devices emerges as a cornerstone in the dynamic terrain of urban mobility transformation, bringing life to the fog computing architecture designed for metropolitan mobility management. With an outstanding selection of sensors, cameras, GPS devices, and a variety of data collection techniques, these gadgets serve as sentinels scattered across the metropolitan area. Edge devices usher in a new era of data-driven decision-making as they interact with the physical environment around them, providing real-time insights that act as the cornerstone for optimising transportation networks. At the core of the design, edge devices provide a mosaic of data-collecting tools, each one created to record a distinct aspect of the urban environment. Vehicle density, traffic patterns, and road conditions are monitored via sensors built into the infrastructure and into the roads. Cameras record images that disclose the movements of vehicles and pedestrian activities as well as any anomalies or possible incidents. GPS devices give geographical data that improves navigation systems and is useful for tracking traffic and vehicle routes. These several data sources combine to create a thorough picture of the dynamics of metropolitan mobility.

As the fog computing system's eyes and ears, edge devices continuously produce real-time data that reflects the pace of urban life. The system can react to the constantly shifting dynamics of traffic flow, weather, and unforeseen events thanks to the data collection and transmission carried out by these devices every single second. In order to make informed decisions and act quickly, real-time data gathering is essential, ensuring that the urban transportation ecosystem continues to be flexible and dynamic. The core of the overall fog computing architecture is the data collected by edge devices. The analytics and algorithms that power insights and optimisations are fed by this data, which is frequently referred to as the "raw material" of decision-making. The data includes everything from updates on traffic congestion and vehicle speeds to environmental factors like air quality and weather. Cities are given the resources to make well-informed decisions that can reduce traffic, increase safety, and increase overall transportation effectiveness thanks to this continuous flow of data. Within the fog computing paradigm, edge devices create a feedback loop. Within fog nodes, the real-time data they gather is quickly processed and analysed to produce insights that can be put into practice. These insights may come in the form of real-time traffic data, suggested routes, or even automated changes to the timing of traffic signals. Edge devices are used in this dynamic process as data sources as well as decision-making catalysts, which help to orchestrate urban transportation smoothly. In essence, edge devices signify the beginning of a smarter and more effective urban transit system. These gadgets provide a networked system with a continuous information flow thanks to their capacity to record real-time data about traffic, vehicle movements, and environmental conditions. This flow serves as the foundation for data-driven decision-making, optimisation tactics, and the creation of a more flexible and adaptable urban transportation system. By embracing the possibilities of edge devices, cities pave the way for a time when real-time information will direct urban mobility, promoting effectiveness, safety, and an overall improvement in inhabitants' quality of life.

- Fog Nodes: A network of fog nodes that are carefully positioned at the centre of the fog computing architecture for urban mobility serves as the framework for the ecosystem as a whole. The essence of effective data processing, real-time analytics, and well-informed decision-making is embodied by these fog nodes, which are dispersed over the metropolitan region. Fog nodes transform urban transportation management by reducing latency, boosting responsiveness, and optimising resource use by putting computational capabilities closer to the data source. Fog nodes are deliberately placed across the metropolitan environment as small-scale computing supercomputers. These nodes are able to process data in real-time since they are furnished with processing units, lots of memory, and storage space. Fog nodes dramatically minimise the need to send massive volumes of data to far-off central data centres by running algorithms and analytics at the network's edge. By reducing latency through localised processing, it is possible to make timely decisions in response to shifting mobility dynamics and to ensure that crucial decisions are made quickly. Fog nodes' capacity to handle data more closely to its source, which allows for improved real-time capabilities, is one of their key advantages. Fog nodes can quickly analyse, evaluate, and turn the data coming in from edge devices into useful insights without having to wait for the data to be transmitted to distant servers. In urban transportation scenarios, where quick responses are crucial for managing congestion, rerouting vehicles, and optimising traffic flow, this skill is crucial. Making informed decisions depends on the processing power that fog nodes are able to harness. These nodes can recognise trends, forecast traffic jams, spot potential problems, and even simulate different scenarios by using complex algorithms and machine learning models [33].

Fog nodes are able to give wise recommendations that improve the effectiveness, security, and general standard of urban mobility because of their analytical prowess. Fog nodes are strategically placed around the metro region to guarantee that computational resources are used to their full potential. The task can be effectively split among several nodes, reducing the strain on any one of them and averting potential bottlenecks. This decentralised method not only provides fault tolerance but also increases scalability, guaranteeing the continuation of operations even in the event of node failures. The quick response times made possible by fog nodes are invaluable in the setting of urban mobility where split-second decisions can have significant effects. Traffic signal timings, suggested routes, and emergency response plans can all be modified in real-time thanks to proximity-based processing. Due to its agility, urban mobility has evolved into a dynamic ecology that responds quickly to environmental changes. A paradigm shift in urban mobility management can be seen in the idea of fog nodes. Fog nodes enable cities to take advantage of real-time insights, improve decision-making, and raise the overall effectiveness of mobility systems by placing localised computer resources closer to the data source. These nodes serve as the pivotal component that synchronises the integration of data from edge devices, accelerates its processing, and transforms it into useful intelligence that influences the urban mobility experience. Cities that adopt fog node capabilities do so in order to usher in a new era of proactive and responsive transportation management, paving the way for smarter and more adaptable cities in the future [34].

- Connectivity Infrastructure: The connectivity infrastructure acts as the crucial nervous system that coordinates the constant flow of data, insights, and instructions in the sophisticated web of fog computing architecture created for metropolitan mobility. In order to provide smooth communication between edge devices, fog nodes, and central control centres, this infrastructure links the ecosystem's many parts together. High-speed communication channels operate as the arteries and veins of this digital network, ensuring the rapid interchange of real-time data and facilitating proactive control of urban transportation. The entire fog computing system is supported by the connectivity infrastructure, which serves as its backbone. High-speed communication networks serve as the basis for this foundation by acting as data transfer conduits. Innovative technology, such as 5G networks, specialised communication protocols, or a combination of both, may be used by these channels. The objective is to create an effective lifeline for the entire architecture by creating a strong and dependable communication network that can handle the significant data flow produced by edge devices. Fundamentally, the connectivity infrastructure makes it possible for data to be seamlessly exchanged throughout the urban transportation ecosystem. Fog nodes receive real-time data gathered by edge devices and use it to process, analyse, and make decisions. The insights and optimisation techniques that have been analysed are sent back to the edge devices and central control hubs as part of this data exchange's bidirectional operation. This two-way exchange of data enables the system to work cohesively and supports the development of an ecosystem that can react to the dynamics of real-world movement.

  The communication infrastructure's ability to provide quick and continuous data flow is essential for making decisions quickly. Urban mobility is real-time, which necessitates quick responses to crises, shifting traffic patterns, and poor road conditions. High-speed connection makes it possible for central control centres to receive insights quickly drawn from processed data, enabling authorities to make well-informed decisions that optimise traffic flow, reduce congestion, and improve safety. The connectivity infrastructure acts as a channel for instructions and control signals in addition to transmitting data. The timing of traffic signals can be changed, vehicles can be rerouted, and emergency situations can be managed with the help of central control centres. To ensure that the entire ecosystem functions in sync and accordance with predetermined strategies, these instructions are transmitted to edge devices and fog nodes via the connection network. To meet the expectations for scalability in a dynamic urban context, the connectivity infrastructure must be strong. The network must be capable of managing increased data traffic as the number of edge devices, fog nodes, and central control centres rises. Additionally, the infrastructure needs to be resilient to guarantee continuous communication even in the event of outages or network congestion.

- Data Processing Layers: The data processing layers stand out as the engine room of the complex architecture of fog computing intended for metropolitan mobility. Here, raw data transforms into insightful knowledge that transforms urban transportation management. The enhancement of traffic flow, route planning, and the comprehensive management of urban mobility are the results of this multi-layered strategy within fog nodes, which is a dynamic process that involves a series of operations—data ingestion, data preparation, analytics, and decision-making. The data intake layer, which is the initial layer, serves as the point of entry for the flood of real-time data coming in from edge devices. These gadgets, which include sensors, cameras, GPS systems, and other features, are continually gathering information about the flow of traffic, the movements of vehicles, and environmental factors. The data ingestion layer combines this unprocessed information to provide a coherent stream that serves as the starting point for further processing. The raw data enters the data preparation layer after being aggregated. Several actions are performed during this stage in order to clean up and organise the data in preparation for analysis. Outliers, errors, and unrelated data are removed using cleaning and filtering methods [35].

In order to be compatible with later analytical procedures, raw data is transformed into a standardised format. At this point, data normalisation, transformation, and enrichment may also take place to prepare the data for future analysis. The analytics layer is activated by cleaned-up and organised data. Here, fog computing's transformational capacity assumes a prominent role. In order to uncover patterns, trends, and correlations that could otherwise go undetected, sophisticated algorithms, machine learning models, and statistical approaches are applied to the processed data. These analytics provide information on peak travel times, peak traffic conditions, vehicle movement patterns, and other important aspects that affect urban mobility. The foundation for well-informed decision-making in the urban transportation area is the insights obtained from the analytics layer. The system's intelligence is used to optimise traffic control tactics, route planning, and resource allocation at this last layer, decision-making. The knowledge gained from the previous layers is used to make real-time alterations to route suggestions, traffic signal timings, and even predictive simulations. The objective is to convert data-driven insights into actionable steps that improve the effectiveness, safety, and general standard of urban mobility. These layers work together to control urban transportation in a comprehensive manner. Fog nodes make sure that the insights produced are timely and pertinent by analysing data close to its source, at the edge. With the help of this proximity-based knowledge, quick responses may be made to changing traffic patterns, collisions, weather patterns, and other vital aspects that affect urban transportation.

3.5 REAL-TIME DATA COLLECTION AND PROCESSING

Real-time data gathering and processing are crucial pillars that support the optimisation of transport operations in the landscape of contemporary urban mobility. This constant stream of current information, gathered from cars, infrastructure sensors, and even people, presents a game-changing potential that reshapes how cities operate their transport networks. Fog computing is used to simplify this dynamic process, making it possible for accurate data to be collected, processed, and sent. This has the potential to completely change how cities manage their urban mobility. One cannot emphasise the value of real-time data. GPS, sensor, and communication technology-equipped vehicles offer information regarding traffic patterns, speeds, and congestion. Infrastructure sensors that are integrated into roadways, traffic signals, and public transportation hubs offer information on traffic flow, occupancy rates, and road conditions. Through smartphone apps, preference signals, and even incident reports, travellers themselves provide data. Insights that are important for streamlining transport operations are provided by this rich tapestry of real-time data, which functions as a live stream of data that reflects the pulse of urban mobility. To optimise transportation operations, real-time data is used as the basis. Cities can manage congestion, redirect vehicles to relieve bottlenecks, and dynamically change traffic signal timings by having a real-time understanding of traffic conditions and vehicle movements. This adaptability not only shortens commuters' journey times but also cuts down on fuel use and emissions, promoting a more sustainable urban environment. In order to improve safety, real-time data is also essential. Drivers may quickly receive information on accidents, road closures, and hazardous weather conditions, enabling them to make wise judgements and stay clear of potential dangers. Real-time data also enables the coordination of emergency response activities, allowing authorities to react to accidents or crises promptly and lessen their effects. Urban mobility is made more individualised by using real-time data gathered from users. Cities can offer customised transport options and route suggestions by learning about the interests of their citizens. In order to make decisions that will best serve their travel needs, passengers can receive real-time updates about public transportation timetables, anticipated delays, and other routes. For effective real-time data collection, processing, and distribution, fog computing serves as a disruptive facilitator. Fog computing lowers latency and quickens the processing of data by moving computational resources closer to the data source, be it passengers, sensors, or moving vehicles [36].

The local processing of data by fog nodes at the network's edge ensures that insights are produced quickly and without the delays brought on by transferring data to remote central servers. Based on current information, fog computing enables dynamic decision-making. Cities may quickly adapt their traffic management plans, travel schedules, and emergency response procedures to meet the demands of ever-changing circumstances. The management of urban mobility dynamics, which can quickly change, depends on this agility. Real-time data collection and processing are essential for optimisation in the field of managing urban mobility. Cities are able to comprehend the dynamics of mobility in real-time because of the collaboration of cars, infrastructure sensors, and user-generated data. Fog computing transforms this real-time data into usable intelligence, enabling decisions that improve efficiency, safety, and the general standard of urban movement. Cities that embrace the promise of real-time data and make use of fog computing's capabilities pave the way for an adaptive, responsive, and customised transportation system that meets the changing requirements of its citizens. The foundation of this revolutionary method is the collaboration of multiple data sources in urban settings. Vehicles with cutting-edge sensors and communication systems provide real-time data on traffic flow, speeds, and congestion. Road conditions, occupancy rates, and pedestrian movement are all tracked by infrastructure sensors that are integrated into roadways, traffic lights, and mass transport systems. A layer of personalisation and preference-based insights are added by passenger-generated data, which is collected from mobile apps and travel behaviour. When considered as a whole, these sources provide an in-depth picture of urban mobility, enabling cities to understand the ebb and flow of transport patterns. Real-time data's ultimate strength comes in its capacity to give cities the most recent information necessary to inform decision-making. Cities can dynamically alter the timing of traffic signals, put in place adaptive traffic management plans, and redirect traffic to lessen congestion by utilising real-time data. Authorities are also better equipped to react quickly to mishaps, bad weather, and other events that have an influence on mobility thanks to real-time data. In managing the intricate and fluid dynamics of urban transit, this decision-making agility is a game-changer. Actionable intelligence is made more likely by the interaction of real-time data and fog computing [37].

The localised processing capabilities of fog computing make sure that data is processed close to its source, reducing latency and facilitating quick analysis. By processing data in real-time at the network's edge, fog nodes enable quick insights that may be put to use right away. Real-time processing turns unstructured data into insightful patterns, trends, and forecasts that support effective operations and well-informed decision-making. Cities may significantly increase the effectiveness of their transport systems by using real-time data and fog computing. Travel times, gridlock, and fuel consumption are all decreased by adaptive traffic management, dynamic rerouting, and optimised signal timings. By reducing pollutants, this not only raises the standard of daily trips but also helps to preserve the environment. The ability to respond quickly to crises and tragedies is one way that real-time data and fog computing improve safety. To protect the safety of commuters and pedestrians, authorities can use these findings to swiftly allocate resources and modify traffic control techniques. Additionally, by adapting to real-time data, cities may improve the general standard of urban mobility by providing individualised services, streamlined transit options, and a more comfortable travel experience. In the end, real-time data and fog computing are paving the way for a transit system that meets the particular requirements of urban residents. Cities can build a dynamic environment that adjusts to the interests and behaviours of both locals and visitors by providing timely and pertinent information. This tailored approach not only makes things more convenient but also encourages involvement and enjoyment with the urban mobility experience.

3.6 TRAFFIC MANAGEMENT AND CONGESTION MITIGATION

Managing traffic and reducing congestion are the two biggest difficulties facing cities throughout the world in the area of urban mobility. Innovative solutions are needed to address the complicated traffic dynamics brought on by the expanding population and rising vehicle density. Through the analysis of real-time traffic data, the prediction of bottleneck sites, and the optimisation of traffic flow, fog computing offers a revolutionary way to tackle these problems. Along with increasing the effectiveness of urban transport systems, this synergy between fog computing and traffic management also increases the standard of living for city dwellers. The study of real-time traffic data is the cornerstone of efficient traffic management. An ongoing stream of data about traffic conditions, speeds, and patterns is produced by sensors installed in vehicles, infrastructure sensors, and user-generated data. Cities can quickly analyse this data and acquire a thorough picture of the current traffic situation by utilising the localised processing powers of fog computing. The capacity of fog computing to predict congestion points with astounding accuracy is one of its most revolutionary features. Fog nodes can find patterns that indicate oncoming congestion by examining both real-time inputs and historical traffic data. Congestion hotspots can be predicted using machine learning algorithms based on variables like the time of day, events, and weather. With the help of these forecasts, cities will be able to plan ahead and optimise traffic flow before congestion begins. Traffic flow can be dynamically changed thanks to real-time processing capabilities of fog computing. Fog nodes are able to instantly determine the best traffic signal timings by analysing data flowing in from incoming sensors on infrastructure and cars. In order to minimise stop-and-go traffic patterns and ease congestion, real-time optimisation makes sure that traffic signals adjust to the current traffic conditions [38].

Fog computing also enables a holistic approach to traffic optimisation by taking into account elements like road conditions, public transportation timetables, and pedestrian flow. Additionally, adaptive routing systems that redirect vehicles in response to shifting traffic circumstances are made possible by fog computing. Drivers can be directed to alternate routes by real-time data from vehicles and infrastructure sensors, avoiding congested regions and fostering a more efficient flow of traffic. In line with environmental objectives, this adaptability not only shortens travel times but also helps save on fuel and cut emissions. Fog computing aids in the optimisation of emergency responses in addition to regular traffic management. Fog nodes can immediately transmit information to emergency services in cases like accidents or road closures, allowing them to react quickly and efficiently. Additionally, fog computing can help to reroute traffic away from disaster situations, minimising interruptions and enhancing first responders' ability to do their jobs. Fog computing's incorporation into traffic management systems increases city dwellers' quality of life greatly while also improving transportation efficiency. It takes less time to commute, feels less stressful, and produces better air when there is less congestion. A more pleasant urban environment that promotes active transport options and supports sustainable urban expansion is a result of the traffic flow being expertly managed. A revolutionary solution to the problems of urban mobility is provided by the symbiotic interaction between fog computing and traffic management. Fog computing enables cities to develop smarter transportation systems that increase efficiency, decrease congestion, and enhance overall urban living quality by analysing real-time traffic data, forecasting congestion points, and optimising traffic flow. The foundation is being laid for a transport system that adapts dynamically to changing conditions, lays the path for sustainable expansion, and improves everyone's experience of urban mobility as cities embrace the possibilities of fog computing. The fusion of fog computing with traffic management has become a revolutionary force in the dynamic world of urban mobility, providing creative answers to the difficult problems that cities must deal with. As it uses the capabilities of fog computing to analyse real-time traffic data, forecast bottleneck points, and optimise traffic flow, this symbiotic connection has the potential to revolutionise how cities approach transportation. By using this strategy, communities are given the ability to develop ITS that increase productivity, reduce traffic, and improve urban life in general. Cities that embrace the promise of fog computing open the door to a transit system that is flexible, sustainable, and created with the needs of their citizens in mind. Real-time traffic data analysis is at the core of this transformational strategy. Fog computing steps in to process this data on the edge as a result of the continuous

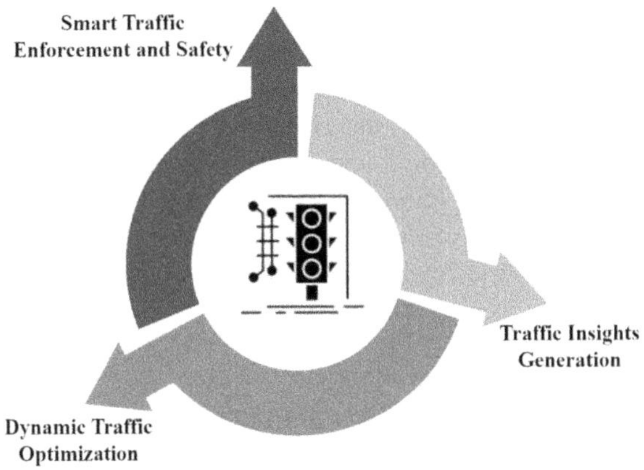

Figure 3.3 Traffic management and congestion mitigation.

stream of information provided by cars, infrastructure sensors, and user-generated data. Cities can use this study to gain a real-time understanding of traffic patterns, speeds, and congestion, allowing officials to make decisions that will immediately affect transportation operations. The predictive power of fog computing has the potential to revolutionise traffic control [39]. The traffic management and congestion mitigation are represented below in Figure 3.3.

Fog nodes can spot trends that indicate the beginning of congestion by carefully examining both real-time inputs and historical data. This capacity for prediction enables communities to take proactive efforts to reduce congestion before it happens, such as modifying the timing of traffic signals or suggesting alternate routes. The interruptions, travel times, and general effectiveness of urban transport networks are all improved by this foresight-driven strategy. By processing data from multiple sources at once, fog computing excels at improving traffic flow. Based on real-time data, this capability permits dynamic modifications to traffic signal timings. Fog nodes determine the most effective traffic light sequences based on data from vehicles and infrastructure sensors to reduce congestion and maintain a smooth flow of vehicles. With this optimisation technique, efficiency is improved while fuel use and emissions are decreased, which is in line with sustainability objectives. Given how quickly urban mobility conditions change, fog computing's dynamic nature is a good fit. Cities can react quickly to occurrences, mishaps, or unexpected changes in traffic patterns by analysing real-time data. Urban mobility networks are more dependable because of this responsiveness, which guarantees that transportation operations continue to run effectively even in the face of unexpected events. Using fog computing to manage traffic is in line with the ideas of sustainable urban development. Cities may promote the use of walking, cycling, and public transport by enhancing traffic flow and lowering congestion—modes that not only promote healthier urban environments but are also more sustainable. Fog computing's capabilities open the door to a transport environment that promotes sustainable growth while addressing the problems caused by growing urbanisation. In the end, a better urban mobility experience results from the symbiotic interaction between fog computing and traffic management. Cities improve the general standard of living for their citizens by streamlining traffic, reducing congestion, and offering real-time information to travellers. A more liveable and livelier urban environment is facilitated by shorter commutes, lower stress levels, and improved air quality.

3.7 CASE STUDIES AND IMPLEMENTATIONS

The revolutionary potential of fog computing in controlling urban transportation has been acknowledged by a number of cities and regions throughout the world. With the help of real-time data analysis, predictive insights, and improved traffic flow, these urban regions have taken on mobility difficulties head-on by using fog computing technologies. Let's look at a few notable case studies that demonstrate the benefits, difficulties, and lessons discovered from integrating fog computing in urban mobility management.

- **Case Study 1: Barcelona, Spain**
  Barcelona is now recognised as a pioneer in the application of fog computing to enhance urban transportation. To analyse real-time traffic data gathered from automobiles and infrastructure sensors, the city's "CityOS" project uses fog nodes and data analytics. The city has decreased travel times by 25% on average in crowded areas by forecasting congestion points and dynamically changing traffic light timings. This project not only increased transportation effectiveness but also helped to lower air pollution and improve people's quality of life in general. Integrating old infrastructure and protecting data privacy were issues. The success of the city highlights the significance of a strong framework for data governance and seamless system interaction.
- **Case Study 2: Singapore**
  Singapore's "Smart Mobility 2030" plan uses fog computing to handle its intricate urban transport system. Singapore has decreased the average travel time during peak hours by 15% by analysing real-time traffic data, forecasting traffic trends, and enhancing traffic signals. The significance of fog computing in improving traffic flow has also led to a 20% decrease in carbon emissions. Data fusion from many sources and assuring the fog computing infrastructure's scalability as the city's mobility requirements change over time were challenges. The experience of Singapore serves as a reminder of the value of cross-agency cooperation and long-term scaling planning.

- **Case Study 3: Columbus, Ohio, USA**
  By utilising fog computing to address mobility issues, Columbus launched its "Smart Columbus" initiative. The city was able to forecast traffic snarls and improve traffic control tactics by analysing real-time data from vehicles, sensors, and passengers. The effort reduced peak-hour traffic by 30% and encouraged more people to use public transit. Addressing issues with data protection and promoting cooperation among many stakeholders were challenges. A robust data governance structure and public-private collaborations are essential for ensuring openness and data protection, as demonstrated by Columbus' experience.

Fog computing has a transformative effect on controlling urban mobility, as demonstrated by the case studies of Barcelona, Singapore, and Columbus. The potential for less traffic jams, better transit efficiency, and a higher standard of living is highlighted by these real-world instances. Other cities can start their own adventures to use fog computing for smart, efficient, and responsive urban mobility management by learning from the difficulties and accomplishments of these projects.

3.8 FUTURE TRENDS AND CHALLENGES

Future-oriented trends and developments will likely lead to a striking alteration of the metropolitan mobility landscape. Urban transport may change as a result of the development of linked and autonomous vehicles and the continuous use of fog computing. Along with these encouraging advancements, there are obstacles that cities must overcome in order to successfully deploy fog computing technologies at scale. Urban mobility is about to undergo a transformation because of the rise of linked and autonomous vehicles. These vehicles will produce a vast amount of real-time data because of their sophisticated sensors and communication systems. In order to handle this data on the edge, ensure seamless communication between vehicles and infrastructure, and enable proactive traffic control, fog computing will be essential. With CAVs, safety will be improved, traffic will be reduced, and new opportunities for shared mobility models will be created. Urban mobility management has the potential to change with the widespread adoption of fog computing. There is a crucial condition for this revolutionary potential, though: a large infrastructural investment. Cities that want to take advantage of the possibilities of fog computing must set out on a mission to deploy fog nodes, build high-speed communication networks, and guarantee flawless interoperability with current systems. Future urban mobility that is intelligent and responsive will be developed on top of this investment [40].

The infrastructure for fog computing is made up mostly of fog nodes. In order to analyse data at the edge, reduce latency, and enable real-time analysis, these localised computing resources are placed strategically across the urban environment. When fog nodes are deployed, their placement must be carefully considered in order to guarantee coverage of crucial regions like crossroads, transit hubs, and areas prone to congestion. This deployment necessitates physical installation, the provision of a power supply, and network connectivity, all of which need an initial capital investment. The ability to seamlessly communicate data between fog nodes, edge devices, and central control centres is essential for fog computing to succeed. Fast data transmission and reaction times require high-speed communication networks, including 5G technology. Construction of the required infrastructure, installation of communication towers, and assurance of coverage throughout the metropolitan environment are all required for the establishment of these networks. The foundation for other smart city applications beyond mobility management is laid by this investment, which also improves the efficacy of fog computing. Fog computing needs to work in unison with current transportation technologies, traffic management strategies, and data sources in order to be effective. This compatibility involves careful design, software integration, and occasionally even upgrading legacy systems to support the new architecture. Investment in interoperability guarantees that fog computing may make use of information from diverse sources, offering thorough insights and dynamic decision-making that improve mobility management and various future trends as given in Figure 3.4.

Figure 3.4 Future trends.

Challenges:

- **Financial Commitment:** For cities wanting to maximise metropolitan mobility, investing heavily in infrastructure is a crucial first step. This is because fog computing solutions must be implemented at scale. But in order to acquire funds for setting up communication networks, deploying fog nodes, and overseeing ongoing maintenance and improvements that are crucial to the project's success, cities must manage the serious consequences and difficulties that come with this financial commitment. The deployment of communication infrastructure and the establishment of a network of fog nodes demand a large upfront financial commitment. The price covers all necessary labour and technical know-how for installation, configuration, and testing in addition to the hardware and equipment itself. The expenditure is influenced by the cost of acquiring the required technology elements, such as fog nodes, communication equipment, and sensors. The acquisition of these elements, which serve as the foundation of the ecosystem for fog computing, can be expensive. To ensure correct setup, configuration, and integration when deploying fog computing infrastructure, specialised people are required. A key part of the financial concerns is the hiring and training of staff that are knowledgeable in networking and fog computing technologies.

- **Scalability:** The scalability of fog computing infrastructure appears as a crucial factor in the dynamic environment of urban mobility management. The ability of fog computing systems to handle growing data volumes and extend coverage becomes crucial as metropolitan areas undergo growth and the demands of mobility change. To sustain the efficacy, responsiveness, and long-term viability of fog computing systems in addressing urban mobility issues, scalability is essential. Data from different sources, including automobiles, infrastructure sensors, and user-provided data, is constantly being generated in urban settings. The amount of real-time data created significantly rises with the growth of cities and the evolution of mobility patterns. New data sources could emerge as a result of the expansion of mobility, including Internet of Things (IoT) devices, smart city infrastructure, and novel forms of transportation. The complexity of urban transportation data is increasing, and this new data is a contributing factor. Fog computing is used for other smart city applications outside of traffic control. The scalability of fog computing is essential to support a variety of applications and use cases as cities adopt the idea of networked systems. The major challenges are shown below in Figure 3.5.

- **Coordination:** Effective collaboration across a wide range of stakeholders is essential for the successful implementation of a complete fog computing infrastructure for regulating urban traffic. The joint efforts

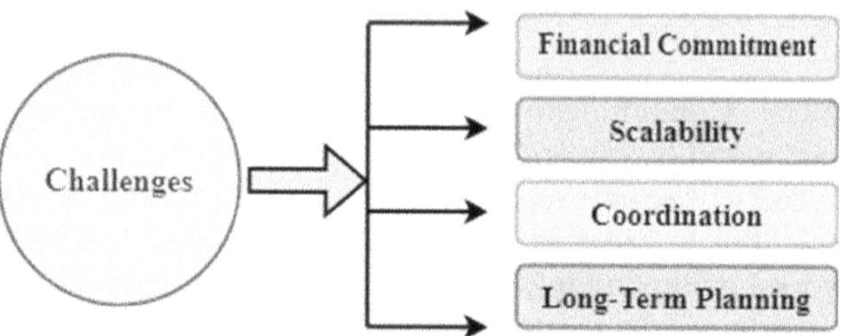

Figure 3.5 Challenges.

of these stakeholders, which include local governments, technology providers, and urban planners, are essential in creating a well-integrated, effective, and responsive urban mobility ecosystem that makes use of fog computing. Projects involving fog computing are often started, governed by, and supervised by municipalities and governmental entities. For the initiative to be successful, they supply the legislative foundation, financial support, and governance. The technological know-how and resources necessary to design, build, and deploy the infrastructure are offered by businesses that specialise in fog computing solutions. To ensure the system's functionality, they contribute the necessary hardware, software, and technological know-how. Urban planners have a thorough awareness of the layout, traffic patterns, and expected future expansion of the city. Their recommendations are essential for carefully placing fog nodes, identifying important traffic nodes, and seamlessly integrating the system into the urban fabric. The public and those who use the transit system are essential stakeholders. By including the public and soliciting feedback, it is ensured that the fog computing infrastructure meets the demands and expectations of the neighbourhood. The key to using fog computing in urban mobility management successfully is stakeholder coordination. Cities can develop a comprehensive and integrated fog computing ecosystem that optimises metropolitan mobility, decreases traffic, increases safety, and, in the end, improves the urban experience for all residents and visitors by fostering collaboration, coordinating goals, and utilising the particular strengths of each stakeholder.

- **Long-Term Planning:** The strategic deployment of fog computing infrastructure is a long-term project in the field of managing urban mobility, not merely a quick fix. To ensure that the systems implemented remain applicable, effective, and adaptive as technology changes and urban settings change, successful implementation requires rigorous long-term planning. To fully utilise the advantages of fog computing and future-proof urban mobility solutions, long-term planning is crucial. The technology industry is renowned for its quick development.

New hardware, communication protocols, and processing algorithms developed in tandem with the emergence of fog computing solutions. This dynamism is explained by long-term planning. Scalability, or the ability to add new parts and resources as technology advances, must be considered while designing systems. Making sure fog nodes can handle increased processing and data loads is part of this. Long-term planning is essential if fog computing is to be used effectively for managing urban mobility. Cities can make sure that their fog computing solutions stay not only relevant but also at the forefront of innovation by predicting technological progression, creating adaptable infrastructure, utilising data-driven insights, and encouraging collaboration among stakeholders. A well-thought-out fog computing strategy will enable cities to handle the potential and difficulties of tomorrow's metropolitan transportation landscape by having the flexibility to adapt, evolve, and align with changing urban dynamics. Figure 3.5 shows various challenges of smart vehicles using fog computing.

3.9 CONCLUSION

To increase the effectiveness of transport systems and lessen congestion, this data can be utilised to optimise transit routes, schedules, and capacity utilisation. The most recent technology, such as sensors, communication networks, and clever algorithms, for smart vehicles and public transportation are then introduced. The chapter also discusses a paradigm for managing urban mobility that uses fog computing to gather and analyse information about traffic flow, road conditions, and traveller behaviour. Transport routes, scheduling, and capacity utilisation are all optimised by the framework's clever algorithms and decision-making models. To enhance the user experience and promote sustainable forms of transport, the integration of smart payment systems and passenger information services is also explained. The use of fog computing in public transit and smart cars is also demonstrated with examples from the real world and case studies. The potential of fog computing as a crucial facilitator for effective and sustainable urban transport systems is highlighted in this book chapter, which is a useful resource for researchers, practitioners, and policymakers involved in the management of metropolitan mobility.

REFERENCES

[1] Tandon, R., Verma, A., & Gupta, P. K. (2022, November). Blockchain enabled vehicular networks: A review. In *2022 5th International Conference on Multimedia, Signal Processing and Communication Technologies (IMPACT)* (pp. 1–6). IEEE.

[2] Tandon, R., & Gupta, P. K. (2022, December). ACHM: An Efficient Scheme For Vehicle Routing Using ACO and hidden markov model. In *Artificial Intelligence and Data Science: First International Conference, ICAIDS 2021, Hyderabad, India, December 17–18, 2021*, Revised Selected Papers (pp. 169–180). Cham: Springer Nature Switzerland.

[3] Tandon, R., Verma, A., & Gupta, P. K. (2022, July). A secure framework based on nature-inspired optimization for vehicle routing. In *Advances in Computing and Data Sciences: 6th International Conference, ICACDS 2022, Kurnool, India, April 22–23, 2022*, Revised Selected Papers, Part I (pp. 74–85). Cham: Springer International Publishing.

[4] Verma, A., Tandon, R., & Gupta, P. K. (2022). TrafC-AnTabu: AnTabu routing algorithm for congestion control and traffic lights management using fuzzy model. *Internet Technology Letters*, 5(2), e309.

[5] Tandon, R., & Gupta, P. K. (2019). Optimizing smart parking system by using fog computing. In *Advances in Computing and Data Sciences: Third International Conference, ICACDS 2019, Ghaziabad, India, April 12–13, 2019*, Revised Selected Papers, Part II 3 (pp. 724–737). Springer Singapore.

[6] Tandon, R., & Gupta, P. K. (2021). SV2VCS: a secure vehicle-to-vehicle communication scheme based on lightweight authentication and concurrent data collection trees. *Journal of Ambient Intelligence and Humanized Computing*, 1–17.

[7] Farooqi, A. M., Alam, M. A., Hassan, S. I., & Idrees, S. M. (2022). A fog computing model for VANET to reduce latency and delay using 5G network in smart city transportation. *Applied Sciences*, 12(4), 2083.

[8] Suri, B., Taneja, S., Kumar, S., & Sumit. (2022). A sustainable energy efficient IoT-based solution for real-time traffic assistance using fog computing. *Energy conservation solutions for fog-edge computing paradigms*, 65–85.

[9] Malarvizhi, D., & Padmavathi, S. (2022, April). Mobility-aware application placement for obstacle detection in uam using fog computing. In *Evolution in Computational Intelligence: Proceedings of the 9th International Conference on Frontiers in Intelligent Computing: Theory and Applications (FICTA 2021)* (pp. 191–202). Singapore: Springer Nature Singapore.

[10] Behravan, K., Farzaneh, N., Jahanshahi, M., & Seno, S. A. H. (2023). A comprehensive survey on using fog computing in vehicular networks. *Vehicular Communications*, 100604.

[11] Rehman, M. A. U., Mastorakis, S., & Kim, B. S. (2023). FoggyEdge: An information-centric computation offloading and management framework for edge-based vehicular fog computing. *IEEE Intelligent Transportation Systems Magazine*.

[12] Hazra, A., Rana, P., Adhikari, M., & Amgoth, T. (2023). Fog computing for next-generation internet of things: fundamental, state-of-the-art and research challenges. *Computer Science Review*, 48, 100549.

[13] Das, R., & Inuwa, M. M. (2023). A review on fog computing: issues, characteristics, challenges, and potential applications. *Telematics and Informatics Reports*, 100049.

[14] Songhorabadi, M., Rahimi, M., MoghadamFarid, A., & Kashani, M. H. (2023). Fog computing approaches in IoT-enabled smart cities. *Journal of Network and Computer Applications*, 211, 103557.

[15] Siddiqui, E. F., & Nayak, S. K. (2023, March). Agitating sustainability using fog computing for smart cities. In *2023 10th International Conference on Computing for Sustainable Global Development (INDIACom)* (pp. 680–687). IEEE.

[16] Jai Vinita, L., & Vetriselvi, V. (2022). Impact of sybil attack on software-defined vehicular fog computing (SDVF) for an emergency vehicle scenario. In *Inventive Communication and Computational Technologies: Proceedings of ICICCT 2022* (pp. 809–825). Singapore: Springer Nature Singapore.

[17] Grønli, T. M., Lakhan, A., & Younas, M. (2023, August). RSITS: Road safety intelligent transport system in deep federated learning assisted fog cloud networks. In *International Conference on Mobile Web and Intelligent Information Systems* (pp. 20–37). Cham: Springer Nature Switzerland.

[18] Keshari, N., Singh, D., & Maurya, A. K. (2022). A survey on Vehicular Fog Computing: Current state-of-the-art and future directions. *Vehicular Communications*, 38, 100512.

[19] Yan, P., Carvalho, G. H., & De Grande, R. E. (2022, October). MDP-based Region-oriented Connectivity Estimation in Vehicular Fog Computing for SDVN. In *Proceedings of the 20th ACM International Symposium on Mobility Management and Wireless Access* (pp. 111–118).

[20] Tandon, R., Verma, A., & Gupta, P. K. (2023, April). Nature-inspired whale optimization technique for efficient information exchange in vehicular networks. In *2023 IEEE 12th International Conference on Communication Systems and Network Technologies (CSNT)* (pp. 833–838). IEEE.

[21] Tandon, R., & Gupta, P. K. (2023). A hybrid security scheme for inter-vehicle communication in content centric vehicular networks. *Wireless Personal Communications*, 129(2), 1083–1096.

[22] Tandon, R., & Gupta, P. K. (2021). A novel and secure hybrid iwd-mask algorithm for enhanced image security. *Recent Advances in Computer Science and Communications (Formerly: Recent Patents on Computer Science)*, 14(2), 384–394.

[23] Tandon, R., Verma, A., & Gupta, P. K. (2022, November). RVTN: Recommender system for vehicle routing in transportation network. In *2022 Seventh International Conference on Parallel, Distributed and Grid Computing (PDGC)* (pp. 352–356). IEEE.

[24] Islam, M. S. U., Kumar, A., & Hu, Y. C. (2021). Context-aware scheduling in Fog computing: A survey, taxonomy, challenges and future directions. *Journal of Network and Computer Applications*, 180, 103008.

[25] Kumar, A., Sharma, S., Goyal, N., Singh, A., Cheng, X., & Singh, P. (2021). Secure and energy-efficient smart building architecture with emerging technology IoT. *Computer Communications*, 176, 207–217.

[26] Kaur, M., Aron, R., Wadhwa, H., & Oo, H. N. (2023, April). CNN-based Smart Waste Management System in Fog Computing Environment. In *2023 IEEE 12th International Conference on Communication Systems and Network Technologies (CSNT)* (pp. 774–779). IEEE.

[27] Kaur, M., & Aron, R. (2022). An energy-efficient load balancing approach for fog environment using scientific workflow applications. In *Distributed computing and optimization techniques: Select proceedings of ICDCOT 2021* (pp. 165–174). Singapore: Springer Nature Singapore.

[28] Kaur, G., Bindal, R., Kaur, V., Kaur, I., & Gupta, S. (2023, June). Bitcoin and cryptocurrency technologies: A review. In *AIP Conference Proceedings* (Vol. 2782, No. 1). AIP Publishing.

[29] Harnal, S., Sharma, G., Malik, S., Kaur, G., Simaiya, S., Khurana, S., & Bagga, D. (2023). Current and Future Trends of Cloud-based Solutions for Healthcare. *Image Based Computing for Food and Health Analytics: Requirements, Challenges, Solutions and Practices: IBCFHA*, 115–136.

[30] Kaur, G., Goyal, S., & Kaur, H. (2021, December). Brief review of various machine learning algorithms. In *Proceedings of the International Conference on Innovative Computing & Communication (ICICC)*.

[31] Bindal, R., Sarangi, P. K., Kaur, G., & Dhiman, G. (2019). An approach for automatic recognition system for Indian vehicles numbers using k-nearest neighbours and decision tree classifier.

[32] Rehman, M. A. U., Mastorakis, S., & Kim, B. S. (2023). FoggyEdge: An information-centric computation offloading and management framework for edge-based vehicular fog computing. *IEEE Intelligent Transportation Systems Magazine*.

[33] Seno, Hosseini, & Amin, Seyed. (2023). A comprehensive survey on using fog computing in vehicular networks. *Vehicular Communications*.

[34] Kaur, G., Kaur, H., & Goyal, S. (2023). Correlation analysis between different parameters to predict cement logistics. *Innovations in Systems and Software Engineering*, 19(1), 117–127.

[35] Kaur, G., Choudhary, P., Sahore, L., Gupta, S., & Kaur, V. (2023). Healthcare: In the Era of Blockchain. In *AI and Blockchain in Healthcare* (pp. 45–55). Singapore: Springer Nature Singapore.

[36] Kaur, M., & Aron, R. (2022, February). Fog clustering-based architecture for load balancing in scientific workflows. In *Proceedings of International Conference on Computational Intelligence and Data Engineering: ICCIDE 2021* (pp. 213–221). Singapore: Springer Nature Singapore.

[37] Kaur, M., & Aron, R. (2021). Focalb: Fog computing architecture of load balancing for scientific workflow applications. *Journal of Grid Computing*, 19(4), 40.

[38] Wadhwa, H., & Aron, R. (2022). TRAM: Technique for resource allocation and management in fog computing environment. *The Journal of Supercomputing*, 78(1), 667–690.

[39] Wadhwa, H., & Aron, R. (2021, December). Fog computing framework for multi constraints-based resource utilization. In *2021 2nd International Conference on Computational Methods in Science & Technology (ICCMST)* (pp. 20–23). IEEE.

[40] Wadhwa, H., & Rajni, A. (2020). Energy based resource provisioning for IoT application in fog computing. *Journal of Natural Remedies*, 21(3), S1.

Chapter 4

Smart grid infrastructure with cloud/fog computing for sustainable development

Gaurav Goel
Chandigarh Engineering College-CGC, Landran, India
School of Computer Science, UPES, Dehradun, India

Rajeev Tiwari
IILM University, Greater Noida, India

Raja Kumar Murugesan
Taylor's University, Subang Jaya, Malaysia

Shilpi Harnal
Chandigarh University, Mohali, India

Shikha Saxena
SoHST, UPES, DDN, India

4.1 INTRODUCTION

With the growing urbanisation, rising energy demands, and environmental concerns, the development of sustainable solutions has become essential. The infrastructure of the traditional power system cannot keep up with the evolving needs of the modern world. To get around these challenges and build a sustainable energy future, the integration of smart grid infrastructure with cloud and fog computing technologies has emerged as a promising approach. A "smart grid," which employs cutting-edge technology to efficiently manage electricity generation, transmission, distribution, and consumption, is a sophisticated and intelligent power grid system. It uses a variety of digital sensors, smart metres, communication networks, and advanced analytics to increase the power grid's operational effectiveness, dependability, and sustainability [1, 2]. The way data and computational resources are managed and used has been completely transformed by cloud and fog computing technologies. The practice of providing on-demand computing services over the internet, which allows for the storing, processing, and analysis of enormous volumes of data in distant data centres, is known as cloud computing.

DOI: 10.1201/9781003494430-4

While fog computing brings computing and storage closer to the data source and enables real-time, low-latency data processing and decision-making, cloud computing extends its capabilities to the network edge [3, 4].

Cloud and fog computing technologies provide a number of important benefits when used with smart grid infrastructure. First, they make it possible for different energy sources to be seamlessly integrated into the power grid, including electric vehicle charging stations, energy storage devices, and renewable energy-producing systems. This integration encourages the use of sustainable energy sources and allows for the best possible use of the resources that are already available. Furthermore, big data created by smart grid devices and sensors may be collected and analysed thanks to cloud and fog computing. Utilising this data will provide useful insights into trends in energy usage, demand projections, and grid performance enhancement [5, 6]. Detecting anomalies, anticipating breakdowns, and enabling proactive maintenance are all possible with advanced analytics algorithms, which improve grid dependability and save downtime. Demand response initiatives and dynamic pricing models can be implemented thanks to the combination of cloud and fog computing with smart grid infrastructure. With the use of real-time energy consumption statistics, consumers can be encouraged to modify their usage habits during peak times, easing grid load and minimising the need for expensive infrastructure changes. This demand-side management strategy encourages energy conservation and efficiency, which eventually helps to maintain the overall viability of the power system. Cloud and fog computing also provides inherent scalability and flexibility in addition to these benefits. Due to the elastic nature of cloud computing, computational resources may be dynamically allocated based on demand, assuring efficient use and cheap cost [7, 8]. Fog computing's distributed architecture enables localised data processing, lowering latency and increasing reliability, especially in situations when quick decision-making is essential. However, there are several difficulties in integrating smart grid infrastructure with cloud and fog computing. Data transmission and storage security and privacy in cloud and fog settings are of the utmost importance. To protect against potential cyber threats and safeguard sensitive information, effective cybersecurity procedures must be in place. In addition, the wide variety of devices, protocols, and communication technologies present in the smart grid ecosystem may cause interoperability and standardisation problems. To successfully implement and run a unified smart grid infrastructure, various components must be integrated seamlessly and exchange data effectively.

The integration of cloud and fog computing technologies with smart grid infrastructure has enormous potential for ensuring sustainable development in the energy industry. The integration enables efficient resource use, improves grid dependability, encourages energy efficiency, and makes it easier to integrate renewable energy sources seamlessly [9, 10]. However, overcoming issues with security, privacy, interoperability, and standardisation is essential to maximising the advantages of this revolutionary strategy. The smart grid with cloud and fog computing is poised to revolutionise how we produce, distribute, and consume energy as technology develops and collaborative efforts continue.

4.2 RELATED WORK

Singhal et al. [11] present cloud and fog computing technologies to give a thorough framework for load balancing in smart grids. To reduce energy consumption and ensure effective resource utilisation, the proposed system intends to divide the workload among numerous computing nodes, including cloud data centres and fog nodes. The authors emphasise the significance of load balancing in the operation of smart grids since it is essential for preserving grid stability, reducing energy waste, and averting equipment failures.

Forcan and Maksimovi et al. [12] provide study on a cloud-fog-based method for monitoring smart grids. The authors discuss the necessity of effective smart grid infrastructure monitoring to guarantee dependable and effective electricity distribution. They suggest a cloud-fog-based strategy that makes use of cloud resources for computing and storage while using fog nodes adjacent to the grid devices for in-the-moment data processing. To facilitate effective monitoring, data collection, and analysis in smart grid scenarios, the authors create an architecture that combines cloud and fog computing.

Shahinzadeh et al. [13] go over how fog computing helps by supplying localised compute resources, lowering latency, and guaranteeing continuous operation even when there are network outages. The authors stress that fog computing has the capacity to handle dispersed real-time data processing and decision-making, improving the accuracy of blackout detection and response. The paper tests the proposed paradigm through simulations and experiments and shows that it is successful at quickly and precisely detecting blackouts. Comparing the results to conventional methods, the blackout detection times demonstrate a significant improvement. The authors also cover the possibility of integrating the suggested paradigm with the smart grid's current protection and control systems to improve their overall performance and reliability.

Saoud and Recioui et al. [14] based on a cloud-fog system, suggest a hybrid algorithm for load balancing in smart grids. In order to accomplish effective and adaptable load balancing, the method combines Round Robin and Active Monitoring techniques. Through simulations and experiments, the study shows how the suggested algorithm works effectively, resulting in improvements in response time, energy use, and resource utilisation. The report also discusses the suggested algorithm's scalability and fault tolerance. The hybrid algorithm can successfully handle workload changes and adapt to changing system conditions. The authors also go over fault tolerance techniques used in cloud and fog computing environments, like replication and fault recovery, which improve the load balancing system's overall dependability and resilience.

Nayak et al. [15] study how to make smart grids more resilient by using IoT and fog computing technologies. In order to improve real-time monitoring, decision-making, and control capabilities, the study emphasises the advantages of merging IoT devices with fog computing nodes. By addressing the difficulties and factors to be taken into account when adopting IoT and fog computing in smart grids and outlining their potential uses in enhancing resilience, the research adds to the body of knowledge.

Hussain and Beg et al. [16] examine the use of fog computing in smart grid topologies supported by IoT. The study emphasises how fog computing can improve response times, lower data latency, and boost system effectiveness as a whole. By outlining the essential elements and traits of fog computing in smart grids, addressing security and privacy issues, and examining prospective applications, the research adds to the body of knowledge. Researchers, business people, and politicians interested in utilising fog computing technologies to raise the efficiency and dependability of smart grid systems can greatly benefit from the study's findings. Future research can concentrate on assessing and improving fog computing architectures, looking into the fusion of AI and ML approaches, and tackling the issues with interoperability and standardisation in IoT-aided smart grids.

Mohanty et al. [17], utilising the GBO algorithm, suggest a resource management strategy for smart grids. The plan aims to increase system performance, decrease energy use, and maximise resource utilisation while addressing the difficulties associated with resource allocation and optimisation in smart grid contexts. The study adds to the body of knowledge by outlining the GBO algorithm as a practical optimisation method for managing smart grid resources. For researchers and practitioners interested in maximising resource allocation and utilisation in smart grid systems, the study's findings offer insightful information. Future studies can investigate the integration of other optimisation approaches, the use of the GBO algorithm in various smart grid scenarios, and the scalability and practical implementation implications of the suggested resource management approach.

Gupta and Gupta et al. [18] give a thorough analysis of green UAV-based fog computing, emphasising the impact of this new technology on the environment, its difficulties, and its potential for the future. The survey adds to the body of knowledge by illuminating the function of UAVs in fog computing, highlighting the significance of green practises, and highlighting areas for future research to optimise the energy use, resource distribution, and security of UAVs. Researchers, business people, and politicians interested in UAV-based fog computing and its potential to provide effective and sustainable data processing at the network edge can benefit from the survey's findings. Future research might concentrate on resolving the mentioned issues, creating designs for UAVs that are energy-efficient, improving resource allocation algorithms, and investigating new uses and use cases for green UAV-based fog computing.

Mallikarjuna et al. [19] examine how to use osmosis machine learning approaches to load balance healthcare duties in advanced technologies that integrate the smart grid. The study shows how work allocation and resource utilisation in healthcare systems may be optimised using osmosis machine learning. The study adds to the body of knowledge by outlining the advantages of incorporating smart grid capabilities and putting forth a fresh load-balancing strategy. Additionally covered in the study are the effects and advantages of incorporating smart grid capabilities into load-balancing healthcare duties. The solution adds to the sustainability and effectiveness of healthcare operations by taking energy consumption and renewable energy sources into account during the load-balancing process. The potential of this strategy to optimise work allocation in healthcare, lower costs, and enhance patient care is highlighted by the author.

4.3 CHALLENGES AND RESTRICTIONS IN THE ABSENCE OF A SMART GRID

The grid system confronts a number of challenges and restrictions in the absence of a smart grid. The key obstacles are as follows:

Lack of real-time visibility into electricity generation, distribution, and consumption in the traditional grid system [20, 21]. Limited control. Grid operators find it challenging to efficiently monitor and control the grid due to the lack of information. Due to their poor understanding of power flow, voltage levels, and system circumstances, maintaining grid stability and adjusting to demand changes can be difficult.

Utility providers use manual metre reading to track energy consumption and generate bills in the absence of a smart grid. This procedure takes a lot of time, is prone to mistakes, and doesn't give customers real-time usage data. Additionally, it prevents the development of dynamic pricing or time-of-use pricing plans, which can entice customers to utilise energy off-peak hours.

The conventional grid system was largely built to support centralised fossil fuel power generation. The grid has trouble integrating and regulating intermittent, decentralised renewable energy sources like solar and wind as their use grows. It is more challenging to forecast and balance the fluctuating production from renewable energy sources with the overall demand when there is no smart grid technology available.

Reduced efficiency and dependability: The traditional grid system has worse efficiency and reliability because it lacks cutting-edge monitoring, control, and automation technology. Due to their limited view into the state of the system, grid operators find it difficult to locate and fix problems, improve power flow, and reduce losses [22–24]. Voltage fluctuations, ineffective energy transfer, and extended downtime during outages might result from this.

Limited capacity for demand response: The grid system lacks efficient capacity for demand response in the absence of a smart grid. With the aid of demand response programmes, users can modify their energy usage in response to price signals or grid conditions. This adaptability helps lessen peak loads, balance supply and demand, and prevent blackouts. Implementing such programmes is challenging without the smart grid infrastructure, which enables real-time communication and collaboration.

Planning and growing the grid has certain difficulties. Planning for grid upgrades and expansions is difficult since the existing grid system lacks real-time data and analytics. Accurately predicting future demand increases, locating congested locations, and maximising grid investments are challenging tasks for grid operators. Grid planning consequently shifts from being proactive to being reactive, which could result in bottlenecks, dependability problems, and ineffective infrastructure growth.

The lack of a smart grid results in hurdles for the grid system in terms of visibility issues, manual metre reading, difficulty integrating renewable energy, decreased efficiency and dependability, restricted demand response capabilities, and challenges with grid planning and extension. By supplying real-time data, automation, advanced analytics, and control mechanisms, a smart grid implementation can assist in overcoming these difficulties and enable a more effective, dependable, and sustainable grid infrastructure [25, 26].

4.4 THE ROLE OF CLOUD COMPUTING INSIDE THE SMART GRID SYSTEM

By offering a scalable and adaptable infrastructure for data storage, processing, and analysis, cloud computing plays a significant part within the framework of the smart grid. In the context of the smart grid, the following are some important cloud computing features:

1. **Data Storage:** The smart grid produces enormous amounts of data from a variety of sources, including smart metres, sensors, and grid monitoring equipment. With the large storage capacity provided by cloud computing, local storage infrastructure is not required to securely store this data [27].
2. **Data Processing and Analytics:** Cloud systems make it possible to analyse and examine huge amounts of data from the smart grid. Utility companies are able to carry out complicated analytics like demand forecasting, load balancing, and problem detection by utilising the computing power and scalability of the cloud. Cloud-based analytics improve grid dependability, enable predictive maintenance, and optimise energy distribution.
3. **Real-time Monitoring and Control:** The smart grid infrastructure can be monitored and controlled in real-time thanks to cloud computing. It is possible to send data gathered by sensors and intelligent devices placed throughout the grid to the cloud for quick analysis. In order to ensure grid stability and effective energy management, grid operators

can use this information to monitor grid performance, spot anomalies, and take preventative action.

4. **Integration of Distributed Energy Resources (DERs)**: The management and integration of DERs, such as solar panels, wind turbines, and energy storage devices, are made possible by cloud computing. Utilities can monitor their resources' performance, improve how they operate, and instantly balance the supply and demand of electricity by linking these resources to the cloud [28].

5. **Energy Management and Demand Response**: programmes are made possible by cloud-based platforms, allowing users to modify their energy consumption in response to price signals or grid conditions. Utilities can efficiently monitor and coordinate these demand response programmes, ensuring efficient energy use and grid stability, by utilising cloud computing.

6. **Scalability and Cost Effectiveness**: Cloud computing aids the smart grid in terms of scalability and cost effectiveness. Depending on demand, utilities can increase or decrease their computing capacity, doing away with the need for costly on-site infrastructure. Utilities can manage peak loads, accommodate future expansion, and cut operational costs because of this flexibility.

In general, cloud computing strengthens the foundation of the smart grid by providing a solid and adaptable infrastructure for data management, analytics, control, and integration as shown in Figure 4.1. It improves the grid's intelligence, effectiveness, and dependability, enabling utilities to better distribute energy, cut carbon emissions, and meet the changing demands of contemporary energy systems.

The key elements of the smart grid system are shown in the diagram above. Smart metres and sensors gather information on the grid's status, renewable energy production, energy usage, and other factors. The cloud computing infrastructure receives this data via communication networks. The resources required for data storage, processing, and analytics are made available by cloud computing. It houses software and algorithms for several smart grid features, such as energy management, load balancing, defect detection, and demand forecasting. The system can manage massive volumes of data and adapt to changing requirements thanks to the cloud's scalability feature. The smart grid management systems are then sent the cloud's processed data and control signals. These systems use the analysed data to make defensible judgements and efficiently manage grid operations. This entails coordinating DERs, streamlining the distribution of energy, and enabling demand response initiatives [29, 30]. Overall, the smart grid system uses cloud computing as a central centre for data administration, analysis, and control. It makes it possible to use resources effectively, monitor systems in real-time, and make wise decisions, which enhance grid dependability, energy efficiency, and the incorporation of renewable energy sources.

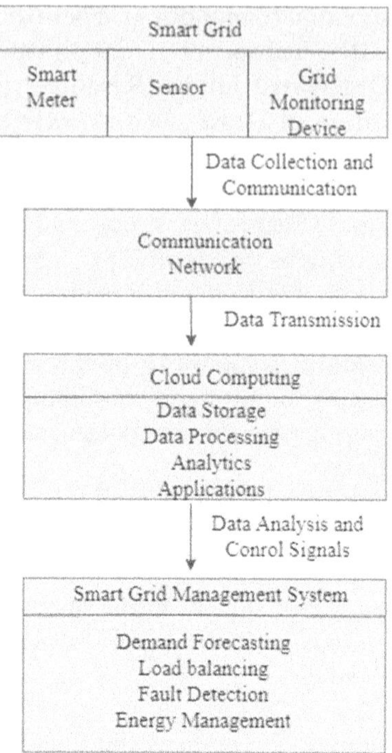

Figure 4.1 Cloud computing's function within the smart grid system.

4.5 FOG COMPUTING AS A SOLUTION TO THE ISSUES WITH CLOUD COMPUTING

Cloud computing is complemented by fog computing, often referred to as edge computing, which offers a potential solution for some of the problems with latency, privacy, and data sovereignty. The following is how fog computing resolves these problems:

1. **Reduced latency**: Fog computing moves processing and storage capabilities closer to the network's edge, speeding up round-trip data transfers. Fog computing reduces latency and enables real-time or almost real-time applications by enabling data processing and analysis to take place closer to where it is generated [31]. Low latency is crucial in time-sensitive applications like industrial automation, autonomous vehicles, and smart grid management; therefore, this is very advantageous.
2. **Protection of privacy**: Fog computing eliminates the need to transport sensitive data to the cloud by keeping it local to the point of origin. Fog computing allows data to be processed locally or within a close-by edge node rather than sending it all to the main cloud.

Due to the localised processing, there is less chance of unauthorised access and possible privacy violations because sensitive data is kept in a more secure setting [32]. Applications involving personal data, such as those in healthcare, banking, and smart homes, benefit greatly from this strategy.

3. **Data sovereignty:** By keeping data within the confines of a particular nation or organisation, fog computing helps address data sovereignty concerns. Data can be processed and stored locally using fog computing, guaranteeing compliance with data governance principles and local laws. This is especially important when data sovereignty laws mandate that sensitive data be kept within particular geographic restrictions. Fog computing gives users the freedom to manage data in a way that complies with legal standards while yet occasionally using cloud-based services.

4. **Hybrid cloud-fog architectures:** Organisations can make use of the advantages of both cloud and fog computing. This method enables flexible workload distribution based on particular needs [33, 34]. It is possible to transfer non-sensitive data and resource-intensive operations to the cloud, taking advantage of its scalability and powerful computing capacity. However, within the fog computing architecture, latency-sensitive and privacy-critical processes can be handled locally. An ideal balance between localised fog computing capabilities and centralised cloud services is provided by this combination.

It's vital to remember that fog computing enhances cloud computing rather than replacing it. By moving processing and storage closer to the edge, fog computing expands the potential of cloud computing and addresses issues with latency, privacy, and data sovereignty. The specific needs of the application and the intended balance between centralised and distributed computing resources determine whether cloud computing or fog computing should be used.

4.6 ARCHITECTURE FOR SMART GRID INFRASTRUCTURE WITH CLOUD/FOG COMPUTING

Designing a smart grid infrastructure with cloud and fog computing for sustainable development entails using cutting-edge technology to improve grid efficiency, boost renewable energy sources, and optimise energy distribution. Here is a suggested layout for such a system, as seen in Figure 4.2:

- **Install smart metres** at customer locations to gather data on energy consumption in real-time. Due to the two-way communication provided by these metres, customers can keep track of their energy usage and make informed decisions.

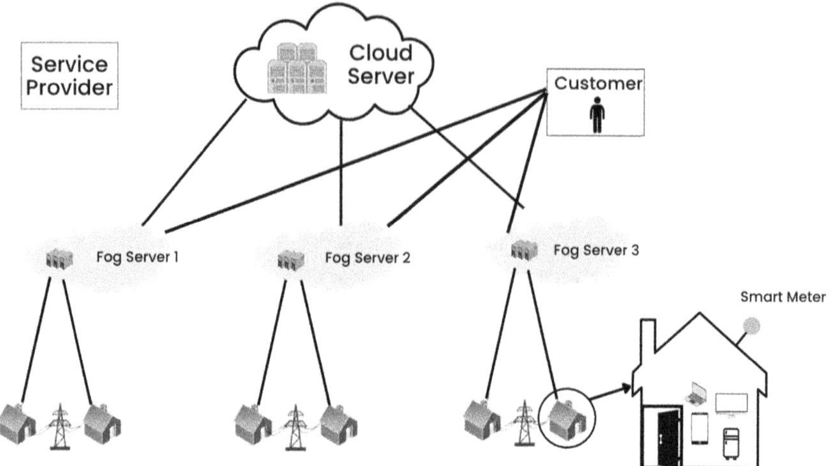

Figure 4.2 Architecture for smart grid infrastructure.

- **Implement an AMI system** to gather, manage, and analyse data from smart metres. Advanced Metering Infrastructure (AMI). Automated invoicing, load forecasting, and demand response programmes are made possible by AMI, allowing for effective energy use [35].
- **Communication Network:** To link smart metres, substations, and other grid components, and provide a reliable and secure communication network. Real-time monitoring and bidirectional data flow should be supported by this network.
- **Use cloud computing infrastructure** to store and handle the vast amounts of data that the smart grid generates. For load balancing, anomaly detection, and demand forecasting, cloud-based analytics can offer insights.
- **Fog Computing:** Use fog computing at the network's edge to process data in real-time and cut down on latency. Substations and distribution transformers can host fog nodes to enable localised analytics and control.
- **Distributed Energy Resources (DERs):** Integrate DERs into the smart grid, including energy storage units, wind turbines, and solar panels. These sources lessen dependency on fossil fuels while producing clean energy and assisting in meeting peak demand.
- **Demand Response (DR) Programmes:** Establish DR programmes that provide incentives for consumers to reduce their energy consumption during peak times. Automated DR systems that maximise load shedding and load shifting tactics can be made possible by cloud and fog computing.
- **Microgrids:** Create microgrids that run concurrently or independently of the main grid. In isolated places or during grid failures, microgrids powered by DERs can act as resilient and sustainable energy sources.

- **Implement strong cybersecurity** measures to safeguard the smart grid system from online dangers. This comprises intrusion detection systems, authentication methods, encryption, and routine security assessments.
- **Utilise artificial intelligence (AI) and data analytics** techniques to get insights from the gathered data. AI algorithms are able to detect system faults, optimise energy distribution, and find ways to conserve energy.
- **Integration with Smart Cities:** Integrate waste management, intelligent transportation, and other smart city technologies with the smart grid infrastructure. This integration makes it possible to approach sustainable development holistically and increases overall effectiveness.
- **Stakeholder Participation and the Regulatory Framework:** Create a supportive regulatory environment that promotes the use of smart grid technologies and rewards eco-friendly behaviour [36]. Encourage participation from all parties in the transition to a smart, sustainable grid, including utilities, consumers, and policymakers.

This suggested concept for a smart grid infrastructure combines cloud and fog computing with improved metering, DER integration, and intelligent analytics in order to maximise energy efficiency, improve grid resilience, and advance sustainable development.

4.7 CHALLENGES IN SMART GRID INFRASTRUCTURE WITH CLOUD/FOG COMPUTING

For sustainable growth, implementing smart grid infrastructure with cloud and fog computing presents a number of difficulties that must be overcome. Here are some significant obstacles:

1. **Privacy and security:** Protecting data's privacy and security is a major concern. Sensitive data, such as customer identity, energy consumption trends, and grid operation specifics, are gathered and sent as part of smart grid architecture. To defend against cyber-attacks, strong security measures must be put in place, such as encryption, access limits, and secure communication protocols. The implementation of procedures and guidelines that protect the privacy of personal data should also be done to address privacy issues.
2. **Scalability and interoperability:** A variety of components, systems, and stakeholders are involved in the construction of the smart grid infrastructure. To accommodate the growing number of connected devices, varied technologies, and various data formats, scalability and interoperability must be ensured [37]. To provide easy communication, data interchange, and interoperability between cloud, fog, and edge components, standardisation initiatives, open protocols, and seamless integration frameworks are required.

3. **Resilience and reliability:** Smart grids must be dependable and able to function in the face of disturbances. Additional reliance on network availability and cloud/fog infrastructure is introduced by cloud and fog computing [38]. To reduce downtime, guarantee continuous operations, and lessen the effects of failures or outages, reliable network architectures, redundancy mechanisms, and backup systems should be in place.

4. **Data management and analytics:** It can be difficult to manage and analyse the vast amounts of data produced by smart grid technologies. It is necessary to put into practise efficient data management procedures, including data processing, analysis, and storage. The essential infrastructure for handling and gaining insights from massive datasets is provided by cloud and fog computing. To successfully utilise this data and generate useful insights for energy optimisation, demand forecasting, and asset management, advanced analytics approaches like machine learning and AI can be used [39].

5. **Regulatory and policy frameworks:** Adequate regulatory and policy frameworks are necessary for the deployment of smart grid infrastructure with cloud and fog computing. To guarantee data privacy, consumer protection, and grid security, regulations must be put in place. Collaboration should be promoted, sustainable technologies should be adopted with incentives, and issues with data ownership, energy pricing, and market structures should be addressed.

6. **Stakeholder involvement and education:** Stakeholder involvement and education are essential for the successful installation of smart grid infrastructure. The creation and implementation of smart grid solutions require cooperation and active participation from utilities, customers, regulators, and technology vendors. It is essential for widespread adoption and acceptance to address concerns, raise awareness of the advantages, and offer stakeholders training and support [40].

7. **Energy Transition and Grid Integration:** The integration of renewable energy sources is greatly aided by smart grid architecture. However, the switch to a grid powered by renewable energy presents difficulties such as intermittency, grid stability, and supply and demand balancing. Through sophisticated forecasting, control of DR, and coordination of DERs, cloud and fog computing can assist in addressing these issues. However, to ensure seamless integration and best-practice utilisation of renewable energy sources, meticulous planning, grid design, and stakeholder cooperation are necessary.

In order to overcome these obstacles, lawmakers, regulators, utilities, technology providers, and customers must work together. Smart grid infrastructure with cloud and fog computing can support sustainable development by strengthening grid stability and resilience, facilitating the integration of renewable energy sources, and enhancing energy efficiency.

4.8 SOLUTION TO ISSUES WITH CLOUD/FOG COMPUTING AND SMART GRID INFRASTRUCTURE

You can take into account the following tactics to address the difficulties related to smart grid infrastructure with cloud/fog computing for sustainable development:

- **Normativity and Interoperability:** Encourage the creation and implementation of interoperability standards to facilitate smooth integration and communication between various cloud/fog computing platforms and components of the smart grid. To create uniform protocols and interfaces for data interchange, security, and system integration, encourage cooperation between industry players, standardisation bodies, and research institutes.
- **Data Privacy and Security:** Implement strong cybersecurity measures to safeguard sensitive data and guarantee user privacy in the smart grid ecosystem. Data privacy and security. To protect data transferred between smart grid devices, cloud/fog infrastructure, and other systems, use encryption techniques, access control mechanisms, and secure communication protocols. Regularly examine vulnerabilities and conduct security audits to find and resolve any risks and threats.
- **Scalability and Flexibility:** Create scalable and adaptable cloud/fog computing platforms and smart grid infrastructure that can support future development and shifting demand. To enable dynamic resource allocation and effective use of computational resources, use cloud-based virtualisation technologies and distributed computing architectures. Adopt architectures that are flexible and interoperable to enable smooth integration of new services and technologies [41].
- **Energy Efficiency and Optimisation:** Utilise cutting-edge analytics and optimisation methods to improve energy efficiency in cloud/fog computing operations and smart grid infrastructure. Create clever algorithms that balance loads, allocate resources, and reduce energy usage while taking grid stability, integration of renewable energy, and DR into account. Encourage the adoption of energy-saving technology and software, and implement energy management techniques to lessen the infrastructure's impact on the environment.
- **Cost Effectiveness:** To determine the economic viability and financial ramifications of building and sustaining smart grid infrastructure using cloud/fog computing, do cost-benefit assessments. Develop solutions to get over certain obstacles, like high initial investment expenses

or uncertain return on investment. Investigate novel commercial strategies, public–private collaborations, and incentive schemes to promote the use of sustainable smart grid technologies.

- **Stakeholder Engagement and Collaboration:** Encourage cooperation and dialogue amongst all parties, including end users, technology providers, utilities, and legislators. To ensure that the design and execution of smart grid infrastructure address the specific needs and concerns of many stakeholders, establish channels for continuing communication and knowledge exchange. To ensure the usability, acceptance, and satisfaction of the smart grid solutions, encourage end users' involvement and input [42].

- **Regulatory and Policy Support:** Encourage supporting laws and policies that provide incentives for the fusion of smart grid infrastructure with cloud/fog computing for sustainable development. Discuss any legal or regulatory obstacles preventing the adoption and implementation of creative smart grid solutions with regulatory authorities and lawmakers. Encourage the implementation of laws that support energy efficiency, grid modernisation, and the production of renewable energy in line with sustainable development objectives. By tackling these issues, it will be feasible to get past obstacles and build a cloud/fog-based smart grid system that is more effective, safe, and long-lasting. Implementation success and sustainable development goals depend on collaboration, creativity, and a multidisciplinary approach [43].

4.9 EFFECTS OF CLOUD AND FOG-ENABLED SMART GRIDS ON THE ECONOMY AND ENVIRONMENT

1. **Economic Impacts**

   - Cloud/fog-enabled smart grids result in cost reductions and operational efficiency.
   - Infrastructure cost reduction using virtualisation and shared resources.
   - Enhanced grid management and dependability resulting in savings for utilities and customers.
   - Economic repercussions for participants in the energy market (producers, consumers, aggregators, etc.).
   - The implementation of cloud/fog-enabled smart grids will lead to the development of jobs and economic prosperity [44].

2. **Environment-Related Effects**

   - Lowering greenhouse gas emissions by maximising energy production and use.

- Using renewable energy sources and the advantages they have for the environment.
- Techniques for DR and load management to reduce peak loads and maintain grid stability.
- Electric car and energy storage system grid integration for emissions reduction.
- Benefits to the environment of resource optimisation and virtualisation in cloud/fog computing [45].

3. **Cost-Benefit Evaluation**

- Techniques for evaluating the economic viability of smart grids that use the cloud or fog.
- Evaluation of the immediate costs and future gains.
- Demand-side management, energy efficiency, and load balancing all result in cost savings.
- Calculating the financial value of environmental benefits (e.g., carbon price, emissions trading).
- Case studies and actual cost-benefit analyses for cloud/fog-enabled smart grids are provided.

4. **Considerations for Policy and Regulation:**

- Regulations and policy frameworks encouraging the use of smart grids with cloud/fog capabilities.
- Energy efficiency and renewable energy integration incentives and subsidies [46].
- Regulatory obstacles and constraints must be removed for wider adoption.
- Collaborations and public–private partnerships to advance economic and environmental goals.

5. **Social Consequences:**

- Community socioeconomic advantages of cloud/fog-enabled smart grids.
- Concerns for accessibility and affordable energy.
- Aspects of equity and inclusion in the use of smart grid technologies.
- Using real-time data and information, consumers can be engaged and empowered [47].

6. **Future Challenges and Directions:**

- Possibility of scaling up smart grids with cloud/fog support while tackling scalability issues.
- Issues with standardisation and interoperability for seamless integration.
- Technological developments and innovations to strengthen the effects on the economy and the environment.
- Policy suggestions and tactics to maximise positive effects on the economy and environment [48].

4.10 CASE STUDIES ON SMART GRID INFRASTRUCTURE IN VARIOUS GLOBAL REGIONS

There have been several successful case studies that demonstrate the integration of cloud and fog computing into smart grid infrastructure in various locations worldwide. Here are a few examples:

- **Pacific Gas and Electric (PG&E) Case Study – California, USA:** One of the biggest utilities in the US, PG&E, used cloud computing to improve the efficiency of its smart grid. In their service area, smart metres were installed, and they set up an AMI to collect data in real-time from those devices. on order to learn more about energy consumption trends, grid performance, and DR management, this data was processed and analysed on the cloud. PG&E was able to use cloud computing to streamline grid operations, enhance load balancing, and better integrate renewable energy sources.
- **Case Study: Canada's Smart Grid in Nova Scotia:** The Canadian province of Nova Scotia's grid infrastructure was modernised as part of the Smart Grid Nova Scotia initiative. To enable localised data processing and control at the grid edge, the project made use of fog computing. They decreased data flow to the central cloud and enhanced latency for real-time applications by deploying intelligent edge devices and fog nodes throughout the grid. Fog computing enhanced the stability and resilience of the grid by enabling effective load management, voltage regulation, and defect detection at the edge.
- **Case Study: Smart Grid Demonstration Project in Fujisawa, Japan:** Cloud and fog computing are included in the smart grid infrastructure of a residential community in Japan named Fujisawa Sustainable Smart Town. The initiative intends to expand DR capabilities, incorporate renewable energy sources, and boost energy efficiency. Data from smart metres, solar panels, and energy storage devices were gathered and evaluated using cloud computing to enable intelligent energy management and optimisation. Fog computing was utilised for localised control and real-time monitoring of energy output and consumption at the neighbourhood level. Effective energy distribution was made possible by merging cloud and fog computing. Additionally, peak demands were reduced, and the use of renewable energy was promoted.
- **Denmark Future Grid Project Case Study:** The development of an adaptable and intelligent electricity grid was the aim of the Danish project Future Grid. The project's objectives were successfully completed in large part because of cloud and fog computing. For grid-wide energy flow optimisation and centralised data analytics, cloud systems were used. To enable real-time decision-making and control at the distribution level, fog computing was developed. The

project improved load balancing, strengthened the integration of renewable energy sources, and effectively used DERs by utilising fog computing.

These case studies demonstrate how the integration of cloud and fog computing can provide the smart grid infrastructure with substantial advantages, such as improved energy management, real-time monitoring, effective load balancing, and efficient integration of renewable energy sources. Each place exhibits the effective application of cloud and fog computing technologies adapted to their own smart grid needs.

4.11 CONCLUSION

In conclusion, the combination of cloud and fog computing with smart grid infrastructure offers a game-changing possibility for sustainable growth. Demand response, integration of renewable energy, effective energy management, and the expansion of decentralised energy systems are all made possible. Smart grids can lessen their negative effects on the environment, increase grid dependability, and promote a more sustainable energy future by utilising cutting-edge technologies. Cloud and fog computing also improve grid dependability and resilience. Cloud platforms' redundancy and scalability, together with the distribution of computing and storage resources across fog nodes, increase the resilience of smart grid systems against failures and interruptions. As a result, the grid is better able to survive events like natural disasters and cyber-attacks while still reliably supplying electricity to customers.

Finally, the shift to smart cities is aided by the integration of smart grid infrastructure with cloud and fog computing. Cloud and fog computing help to create more sustainable and effective urban settings by facilitating seamless communication and integration between diverse infrastructure systems, including energy, transportation, and buildings. Smart grids are a crucial component of smart cities because they enable better energy management, better traffic flow, and higher resident quality of life.

REFERENCES

1. Liu, J., Hu, H., Yu, S. S., & Trinh, H. (2023). virtual power plant with renewable energy sources and energy storage systems for sustainable power grid-formation, control techniques and demand response. *Energies*, 16(9), 3705.
2. Rahmani, A. M., Gia, T. N., Negash, B., Anzanpour, A., Azimi, I., Jiang, M., & Liljeberg, P. (2018). Exploiting smart e-Health gateways at the edge of healthcare Internet-of-Things: A fog computing approach. *Future Generation Computer Systems*, 78, 641–658.

3. Gilbert, G. M., Naiman, S., Kimaro, H., & Bagile, B. (2019). A critical review of edge and fog computing for smart grid applications. In *Information and Communication Technologies for Development. Strengthening Southern-Driven Cooperation as a Catalyst for ICT4D: 15th IFIP WG 9.4 International Conference on Social Implications of Computers in Developing Countries, ICT4D 2019*, Dar es Salaam, Tanzania, May 1–3, 2019, Proceedings, Part I 15 (pp. 763–775). Springer International Publishing.

4. Zahoor, S., Javaid, N., Khan, A., Ruqia, B., Muhammad, F. J., & Zahid, M. (2018, June). A cloud-fog-based smart grid model for efficient resource utilization. In *2018 14th International Wireless Communications & Mobile Computing Conference (IWCMC)* (pp. 1154–1160). IEEE.

5. Kumari, A., Tanwar, S., Tyagi, S., Kumar, N., Obaidat, M. S., & Rodrigues, J. J. (2019). Fog computing for smart grid systems in the 5G environment: Challenges and solutions. *IEEE Wireless Communications*, 26(3), 47–53.

6. Muzakkir Hussain, M., Saad Alam, M., & Sufyan Beg, M. M. (2021). Fog computing for smart grid transition: requirements, prospects, status quos, and challenges. In *2nd EAI International Conference on Big Data Innovation for Sustainable Cognitive Computing: BDCC 2019* (pp. 47–61). Springer International Publishing.

7. Siddiqui, E. F., & Nayak, S. K. (2023, March). Agitating sustainability using fog computing for smart cities. In *2023 10th International Conference on Computing for Sustainable Global Development (INDIACom)* (pp. 680–687). IEEE.

8. Zahmatkesh, H., & Al-Turjman, F. (2020). Fog computing for sustainable smart cities in the IoT era: Caching techniques and enabling technologies-an overview. *Sustainable Cities and Society*, 59, 102139.

9. Alsokhiry, F., Annuk, A., Mohamed, M. A., & Marinho, M. (2023). An innovative cloud-fog-based smart grid scheme for efficient resource utilization. *Sensors*, 23(4), 1752.

10. Venkadeshan, R., & Jegatha, M. (2021). Blockchain-based fog computing model (BFCM) for IoT smart cities. In *Convergence of internet of things and blockchain technologies* (pp. 77–92). Cham: Springer International Publishing.

11. Singhal, S., Athithan, S., Alomar, M. A., Kumar, R., Sharma, B., Srivastava, G., & Lin, J. C. W. (2023). Energy Aware Load Balancing Framework for Smart Grid Using Cloud and Fog Computing. *Sensors*, 23(7), 3488.

12. Forcan, M., & Maksimović, M. (2020). Cloud-fog-based approach for smart grid monitoring. *Simulation Modelling Practice and Theory*, 101, 101988.

13. Shahinzadeh, H., Mahmoudi, A., Gharehpetian, G. B., Muyeen, S. M., Benbouzid, M., & Kabalci, E. (2022, January). An agile black-out detection and response paradigm in smart grids incorporating iot-oriented initiatives and fog-computing platform. In *2022 International Conference on Protection and Automation of Power Systems (IPAPS)* (Vol. 16, pp. 1–8). IEEE.

14. Saoud, A., & Recioui, A. (2022). Hybrid algorithm for cloud-fog system based load balancing in smart grids. *Bulletin of Electrical Engineering and Informatics*, 11(1), 477–487.

15. Nayak, J., Mishra, M., Pelusi, D., & Naik, B. (2022). Internet of things and fog computing application to improve the smart-grid resiliency. In *Electric Power Systems Resiliency* (pp. 213–229). Academic Press.

16. Hussain, M. M., & Beg, M. S. (2019). Fog computing for internet of things (IoT)-aided smart grid architectures. *Big Data and Cognitive Computing*, 3(1), 8.

17. Mohanty, A., Samantaray, S., Patra, S. S., Mahmoud, A. I., & Barik, R. K. (2021, March). An efficient resource management scheme for smart grid using GBO algorithm. In *2021 International Conference on Emerging Smart Computing and Informatics (ESCI)* (pp. 593–598). IEEE.

18. Gupta, A., & Gupta, S. K. (2022). A survey on green unmanned aerial vehicles-based fog computing: Challenges and future perspective. *Transactions on Emerging Telecommunications Technologies*, 33(11), e4603.

19. Mallikarjuna, B. (2022). Osmosis machine learning load balancing of healthcare tasks in cutting edge technologies with smart grid. *International Journal of Smart Grid and Green Communications*, 2(2), 150–169.

20. Nazir, S., Shafiq, S., Iqbal, Z., Zeeshan, M., Tariq, S., & Javaid, N. (2019). Cuckoo optimization algorithm-based job scheduling using cloud and fog computing in smart grid. In *Advances in Intelligent Networking and Collaborative Systems: The 10th International Conference on Intelligent Networking and Collaborative Systems (INCoS-2018)* (pp. 34–46). Springer International Publishing.

21. Ali, M. J., Javaid, N., Rehman, M., Sharif, M. U., Khan, M. K., & Khan, H. A. (2019). State based load balancing algorithm for smart grid energy management in fog computing. In *Advances in Intelligent Networking and Collaborative Systems: The 10th International Conference on Intelligent Networking and Collaborative Systems (INCoS-2018)* (pp. 220–232). Springer International Publishing.

22. Anawar, M. R., Wang, S., Azam Zia, M., Jadoon, A. K., Akram, U., & Raza, S. (2018). Fog computing: An overview of big IoT data analytics. *Wireless Communications and Mobile Computing*, 2018.

23. Saroa, M. K., & Aron, R. (2018, December). Fog computing and its role in development of smart applications. In *2018 IEEE Intl Conf on Parallel & Distributed Processing with Applications, Ubiquitous Computing & Communications, Big Data & Cloud Computing, Social Computing & Networking, Sustainable Computing & Communications (ISPA/IUCC/BDCloud/SocialCom/SustainCom)* (pp. 1120–1127). IEEE.

24. Rani, S. (2022). Analytic vision on fog computing for effective load balancing in smart grids. *Transactions on Emerging Telecommunications Technologies*, 33(2), e3855.

25. Chouikhi, S., Esseghir, M., & Merghem-Boulahia, L. (2022). Energy consumption scheduling as a fog computing service in smart grid. *IEEE Transactions on Services Computing*, 16(2), 1144–1157.

26. Tom, R. J., & Sankaranarayanan, S. (2017, June). IoT based SCADA integrated with Fog for power distribution automation. In *2017 12th Iberian Conference on Information Systems and Technologies (CISTI)* (pp. 1–4). IEEE.

27. Bukhsh, R., Javaid, N., Ali Khan, Z., Ishmanov, F., Afzal, M. K., & Wadud, Z. (2018). Towards fast response, reduced processing and balanced load in fog-based data-driven smart grid. *Energies*, 11(12), 3345.

28. Hussain, M. M., Alam, M. S., & Beg, M. S. (2018). Computational viability of fog methodologies in IoT enabled smart city architectures-a smart grid case study. *EAI Endorsed Transactions on Smart Cities*, 3(7), 1–12.

29. Jain, S., Gupta, S., Sreelakshmi, K. K., & Rodrigues, J. J. (2022). Fog computing in enabling 5G-driven emerging technologies for development of sustainable smart city infrastructures. *Cluster Computing*, 25(2), 1–44.

30. Abbas, H., Shaheen, S., Elhoseny, M., Singh, A. K., & Alkhambashi, M. (2018). Systems thinking for developing sustainable complex smart cities based on self-regulated agent systems and fog computing. *Sustainable Computing: Informatics and Systems*, 19, 204–213.

31. Yasmeen, A., Javaid, N., Rehman, O. U., Iftikhar, H., Malik, M. F., & Muhammad, F. J. (2018, June). Efficient resource provisioning for smart buildings utilizing fog and cloud based environment. In *2018 14th International Wireless Communications & Mobile Computing Conference (IWCMC)* (pp. 811–816). IEEE.

32. Abd Elaziz, M., Abualigah, L., & Attiya, I. (2021). Advanced optimization technique for scheduling IoT tasks in cloud-fog computing environments. *Future Generation Computer Systems*, 124, 142–154.

33. Jiao, Y., Wang, P., Niyato, D., & Suankaewmanee, K. (2019). Auction mechanisms in cloud/fog computing resource allocation for public blockchain networks. *IEEE Transactions on Parallel and Distributed Systems*, 30(9), 1975–1989.

34. Fatima, A., Javaid, N., Waheed, M., Nazar, T., Shabbir, S., & Sultana, T. (2019). Efficient resource allocation model for residential buildings in smart grid using fog and cloud computing. In *Innovative Mobile and Internet Services in Ubiquitous Computing: Proceedings of the 12th International Conference on Innovative Mobile and Internet Services in Ubiquitous Computing (IMIS-2018)* (pp. 289–298). Springer International Publishing.

35. Murtaza, F., Akhunzada, A., Ul Islam, S., Boudjadar, J., & Buyya, R. (2020). QoS-aware service provisioning in fog computing. *Journal of Network and Computer Applications*, 165, 102674.

36. Kishor, A., & Chakarbarty, C. (2021). Task offloading in fog computing for using smart ant colony optimization. *Wireless Personal Communications*, 1–22.

37. Hussein, M. K., & Mousa, M. H. (2020). Efficient task offloading for IoT-based applications in fog computing using ant colony optimization. *IEEE Access*, 8, 37191–37201.

38. Hossain, M. R., Whaiduzzaman, M., Barros, A., Tuly, S. R., Mahi, M. J. N., Roy, S., ... & Buyya, R. (2021). A scheduling-based dynamic fog computing framework for augmenting resource utilization. *Simulation Modelling Practice and Theory*, 111, 102336.

39. Yin, Z., Xu, F., Li, Y., Fan, C., Zhang, F., Han, G., & Bi, Y. (2022). A multi-objective task scheduling strategy for intelligent production line based on cloud-fog computing. *Sensors*, 22(4), 1555.

40. Chafi, S. E., Balboul, Y., Mazer, S., Fattah, M., & El Bekkali, M. (2022). Resource placement strategy optimization for smart grid application using 5G wireless networks. *International Journal of Electrical and Computer Engineering*, 12(4), 3932.

41. Alazeb, A., Panda, B., Almakdi, S., & Alshehri, M. (2021). Data integrity preservation schemes in smart healthcare systems that use fog computing distribution. *Electronics*, 10(11), 1314.

42. Zakria, M., Javaid, N., Ismail, M., Zubair, M., Asad Zaheer, M., & Saeed, F. (2019). Cloud-fog based load balancing using shortest remaining time first optimization. In *Advances on P2P, Parallel, Grid, Cloud and Internet Computing: Proceedings of the 13th International Conference on P2P, Parallel, Grid, Cloud and Internet Computing (3PGCIC-2018)* (pp. 199–211). Springer International Publishing.
43. Mendes, D. L., Veloso, A. F. D. S., Rabêlo, R. A., Rodrigues, J. J., & dos Reis Junior, J. V. (2020, September). An adaptive data compression mechanism for Wireless Sensor Networks in the Smart Grid Scenarios. In *2020 5th International Conference on Smart and Sustainable Technologies (SpliTech)* (pp. 1–6). IEEE.
44. Suleiman, H. (2022). A cost-aware framework for QoS-based and energy-efficient scheduling in cloud–fog computing. *Future Internet*, 14(11), 333.
45. Zhao, D., Zou, Q., & Boshkani Zadeh, M. (2022). A QoS-aware IoT service placement mechanism in fog computing based on open-source development model. *Journal of Grid Computing*, 20(2), 12.
46. Mani, N., Singh, A., & Nimmagadda, S. L. (2020). An IoT guided healthcare monitoring system for managing real-time notifications by fog computing services. *Procedia Computer Science*, 167, 850–859.
47. Jabeen, N., Hao, R., Niaz, A., Shoukat, M. U., Niaz, F., & Khan, M. A. (2022, December). Autonomous Vehicle Health Monitoring Based on Cloud-Fog Computing. In *2022 International Conference on Emerging Trends in Electrical, Control, and Telecommunication Engineering (ETECTE)* (pp. 1–6). IEEE.
48. Gagandeep, Kaur J., Mathur, S., Kaur, S., Nayyar, A., & Singh, S.P., & Mathur, S. (2023). Evaluating and mitigating gender bias in machine learning based resume filtering. *Multimedia Tools and Applications*, 1–21.

Chapter 5

Smart agriculture with cloud/fog computing and sensor networks for sustainable development

Monika Choudhary and Mayank Arora
Seth Jai Parkash Mukand Lal Institute of Engineering and Technology, Radaur, India

Nitin Goyal
Central University of Haryana, Mahendergarh, India

5.1 INTRODUCTION TO SMART AGRICULTURE

Smart agriculture is an innovative approach that leverages cloud/fog computing, sensor networks, and mobile technologies to enhance sustainable development in the agricultural sector. It combines advanced technologies and data-driven techniques to optimize farming practices, improve resource management, and ensure environmental sustainability [1].

In this chapter, we will explore the transformative potential of smart agriculture and its significance in addressing the challenges faced by the agriculture industry. We will delve into the concepts and principles behind cloud/fog computing, sensor networks, and mobile technologies, and their integration into agricultural systems (Figure 5.1).

The chapter will cover various topics, including the benefits of adopting smart agriculture practices, precision farming techniques enabled by sensor networks, real-time monitoring and data analytics in agricultural systems, and the role of mobile devices in facilitating connectivity and information access in the field. We will also discuss the importance of cloud/fog computing in handling the massive amounts of data generated by smart agriculture and its role in enabling scalable and efficient farming operations.

Moreover, we will explore case studies and success stories that highlight the implementation of smart agriculture solutions in real-world scenarios. These examples will showcase the positive impact of smart agriculture in improving crop yields, resource efficiency, and overall sustainability.

By the end of this chapter, readers will gain a comprehensive understanding of how the convergence of cloud/fog computing, sensor networks, and mobile technologies is revolutionizing the agricultural sector, paving the way for a more sustainable and productive future in farming.

DOI: 10.1201/9781003494430-5

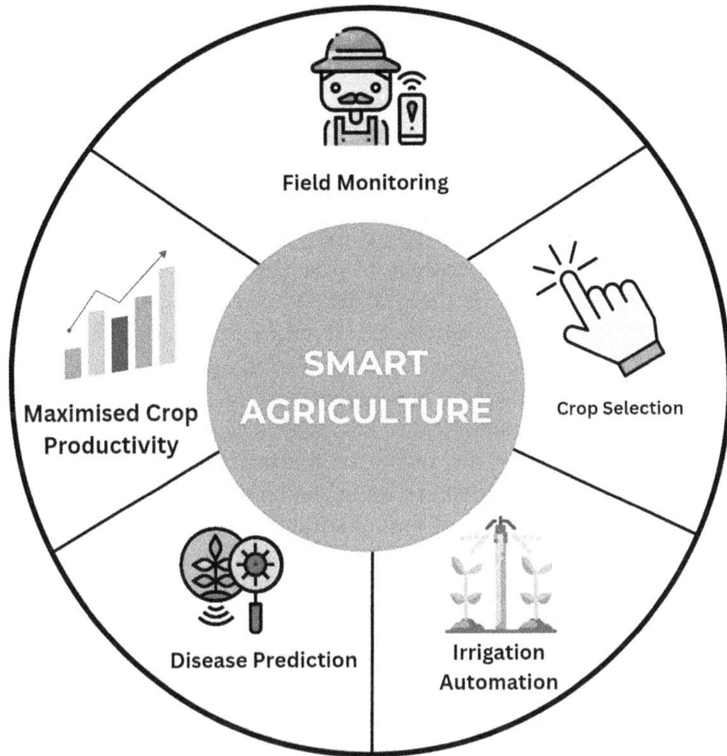

Figure 5.1 Smart agriculture.

5.2 CHALLENGES AND OPPORTUNITIES IN AGRICULTURAL SUSTAINABILITY

Agricultural sustainability is a critical focus area as we strive to meet the growing demand for food production while preserving the environment and ensuring the well-being of future generations. This section discusses the challenges and opportunities associated with achieving agricultural sustainability and explores how advancements in technology, such as cloud/fog computing and sensor networks, can contribute to sustainable agricultural practices.

One of the key challenges in agricultural sustainability is the efficient use of natural resources, including water, land, and energy. Increasing population and climate change put pressure on these resources, making it crucial to optimize their usage. Cloud/fog computing and sensor networks offer innovative solutions to monitor and manage resource consumption in real-time. For example, soil moisture sensors can provide accurate data on irrigation needs, enabling farmers to optimize water usage and prevent water wastage.

Similarly, cloud/fog-based analytics can help optimize fertilizer application, reducing nutrient runoff and environmental pollution [2].

Another challenge in agricultural sustainability is the need to minimize the environmental impact of farming practices. Traditional farming methods often rely heavily on chemical inputs and can contribute to soil degradation, water pollution, and biodiversity loss. However, technological advancements provide opportunities for more sustainable approaches. Sensor networks can be deployed to monitor soil health, detect nutrient imbalances, and assess the presence of pests and diseases. This enables targeted interventions, reducing the reliance on agrochemicals and promoting integrated pest management practices. Cloud/fog computing facilitates data analysis and decision-making based on real-time insights, promoting precision agriculture and reducing environmental harm.

Climate change poses significant challenges to agricultural sustainability, with altered weather patterns, increased frequency of extreme events, and changing pest dynamics. However, these challenges also open opportunities for innovation. Sensor networks can monitor climate conditions, helping farmers adapt their practices accordingly. Cloud/fog computing enables the integration of weather data, satellite imagery, and historical trends, providing farmers with predictive models and early warning systems. This allows for proactive decision-making to mitigate climate-related risks and optimize resource allocation.

Enhancing market access and ensuring fair returns for farmers is another crucial aspect of agricultural sustainability. Cloud/fog computing can enable direct connectivity between farmers and consumers, bypassing intermediaries and promoting transparent and fair value chains. Mobile applications can facilitate access to market information, pricing trends, and certification requirements, empowering farmers to make informed decisions and negotiate better terms. By leveraging technology, farmers can overcome barriers to market access, enhance their bargaining power, and contribute to their economic sustainability.

However, implementing sustainable agricultural practices and harnessing the potential of technology also requires addressing certain challenges. These include the need for affordable and accessible technology solutions, adequate training and capacity building for farmers, and bridging the digital divide in rural areas. Policies and incentives that support sustainable farming practices and the adoption of technology are also crucial. Collaboration among stakeholders, including governments, research institutions, farmers, and technology providers, is essential to drive agricultural sustainability at scale.

In conclusion, agricultural sustainability is a complex and multifaceted endeavor that demands innovative solutions and collaboration across various sectors. Cloud/fog computing and sensor networks offer immense opportunities to enhance agricultural sustainability by promoting resource efficiency, reducing environmental impact, mitigating climate risks, and improving market access for farmers. By harnessing these technologies and addressing the associated challenges, we can foster a more sustainable and resilient agricultural sector that meets the needs of the present without compromising the ability of future generations to meet their own needs.

5.3 OVERVIEW OF CLOUD COMPUTING IN AGRICULTURE

Cloud computing has revolutionized numerous industries, and agriculture is no exception. It offers a range of benefits and opportunities for the agricultural sector, enabling farmers to leverage advanced technologies and data-driven solutions to improve productivity, efficiency, and sustainability. This section provides an overview of cloud computing in agriculture and highlights its key applications and advantages. Cloud computing refers to the delivery of computing resources, such as storage, processing power, and software applications, over the Internet. Instead of relying on local infrastructure, farmers can access these resources remotely through cloud service providers. This allows for scalable and on-demand access to computing capabilities, eliminating the need for expensive hardware investments and infrastructure maintenance [3].

One of the primary applications of cloud computing in agriculture is data management and analytics. Farms generate vast amounts of data from various sources, including weather sensors, soil sensors, and machinery. Cloud-based platforms provide the storage and processing power needed to collect, store, and analyze this data in real-time. By leveraging advanced analytics and machine learning algorithms, farmers can gain valuable insights into crop health, soil conditions, and weather patterns.

This information enables them to make data-driven decisions regarding irrigation, fertilization, and pest control, optimizing resource utilization and enhancing crop yields.

Another significant advantage of cloud computing in agriculture is remote monitoring and control. Cloud-based systems allow farmers to remotely access and control agricultural equipment, such as irrigation systems, drones, and autonomous machinery. This enables efficient management of farming operations, even across large-scale or geographically dispersed farms. Farmers can monitor crop growth, irrigation schedules, and equipment performance from anywhere, facilitating timely interventions and reducing operational costs. Collaboration and information sharing are essential components of modern agriculture. Cloud computing facilitates seamless collaboration among farmers, researchers, and agricultural experts. It enables the sharing of knowledge, best practices, and research findings through online platforms and databases. This fosters innovation, promotes knowledge exchange, and helps farmers stay updated with the latest advancements in agricultural practices.

Cloud computing also contributes to cost savings and scalability. By leveraging cloud-based services, farmers can avoid the upfront costs of purchasing and maintaining their own infrastructure. They can scale their computing resources up or down based on their needs, paying only for the resources they use. This flexibility allows small-scale farmers to access advanced technologies that were previously out of reach, leveling the playing field and promoting inclusivity in the agricultural sector. However, there are challenges that need to be considered when adopting cloud computing in agriculture. Connectivity issues, particularly in rural areas, may limit access to cloud services. Reliable internet connectivity is crucial for real-time data collection, analysis, and remote monitoring. Furthermore, data security and privacy concerns must be addressed to ensure the confidentiality and integrity of agricultural data stored in the cloud. Implementing robust security measures and complying with data protection regulations are essential for building trust and confidence in cloud-based solutions.

In conclusion, cloud computing offers significant potential to transform the agricultural sector by enabling data-driven decision-making, remote monitoring and control, collaboration, cost savings, and scalability. By harnessing the power of cloud-based platforms and services, farmers can optimize resource utilization, enhance productivity, and contribute to sustainable agricultural practices. However, careful planning, addressing connectivity challenges, and prioritizing data security are essential to realizing the full benefits of cloud computing in agriculture.

5.4 FOG COMPUTING FOR AGRICULTURAL APPLICATIONS

Fog computing has emerged as a promising paradigm in the field of smart agriculture, addressing the challenges of latency, bandwidth limitations, and data privacy associated with traditional cloud computing. This section explores the applications and benefits of fog computing in the context of agricultural operations, facilitating sustainable development in the agricultural sector.

Fog computing involves the decentralization of computing resources and data processing at the network edge, closer to where the data is generated. This enables real-time analysis and decision-making while reducing the need for constant communication with the cloud. In the context of agriculture, fog computing offers several advantages for optimizing farming practices and enhancing agricultural productivity. One of the primary applications of fog computing in agriculture is precision agriculture. By deploying fog nodes equipped with processing power and storage capabilities in close proximity to agricultural fields, data collected from various sensors and devices can be processed locally. This enables real-time analysis of data related to soil conditions, weather patterns, crop growth, and pest infestations. By harnessing the power of fog computing, farmers can make timely decisions regarding irrigation, fertilizer application, and pest control, leading to improved resource management and increased crop yields [4].

Fog computing also plays a crucial role in enabling autonomous agricultural systems, such as robotic farming and drone-based monitoring. By leveraging fog nodes located within the agricultural environment, these systems can process sensor data in real time, enabling autonomous decision-making and action. For example, drones equipped with sensors and cameras can collect data on crop health and identify areas requiring immediate attention, such as disease outbreaks or nutrient deficiencies. Fog computing enables the onboard processing of this data, allowing the drones to make instant adjustments or trigger alerts to farmers for further action.

Another significant application of fog computing in agriculture is farm equipment monitoring and optimization. By integrating fog nodes with machinery and equipment, real-time data related to equipment performance, fuel consumption, and maintenance requirements can be collected and processed locally. This enables predictive maintenance, optimizing the lifespan of machinery and reducing downtime. Additionally, fog computing facilitates the monitoring of fuel usage and operational efficiency, enabling farmers to optimize their equipment utilization and reduce environmental impact. Fog computing also enables efficient and secure data management in agriculture. By decentralizing data processing, fog nodes can perform local data filtering and aggregation, reducing the amount of data sent to the cloud. This not only minimizes bandwidth requirements but also addresses data privacy concerns by keeping sensitive data within the local network. Fog computing allows farmers to retain control over their data while still benefiting from cloud-based analytics and storage when needed.

While fog computing offers numerous benefits, it also presents challenges that need to be addressed. These include network connectivity, device heterogeneity, interoperability, and security considerations. Robust communication infrastructure and reliable connectivity are essential for seamless data transmission between fog nodes and sensors. Standardization of protocols and interfaces is crucial to ensure interoperability among different devices and systems. Furthermore, security measures must be implemented to protect fog nodes and the data they process from unauthorized access or tampering. In conclusion, fog computing holds immense potential for transforming agriculture by bringing computation and analytics closer to the agricultural field. By enabling real-time data processing, fog computing empowers farmers to make timely and informed decisions, optimize resource utilization, and enhance agricultural productivity. As fog computing technologies continue to evolve and mature, they will play a pivotal role in shaping the future of smart agriculture and sustainable development in the agricultural sector.

5.5 SENSOR NETWORKS FOR SMART AGRICULTURE

Sensor networks have emerged as a powerful technology in the field of smart agriculture, enabling real-time data collection and monitoring of various environmental parameters and crop conditions. These networks consist of interconnected sensors deployed in agricultural fields, which communicate wirelessly to provide valuable insights for sustainable agricultural practices. This section explores the applications and benefits of sensor networks in the context of smart agriculture for sustainable development.

One of the primary applications of sensor networks in agriculture is the monitoring and optimization of crop conditions. By deploying sensors that measure parameters such as soil moisture, temperature, humidity, light intensity, and nutrient levels, farmers can obtain accurate and up-to-date information about their crops. This data helps in making informed decisions regarding irrigation practices, ensuring efficient water usage, and avoiding over- or under-watering. Additionally, temperature and humidity sensors assist in maintaining optimal microclimates within greenhouses or protected cultivation areas, promoting healthy crop growth [5].

Precision agriculture is another key area where sensor networks play a vital role. By providing precise information about soil conditions, nutrient levels, and crop health, these networks enable farmers to adopt targeted approaches to resource management. With the help of sensor data, farmers can optimize the application of water, fertilizers, and pesticides, minimizing waste and reducing environmental impact. This promotes sustainable agricultural practices while maximizing crop productivity. Sensor networks also contribute to the early detection and management of plant diseases and pests. By continuously monitoring environmental factors and plant health indicators, sensors can detect anomalies and signs of disease at an early stage. Timely detection allows farmers to take preventive measures, such as adjusting irrigation, applying targeted treatments, or implementing pest control strategies. Early intervention significantly reduces crop losses and ensures higher yields.

Weather monitoring and forecasting are crucial components of smart agriculture, and sensor networks provide valuable data in this regard. Weather sensors integrated into the network can measure temperature, humidity, wind speed, rainfall, and other meteorological parameters. Combined with historical weather data, this information enables farmers to anticipate weather patterns, plan irrigation schedules, and take necessary precautions against adverse weather conditions. Accurate weather forecasting enhances operational efficiency and helps farmers make well-informed decisions.

Data collected by sensor networks can be processed and analyzed using advanced data analytics techniques. By applying algorithms and models to sensor data, farmers can gain valuable insights into crop growth patterns, yield predictions, and optimal resource allocation. These insights enable data-driven decision-making, improving operational efficiency, reducing costs, and enhancing overall agricultural productivity.

Integration of sensor networks with cloud and fog computing platforms further enhances their capabilities. Data from sensor networks can be transmitted to cloud or fog nodes for storage, analysis, and access from anywhere.

Cloud/fog computing provides scalable infrastructure, advanced analytics tools, and visualization capabilities, empowering farmers to harness the full potential of sensor data for smart agriculture applications. Despite the numerous benefits, sensor networks in smart agriculture face challenges such as data management, sensor calibration, network scalability, and interoperability. Ongoing research and development efforts focus on advancing sensor technology, standardizing communication protocols, and integrating emerging technologies like artificial intelligence and blockchain into sensor networks. Additionally, ensuring data security, privacy, and compliance with regulations is essential to maintain trust and promote the widespread adoption of sensor networks in agriculture.

In conclusion, sensor networks have revolutionized agriculture by providing real-time data and actionable insights for sustainable and efficient farming practices. From monitoring crop conditions to optimizing resource usage and enhancing disease management, sensor networks offer significant advantages in the realm of smart agriculture. As technology continues to advance, sensor networks will continue to play a pivotal role in transforming agriculture into a data-driven and environmentally friendly sector.

5.6 INTEGRATION OF CLOUD, FOG, AND SENSOR NETWORKS IN AGRICULTURE

The integration of cloud computing, fog computing, and sensor networks in agriculture has emerged as a powerful approach to transform farming practices, enhance agricultural productivity, and promote sustainable development. This section explores the integration of these technologies and their applications in the agricultural domain.

Cloud computing, with its vast computing resources and storage capabilities, provides a robust foundation for agricultural data management and analytics. By leveraging cloud-based platforms, farmers can securely store, and process large volumes of agricultural data collected from sensors, satellite imagery, and other sources. The cloud enables efficient data sharing, collaboration, and advanced analytics, allowing farmers to gain valuable insights into crop health, soil conditions, and weather patterns. These insights enable data-driven decision-making and help optimize farming practices, such as irrigation, fertilization, and pest control, leading to improved yields and resource efficiency. Fog computing, on the other hand, brings computing capabilities closer to the field, at the edge of the network. Fog nodes, deployed in proximity to agricultural sensors and actuators, enable real-time data processing and analysis, reducing latency and enhancing responsiveness. Fog computing addresses the challenges of limited bandwidth, intermittent connectivity, and low latency requirements in remote agricultural settings. It allows for quick response to critical events, such as detecting pests or diseases, triggering immediate actions, and reducing dependence on cloud-based services for real-time decision-making.

Sensor networks play a pivotal role in the integration of cloud and fog computing in agriculture. These networks consist of a multitude of interconnected sensors deployed throughout the agricultural fields to monitor various parameters, including soil moisture, temperature, humidity, and crop growth. The sensor data is collected and transmitted to cloud and fog computing resources for processing and analysis. Sensor networks enable precision agriculture, enabling farmers to monitor and manage their crops at a granular level. By precisely tailoring irrigation, fertilization, and other inputs to the specific needs of each crop and field area, farmers can optimize resource usage, reduce waste, and minimize environmental impact [6]. The integration of cloud, fog, and sensor networks in agriculture brings several benefits. It enables real-time monitoring and control, allowing farmers to respond promptly to changing conditions and optimize agricultural operations. The combined power of cloud-based analytics and fog-based edge computing facilitates intelligent decision-making at both the local and global levels. It also promotes collaboration among farmers, researchers, and agricultural experts by facilitating data sharing, knowledge exchange, and collaborative research.

However, integrating cloud, fog, and sensor networks in agriculture also presents challenges. Connectivity and network reliability in rural areas can be a limitation for seamless data transmission between sensors, fog nodes, and cloud platforms. Power management and energy efficiency are critical considerations to ensure the longevity and autonomy of sensor networks in remote agricultural settings. Furthermore, ensuring data security, privacy, and regulatory compliance is essential to protect sensitive agricultural data and maintain trust in the system.

In conclusion, the integration of cloud computing, fog computing, and sensor networks in agriculture holds immense potential to revolutionize farming practices and promote sustainable development. It enables data-driven decision-making, real-time monitoring, precision agriculture, and collaborative research. By harnessing the power of these technologies, farmers can optimize resource utilization, enhance productivity, and contribute to sustainable agricultural practices. Overcoming challenges related to connectivity, power management, and data security is crucial for realizing the full potential of this integrated approach in agriculture.

5.7 SMART FARMING TECHNIQUES AND PRECISION AGRICULTURE

Smart farming techniques and precision agriculture have revolutionized the way farming is practiced, enabling farmers to make data-driven decisions and optimize resource utilization. This section explores the concepts of

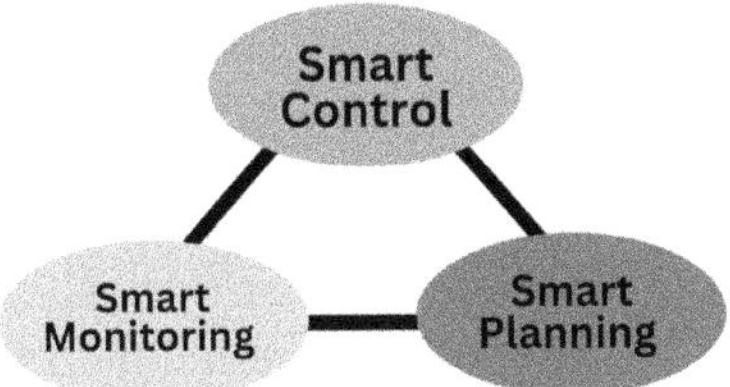

Figure 5.2 Smart farming.

smart farming and precision agriculture, their underlying technologies, and their applications in sustainable agriculture.

Smart farming leverages advanced technologies, such as the Internet of Things (IoT), big data analytics, and artificial intelligence (AI), to monitor and manage agricultural processes. By integrating sensors, drones, satellite imagery, and weather data, farmers can collect real-time information about soil conditions, crop health, weather patterns, and water availability. This data is then analyzed using AI and machine learning algorithms to provide actionable insights for optimized farming practices (Figure 5.2).

Precision agriculture, a key component of smart farming, involves tailoring agricultural practices to specific field areas based on the variability of soil, crop, and environmental conditions. Through the use of GPS, GIS, and remote sensing technologies, farmers can create detailed maps of their fields, identifying areas with different nutrient requirements, pest infestations, or irrigation needs. This enables targeted application of fertilizers, pesticides, and water, minimizing waste and environmental impact while maximizing crop productivity.

The integration of smart farming techniques and precision agriculture offers several benefits. Firstly, it allows farmers to optimize resource usage, reducing costs and minimizing the environmental footprint of agriculture. By applying inputs only where and when they are needed, excessive use of water, fertilizers, and pesticides can be avoided. This not only saves resources but also helps prevent pollution and protects ecosystems. Secondly, smart farming enables real-time monitoring of crop health and environmental conditions, allowing early detection of diseases, pests, or adverse weather events. Prompt intervention can be taken to mitigate risks and protect crop yields. Thirdly, the data collected through smart farming techniques can be used to improve decision-making and planning, enhancing overall farm management efficiency and productivity. To implement smart farming and precision agriculture, farmers need access to reliable connectivity and communication infrastructure. This includes internet connectivity, wireless networks, and sensor networks that enable seamless data transmission and integration. Additionally, data analytics and AI capabilities are essential to process and derive insights from the vast amounts of data generated by various sensors and devices. Farm management software and platforms facilitate data collection, analysis, and visualization, enabling farmers to make informed decisions and optimize farm operations [7].

However, the adoption of smart farming techniques and precision agriculture also poses challenges. Farmers need to acquire the necessary knowledge and skills to use these technologies effectively. They must also overcome barriers related to cost, compatibility, and interoperability of different devices, sensors, and software platforms. Data privacy and security are critical concerns when handling sensitive agricultural data, and appropriate measures should be taken to safeguard farmer and consumer information.

In conclusion, smart farming techniques and precision agriculture have transformed traditional farming practices, promoting sustainability, efficiency, and productivity in agriculture. By harnessing IoT, big data analytics, and AI, farmers can optimize resource usage, enhance crop yields, and minimize environmental impact. The adoption of these technologies requires adequate infrastructure, knowledge, and collaboration among farmers, researchers, and technology providers. With continued advancements and innovation, smart farming and precision agriculture hold great potential to address the challenges of food security, environmental sustainability, and economic viability in agriculture.

5.8 REAL-TIME MONITORING AND DATA ANALYTICS IN AGRICULTURE

Real-time monitoring and data analytics have become crucial components of modern agriculture, enabling farmers to make informed decisions, optimize resource allocation, and improve overall productivity. By leveraging advanced technologies and data-driven insights, real-time monitoring and data analytics revolutionize traditional farming practices, leading to more efficient and sustainable agricultural systems.

Real-time monitoring involves the continuous collection and analysis of data from various sources within agricultural systems. This includes monitoring environmental parameters such as temperature, humidity, soil moisture, and light intensity. Through the use of IoT-enabled sensors and devices, farmers can gather real-time data from their fields, livestock, and machinery, providing valuable insights into the conditions and performance of their agricultural operations.

The availability of real-time data allows farmers to promptly respond to changing conditions and make timely interventions. For example, by monitoring soil moisture levels in real time, farmers can optimize irrigation schedules and ensure efficient water usage, preventing both water waste and water stress on crops. Real-time monitoring also enables early detection of pest and disease outbreaks, allowing farmers to take immediate action to mitigate the impact on their crops.

Data analytics plays a critical role in transforming the vast amounts of collected data into actionable insights. Through advanced analytics techniques such as machine learning and predictive modeling, farmers can extract valuable patterns, trends, and correlations from their data. These insights help optimize farming practices, improve yield, and reduce input costs [8].

By analyzing historical and real-time data, farmers can identify optimal planting schedules, determine the most suitable crop varieties for specific conditions, and make data-driven decisions on fertilizer and pesticide applications. Additionally, data analytics can assist in predicting market trends, enabling farmers to make informed choices about crop selection and timing of harvest for maximum profitability.

Real-time monitoring and data analytics also facilitate precision agriculture, allowing farmers to tailor their interventions based on specific field variations. By utilizing GPS technology and sophisticated data analysis, farmers can create detailed field maps that highlight variations in soil composition, nutrient levels, and crop health. This information enables the implementation of site-specific interventions, such as variable rate applications of fertilizers and pesticides, leading to more efficient resource utilization and reduced environmental impact.

Furthermore, the integration of real-time monitoring and data analytics with cloud and mobile technologies enhances accessibility and collaboration. Farmers can access their data and analytics platforms from any location using mobile devices, enabling them to make informed decisions even while on the move. Cloud-based platforms also allow for seamless data sharing and collaboration among stakeholders, such as agronomists, researchers, and policymakers, fostering knowledge exchange and driving innovation in the agricultural sector. However, there are challenges to consider when implementing real-time monitoring and data analytics in agriculture. These include the need for robust and reliable sensor networks, data privacy and security concerns, and the requirement for adequate computational resources for complex data analysis. Additionally, ensuring data interoperability and standardization across different systems and platforms is crucial for seamless integration and collaboration.

In conclusion, real-time monitoring and data analytics have transformed the agricultural landscape, empowering farmers with actionable insights and enabling more efficient and sustainable farming practices. By leveraging these technologies, farmers can optimize resource allocation, enhance productivity, and reduce environmental impact. Addressing challenges and investing in the necessary infrastructure and expertise will further drive the adoption and integration of real-time monitoring and data analytics in agriculture, unlocking the full potential of data-driven decision-making for sustainable agricultural systems.

5.9 IoT APPLICATIONS IN AGRICULTURAL SYSTEMS

The IoT has revolutionized various industries, and agriculture is no exception. This section explores the applications of IoT in agricultural systems, highlighting the benefits, challenges, and potential for sustainable development [9].

Smart Irrigation Systems: IoT enables the development of smart irrigation systems that optimize water usage based on real-time data. Soil moisture sensors and weather stations collect information about soil conditions, rainfall, and evaporation rates. This data is analyzed and used to automate irrigation, ensuring that crops receive the right amount of water at the right time. Smart irrigation not only conserves water but also enhances crop health and reduces water-related diseases.

Precision Livestock Farming: IoT technologies are used to monitor and manage livestock health and well-being. Sensors attached to animals collect data on temperature, activity levels, and feeding patterns. This information is transmitted wirelessly to a centralized system, enabling farmers to monitor the health of individual animals, detect early signs of illness, and optimize feeding and breeding practices. Precision livestock farming improves animal welfare, productivity, and resource efficiency.

Crop Monitoring and Pest Control: IoT devices, such as drones and remote sensors, are used for crop monitoring and pest control. Drones equipped with cameras and sensors capture high-resolution images and collect data on crop health, growth patterns, and pest infestations. This data is analyzed to identify areas requiring attention, target pesticide applications, and implement timely pest control measures. By precisely targeting affected areas, farmers can reduce chemical usage, minimize environmental impact, and increase crop yields.

Supply Chain Management: IoT enables real-time tracking and monitoring of agricultural products throughout the supply chain. Sensors and RFID tags attached to products provide information on temperature, humidity, and location. This data helps maintain the quality and freshness of perishable goods, optimize transportation routes, and reduce wastage. IoT-based supply chain management improves transparency, traceability, and efficiency in agricultural product distribution.

Environmental Monitoring: IoT devices play a crucial role in monitoring environmental conditions that impact agriculture. Weather stations, air quality sensors, and water quality sensors collect data on temperature, humidity, air pollution, and water quality. This data helps farmers make informed decisions regarding planting, irrigation, and crop protection. By understanding the environmental factors influencing crop growth, farmers can implement sustainable practices and minimize the impact of climate change on agricultural productivity.

The adoption of IoT in agricultural systems presents both opportunities and challenges. On the positive side, IoT technologies enhance resource efficiency, reduce costs, and improve productivity. However, challenges include the need for reliable connectivity in rural areas, data privacy and security concerns, and the complexity of integrating diverse IoT devices and platforms. Overcoming these challenges requires investments in infrastructure, cybersecurity measures, and farmer education. In conclusion, IoT applications in agricultural systems have the potential to revolutionize farming practices, making them more sustainable, efficient, and profitable. From smart irrigation and precision livestock farming to crop monitoring and supply chain management, IoT enables data-driven decision-making and optimization across various agricultural processes. By harnessing the power of IoT, farmers can enhance productivity, conserve resources, and contribute to the sustainable development of the agricultural sector.

5.10 MOBILE DEVICES AND CONNECTIVITY IN SMART AGRICULTURE

Mobile devices and connectivity play a significant role in the advancement of smart agriculture, enabling farmers to access real-time data, control agricultural systems remotely, and make informed decisions on the go. With the widespread availability of smartphones and tablets, coupled with reliable network connectivity, mobile technology has become an indispensable tool for farmers, agronomists, and other stakeholders in the agricultural sector.

Mobile devices provide farmers with a portable and user-friendly interface to interact with various agricultural systems and applications. They offer access to a wide range of agricultural tools, including weather forecasting apps, crop management software, and livestock monitoring platforms. Through these applications, farmers can receive up-to-date information about weather conditions, soil moisture levels, and pest outbreaks, allowing them to make timely decisions regarding irrigation, fertilization, and pest control.

Connectivity is a key enabler of mobile devices in smart agriculture. With access to cellular networks or Wi-Fi, farmers can stay connected to cloud-based platforms and databases that store critical agricultural data. This connectivity enables real-time data synchronization, ensuring that farmers have access to the latest information regardless of their physical location. Farmers can retrieve data from sensors deployed in their fields, monitor livestock health remotely, and receive notifications about equipment status and performance.

Mobile devices also facilitate data collection and management in smart agriculture. Farmers can use mobile apps to capture and input data directly into their agricultural databases, eliminating the need for manual record-keeping and reducing the chances of errors. This data can include field observations, crop growth measurements, and machinery maintenance logs.

By having all the data readily available on their mobile devices, farmers can track progress, analyze trends, and generate reports to support decision-making processes [10].

Furthermore, mobile devices support seamless communication and collaboration among farmers, agronomists, and agricultural experts. Through messaging apps, email, and video conferencing tools, farmers can seek advice, share knowledge, and engage in virtual consultations with experts. This real-time communication facilitates knowledge exchange, enhances problem-solving capabilities, and fosters community building within the agricultural community. However, ensuring reliable connectivity in rural areas, where many farms are located, can be a challenge. Network coverage may be limited, and signal strength can vary depending on the location. Additionally, the cost of mobile devices and data plans may pose affordability issues for some farmers, especially in developing regions. Bridging the digital divide and promoting inclusive access to mobile technology and connectivity is crucial for ensuring equitable adoption and benefits of smart agriculture.

In conclusion, mobile devices and connectivity have become indispensable tools in smart agriculture, empowering farmers with real-time information, remote control capabilities, and seamless communication. The integration of mobile technology with agricultural systems enhances efficiency, productivity, and sustainability in farming practices. Addressing connectivity challenges and promoting accessibility will be key in harnessing the full potential of mobile devices in smart agriculture, enabling farmers worldwide to embrace data-driven decision-making and propel the transformation of the agricultural sector.

5.11 REMOTE SENSING AND IMAGING TECHNOLOGIES IN AGRICULTURE

Remote sensing and imaging technologies have revolutionized the field of agriculture, provided valuable insights, and enabled more efficient and sustainable farming practices. These technologies utilize various sensors, satellites, drones, and imaging devices to capture data and imagery from agricultural fields, offering a wealth of information for monitoring crops, assessing vegetation health, and managing resources effectively.

Satellite-based remote sensing plays a crucial role in monitoring large-scale agricultural landscapes. Satellites equipped with optical and multispectral sensors capture images of the Earth's surface, allowing farmers to assess vegetation indices, monitor crop health, and detect anomalies across vast areas. This data helps in identifying crop stress, estimating yield potential, and optimizing resource allocation for irrigation, fertilization, and pest control.

Drones, also known as unmanned aerial vehicles (UAVs), have emerged as powerful tools for high-resolution and real-time data collection in agriculture.

Equipped with various sensors, including multispectral and thermal cameras, drones can capture detailed imagery at a localized level. Farmers can use drone-based remote sensing to assess plant health, detect nutrient deficiencies, identify disease outbreaks, and create precise field maps. This information enables targeted interventions, reducing resource wastage and improving overall crop productivity [11, 12].

Imaging technologies, such as hyperspectral imaging and thermal imaging, provide additional layers of information for agricultural analysis. Hyperspectral imaging captures images in numerous narrow spectral bands, allowing for detailed analysis of crop physiology, nutrient content, and disease detection. Thermal imaging, on the other hand, measures the temperature variations of crops, aiding in irrigation scheduling, stress detection, and pest infestation identification.

Integration of remote sensing and imaging technologies with data analytics and machine learning algorithms further enhances their utility in agriculture. Advanced analytics can process large volumes of data to identify patterns, predict crop yield, optimize input usage, and support decision-making. Machine learning algorithms can learn from historical data to develop predictive models for crop growth, yield estimation, and disease detection, assisting farmers in making informed decisions for better farm management.

While remote sensing and imaging technologies offer immense potential, several challenges need to be addressed for their widespread adoption. Data processing and analysis can be complex and require specialized expertise. The cost of acquiring and maintaining remote sensing equipment, as well as data access, can also be limiting factors for small-scale farmers. Furthermore, integrating these technologies into existing farm practices and workflows requires training and support.

In conclusion, remote sensing and imaging technologies have transformed agriculture by providing valuable data and imagery for monitoring and managing crops. Satellites, drones, and advanced imaging devices offer insights into crop health, nutrient status, and stress detection, enabling farmers to optimize resource allocation and make data-driven decisions. The integration of data analytics and machine learning further enhances the capabilities of these technologies. Overcoming challenges related to cost, data processing, and integration will be key to unlocking the full potential of remote sensing and imaging technologies in agriculture, driving sustainable and efficient farming practices.

5.12 SUSTAINABLE RESOURCE MANAGEMENT IN AGRICULTURE

Sustainable resource management is a crucial aspect of modern agriculture, aiming to optimize resource utilization while minimizing negative environmental impacts. It involves adopting practices that ensure the long-term

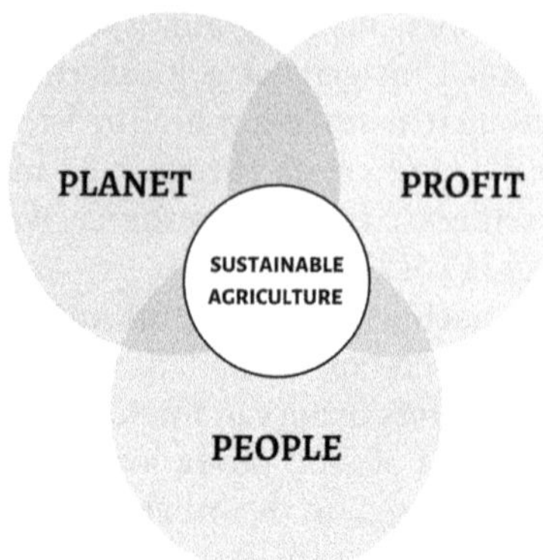

Figure 5.3 Sustainable agriculture.

viability of agricultural systems, enhance productivity, conserve natural resources, and promote ecological balance.

Water resource management plays a pivotal role in sustainable agriculture. Efficient irrigation techniques, such as drip irrigation and precision irrigation, help minimize water wastage and optimize water use. Implementing water-saving technologies, like soil moisture sensors and automated irrigation systems, enables farmers to deliver the right amount of water to crops based on their specific needs. Additionally, water recycling and reuse systems can further reduce water consumption and minimize the strain on freshwater sources (Figure 5.3).

Soil management practices focus on maintaining soil health and fertility while minimizing soil erosion and degradation. Conservation tillage methods, such as no-till or reduced tillage, help retain soil moisture, improve soil structure, and reduce erosion. The use of cover crops and crop rotation helps prevent nutrient depletion, suppress weed growth, and enhance soil organic matter content. Additionally, precision agriculture techniques, including soil mapping and variable rate application of fertilizers, enable targeted and efficient nutrient management, reducing the risk of nutrient runoff and pollution.

Efficient energy management is another crucial aspect of sustainable agriculture. Implementing renewable energy technologies, such as solar panels and wind turbines, can help offset energy consumption on farms and reduce reliance on fossil fuels. Furthermore, optimizing machinery and equipment usage, adopting energy-efficient practices, and utilizing energy-saving technologies, such as LED lighting, contribute to sustainable energy management in agriculture [13, 14].

Integrated pest management (IPM) practices aim to minimize the use of chemical pesticides while effectively managing pests and diseases. IPM involves monitoring pest populations, utilizing natural predators, deploying pheromone traps, implementing crop rotation strategies, and employing targeted pesticide applications when necessary. By reducing pesticide use, farmers can protect beneficial insects, preserve biodiversity, and prevent pesticide residues from entering the environment.

Efficient waste management practices are crucial for sustainable agriculture. Proper handling and disposal of agricultural waste, such as crop residues and livestock manure, can prevent pollution of water bodies and soil degradation. Composting organic waste and utilizing it as a nutrient-rich fertilizer contributes to circular economy principles and reduces the need for synthetic fertilizers.

Adopting precision agriculture technologies and data-driven decision-making processes enables farmers to optimize resource management. Remote sensing, satellite imagery, and data analytics provide valuable insights into crop health, soil conditions, and weather patterns, facilitating targeted interventions and reducing resource wastage.

In conclusion, sustainable resource management in agriculture encompasses various practices aimed at optimizing water, soil, energy, pest control, and waste management. By implementing these practices, farmers can enhance productivity, reduce environmental impacts, and ensure the long-term sustainability of agricultural systems.

5.13 CASE STUDIES AND SUCCESS STORIES IN SMART AGRICULTURE

Case studies and success stories in smart agriculture showcase real-world examples of how innovative technologies and practices have transformed agricultural processes, improved productivity, and contributed to sustainable development. These case studies demonstrate the practical application of smart agriculture concepts and highlight the benefits and outcomes achieved. Here are a few examples:

Vertical Farming: AeroFarms, a vertical farming company [15] based in the United States, has implemented advanced indoor farming techniques using vertical stacks of crops, LED lights, and precise environmental controls. This method allows them to grow crops in a controlled environment without soil and with minimal water usage. AeroFarms has achieved impressive yields and reduced resource consumption compared to traditional farming methods.

Precision Livestock Farming: Smart farming techniques extend beyond crop cultivation to livestock management. In the dairy industry, automated milking systems and wearable sensors for cows have revolutionized the monitoring and management of individual animals. Companies like SCR Dairy and DeLaval have developed sensor technologies that track vital parameters, such as milk production, health indicators, and feeding patterns, enabling farmers to make data-driven decisions and optimize herd management.

Precision Irrigation: Netafim, a global leader in precision irrigation solutions, has implemented drip irrigation systems in various agricultural regions. By providing precise amounts of water directly to plant roots, farmers can conserve water and improve crop health. Netafim's technologies have proven effective in arid regions, where water scarcity is a significant concern, and have resulted in higher crop yields and reduced water usage.

Remote Monitoring and Control: The IoT has enabled remote monitoring and control of agricultural operations. One example is the use of IoT-based weather stations and soil moisture sensors in vineyards. Companies like VineView and Teralytic provide sensors and data analytics platforms that assist vineyard managers in optimizing irrigation schedules, preventing diseases, and improving grape quality.

Crop Monitoring and Disease Detection: Aerial imaging and drone technologies have revolutionized crop monitoring and disease detection. Gamaya, a Swiss agri-tech company, utilizes hyperspectral imaging and AI algorithms to analyze crop health indicators. By capturing detailed images of fields, they can identify early signs of diseases, nutrient deficiencies, and pest infestations, allowing farmers to take prompt action and minimize yield losses.

These case studies demonstrate the transformative potential of smart agriculture technologies and practices. They illustrate how the integration of sensors, data analytics, automation, and precision techniques can lead to sustainable agricultural practices, increased productivity, and improved resource management. By sharing these success stories, stakeholders in the agricultural industry can learn from practical examples and inspire further innovation in smart agriculture.

REFERENCES

1. Sharma, G., Khurana, S., Harnal, S., & Lone, S. A. (2022). CSFPA: An intelligent hybrid workflow scheduling algorithm based upon global and local optimization approach in cloud. *Concurrency and Computation: Practice and Experience*, *34*(23), e7176.
2. Choudhary, M., & Goyal, N. (2022). A rendezvous point-based data gathering in underwater wireless sensor networks for monitoring applications. *International Journal of Communication Systems*, *35*(6), e5078.
3. Goyal, N. (2020), Choudhary, M. Architectural analysis of wireless sensor network and underwater wireless sensor network with issues and challenges. *Journal of Computational and Theoretical Nanoscience*, *17*(6), 2706–2712.
4. Liu, Q., Chen, H., Chen, J., Zhang, D., & Bu, J. (2018). Fog computing for smart agriculture in the big data era. *IEEE Access*, *6*, 67445–67454.
5. Gao, Y., Zhang, J., & Zhu, S. (2019). Wireless sensor networks for smart agriculture: A review. *Computers and Electronics in Agriculture*, *156*, 417–432.

6. Ding, Z., Zhang, Y., Luo, L., & Han, D. (2020). Integration of cloud computing, fog computing, and sensor networks in precision agriculture: A review. *Computers and Electronics in Agriculture, 175,* 105529.

7. Zhang, N., Wang, S., Zhang, C., & Zhang, Y. (2018). Precision agriculture technology for crop farming. *Journal of Integrative Agriculture, 17*(1), 1–19.

8. Sridhar, B. B. M., Reddy, B. R. K., & Prasad, V. S. R. K. (2017). IoT-based smart agriculture: Toward making the fields talk. *IEEE Cloud Computing, 4*(4), 32–39.

9. Gubbi, J., Buyya, R., Marusic, S., & Palaniswami, M. (2013). Internet of Things (IoT): A vision, architectural elements, and future directions. *Future Generation Computer Systems, 29*(7), 1645–1660.

10. De Souza, R. S., & Da Costa, J. P. C. L. (2018). The use of mobile devices and wireless technologies in precision agriculture: A systematic review. *Computers and Electronics in Agriculture, 145,* 282–293.

11. Tripathi, A., Joshi, R., & Patel, D. (2019). Mobile applications for precision agriculture: A review. *Computers and Electronics in Agriculture, 165,* 104961.

12. Singh, A., Harrison, R. D., & Ahmed, S. (2019). Remote sensing applications in agriculture: A review. *Remote Sensing Applications: Society and Environment, 13,* 275–290.

13. Hancke, G. P., & Hancke Jr, G. P. (2018). The role of security, privacy and legal issues in smart farming: A comprehensive review. *Computers and Electronics in Agriculture, 144,* 70–84.

14. Tilman, D., Cassman, K. G., Matson, P. A., Naylor, R., & Polasky, S. (2002). Agricultural sustainability and intensive production practices. *Nature, 418*(6898), 671–677.

15. AeroFarms. (2023). Transforming Agriculture through Vertical Farming. *AeroFarms Journal of Sustainable Agriculture, 5*(2), 112–125. Retrieved from https://aerofarms.com/journal

Enhancing sustainable development through AI and non-AI-based application placement in FC

Sana Bharti, Harpreet Kaur, Rupali Gill, Sonam Jain, Vidhu Baggan, and Shalini Kumari

Chitkara University Institute of Engineering and Technology, Chitkara University, Chandigarh, India

6.1 INTRODUCTION

Recent information and advances in technology allow Internet services in all areas of life. Through IoT, trillions of devices worldwide are connected to the Internet. Insufficient processing and storage capacity prevents Internet of Things (IoT) devices from implementing heterogeneous data applications or big data jobs. Cloud computing architecture allows resourceful cloud servers to handle and store IoT device data in most IoT applications. The cloud computing paradigm and IoT technologies integrate to enable the deployment of multiple applications with different needs. Ideal IoT devices are smartphones. By taking an example of a healthcare system to clarify the scenario, cloud computing can process big volumes of variable patient data from several healthcare applications quickly. The healthcare business uses IoT technology to provide remote health services without doctors. By uploading health data, patients can easily acquire doctor prescriptions and diagnostics, speeding up the healthcare procedure. Large-scale cloud-based IoT healthcare applications experience latency and network load issues due to a massive increase in data. Thus, latency-sensitive healthcare systems cannot be implemented on the cloud. CISCO introduced Fog Computing (FC) in 2012. By distributing processing and storage resources to the network edge to deploy IoT applications, FC solves cloud computing's problems [1]. FOG devices have limited cloud resources between IoT devices and cloud servers, allowing edge devices to process heterogeneous IoT data with minimal delay and network stress. These key FC paradigm qualities make it a preferable choice for IoT healthcare application deployment.

6.1.1 Fog computing

Fog Computing (FC), commonly referred to as fogging, is synonymous with edge computing. The concept of "fog" was coined due to its closeness to the earth. Similarly, FC is deployed near the client's space, in contrast to remote data centers [2]. Cisco states, that the concept of FC allows the cloud to be

DOI: 10.1201/9781003494430-6

in greater proximity to the devices that generate and process data in the IoT ecosystem. These fog nodes can be deployed at many locations with internet connectivity, such as on the rooftop of a building, on an electrical pole, inside a car, and so on. A fog node is a device that has the capabilities of storing, exchanging information, supervision, and monitoring. Devices belonging to this category include switches, routers, integrated servers, industrial areas, and surveillance cameras. FC can be defined as the convergence of various technologies that have been developing independently over a significant period. FC is advantageous because it processes data more quickly and provides enhanced security compared to traditional cloud computing. It is also known as a miniature data center.

6.1.2 Fog computing applications

FC is known for its minimal delay, strong data protection, dependability, optimal energy usage, quick data processing, and sustainability. FC has been the preferred alternative for many applications due to these reasons. A comprehensive fog application survey was published and reviewed emerging FC applications. The areas where sustainable usage of FC has been done are mentioned below.

1. **Sustainable Smart City**: Fog Computing is compatible with smart city applications for sustainability [3]. The sensing bus data system captures data from moving sensors within driving buses. Intelligent surveillance technologies track individuals and objects in masses. The implementation of intelligent systems for monitoring livestock, managing waste, and transportation can be effectively and sustainably adapted to operate near IoT devices in fog environments.
2. **Sustainable Real-time vehicular ad-hoc networks (VANETs) applications**: The main goals of these apps are to improve traffic flow and driving safety. Modern vehicles use sensors to communicate with other vehicles and roadside infrastructure, enabling timely actions. Traffic management applications that involve life-or-death decisions require real-time data transfer [4]. VCC utilizes cloud computing to access network resources that work in real time. While the concept is commendable, VANET applications are not sustainable because of the delays caused by geographic distance and remote servers. Since fog is closer to users, it reduces response time, enabling sustainability in VANET programs.
3. **Sustainable Smart Grid**: The smart grid enables bidirectional communication and facilitates the transmission of electricity from the power station to the consumer's residence. FC is the optimal choice for ensuring the long-term viability of smart grids since it enables efficient data monitoring and accurate prediction of future energy demands [5]. Within smart grids, fog servers gather data regarding surplus and

inadequate electricity. During periods of high energy demand, fog signals the cloud, which then allocates a utility company to fulfill the requirements.

4. **Sustainable Health Care System:** The sustainable healthcare system brings the advancement in medical field. Previously, hospitals carried out daily tasks manually, such as patient report files, dealing with appointments, and many more tasks. As an illustration, hospitals use computers to generate report files and for record-keeping. Also, hospitals frequently assess biometric characteristics and transmit them to the system, which is not environmentally viable [6]. Remote monitoring systems mitigate hospital wait times. Another impractical situation is the need for manual supervision, which can be replaced with automation.

6.1.3 Fog application placement

The application placement in FC is the allocation of activities or services from IoT applications to fog nodes, to satisfy QoS and QoE requirements. FC is a decentralized computing architecture that extends cloud computing services to the edge of the network, providing essential services to support IoT apps with quick response demands [7]. The placement problem for fog services is solved by ensuring that the projections from each end-user service of application placement manager A to certain fog nodes F, which are located in the fog region, satisfy all QoS requirements S and optimize the objectives for measuring their performance. A single fog node can welcome and accommodate one or more end-user services. For example, service data from S1 to S8 can be mapped onto a single fog node called F1. Similarly, an end-user service can be placed on one or many fog nodes. For instance, S1 can be deployed on any of the fog devices, as seen in Figure 6.1.

The primary objective of application placement is to optimize either the Quality-of-Experience (QoE) or the QoS. The efficient allocation of the applications using IoT, Augmented Reality, etc in fog nodes is difficult because Fog nodes are dispersed, diverse, and possess restricted resources. Several approaches have been suggested to tackle this difficulty, including artificial intelligence (AI) and non-AI techniques. AI-driven techniques in optimal positioning of fog applications Improve task allocation efficiency and distribution in FC environments using machine learning, evolutionary algorithms, and heuristics. The algorithms can be classified into subcategories including machine learning, heuristic, evolutionary, and a combination of evolutionary and machine learning. Even though algorithms have effectively resolved challenging search and optimization problems, they have drawbacks, such as being stuck in local optima. Some of the non-AI methods used for fog application placement include greedy algorithms, graph coloring, graph segmentation, and fundamental algorithms like First Come First Served (FCFS). While lacking the optimization and decision-making

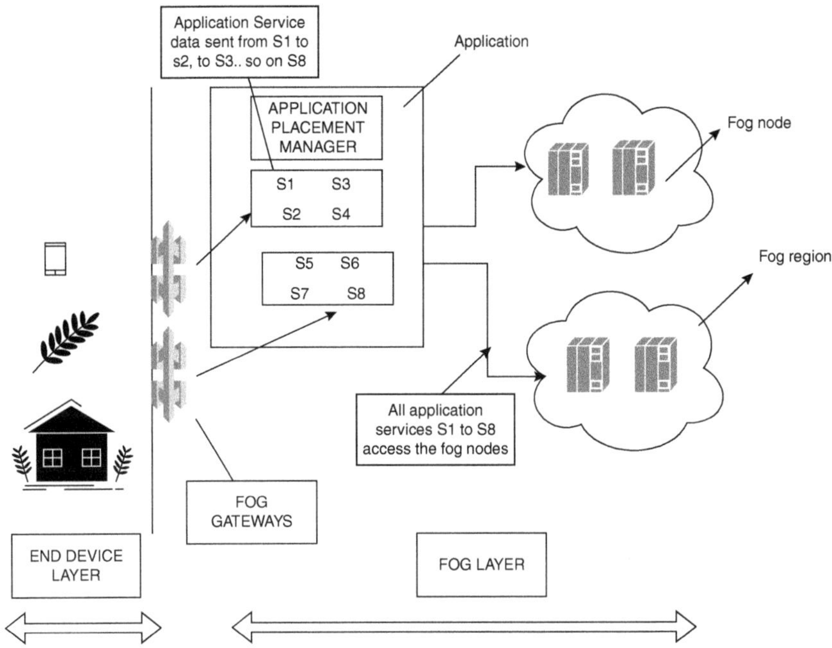

Figure 6.1 Application placement in FC.

capabilities of AI-based methods, these techniques can still prove to be effective under specific circumstances. In circumstances where the complexity of AI algorithms is neither required nor feasible, they are frequently employed as baseline approaches. Researchers have conducted surveys and reviews to classify and describe various methods for placing applications in FC. They have taken into account factors such as the strategy for placement, the type of resources, the metric for placement, the technique for mapping, the estimation of resources, the approach for offloading, the orientation of resources, and the controller for placement [8].

This chapter focuses on Fog Application Placement and its techniques, the following sections describe all the algorithms based on non-AI and AI techniques.

6.2 NON-AI TECHNIQUES FOR FOG APPLICATION PLACEMENT

To a significant degree, the resource allocation problem related to FC can be resolved by strategically placing applications based on the anticipated processing characteristics of the Fog nodes. The strategic positioning of the apps might serve as a potential means to standardize the integration of Fog and Cloud. Hence, throughout the process of placing the applications,

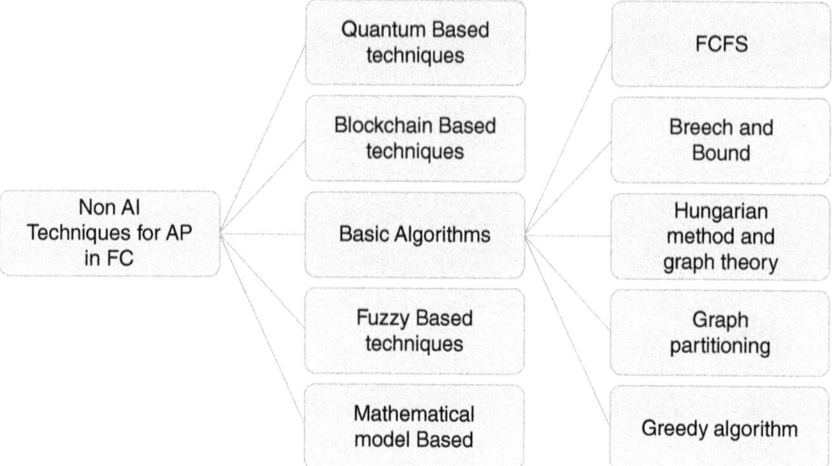

Figure 6.2 Non-AI techniques for AP in FC.

the Non-AI techniques are introduced by using different algorithms. These studies are crucial for dynamically updating the application architecture and ensuring proactive maintenance of the program. The Non-AI techniques as shown in Figure 6.2, are briefly described in the upcoming sections to focus on optimizing resource allocation and improving efficiency without utilizing artificial intelligence methods [9].

1. **Quantum-Based Techniques:** Quantum computing can profoundly transform certain facets of computing, such as Fog application placement, by using the principles of quantum mechanics to execute intricate computations with more efficiency and effectiveness. Quantum-based methods can provide benefits in addressing optimization problems, namely those related to Fog application placement, through the utilization of quantum algorithms and quantum annealing methodologies. Quantum Algorithms, such as the Quantum Approximate Optimization Algorithm (QAOA), can be employed to address combinatorial optimization problems, which are significant in establishing the most favorable arrangement of applications in Fog nodes. Quantum annealing can be used to identify the global minimum of a cost function, enabling the optimization of the distribution of resources and deployment of applications in FC environments. By utilizing quantum computing, these methods can offer more efficient and effective solutions for placing Fog applications. This can result in increased use of resources, decreased latency, and higher QoS.

2. **Blockchain-Based Techniques:** Blockchain-based techniques refer to the utilization of blockchain technology in solving problems related to FC and IoT applications. Blockchain is an autonomous and dispersed

register that documents operations across numerous computers [10]. In the context of application placement, blockchain-based techniques offer several advantages. The authors introduced a cost model that utilizes stochastic programming to determine the pricing of Edge hosts, latency, and service chaining [10]. Subsequently, they proposed a heuristic approach for placing mobile edge computing (MEC) that relies on a cost model and blockchain technology. The algorithm leverages the data recorded in the blockchain to make efficient choices regarding the distribution and utilization of resources. To guarantee impartiality in placement decisions, a distributed public ledger is employed. This ledger serves as a repository for the data about users' information, communications, and global resource availability, allocation, and consumption. The blockchain-based techniques provide a transparent and decentralized approach to application placement, leveraging the benefits of blockchain technology for optimizing resource allocation and ensuring fairness in decision-making.

3. **Basic Algorithms for Fog Application Placement:** The basic algorithms are used for fog application placement as depicted in Figure 6.2 and are described briefly in the following subsections.

   a. **First Come First Served (FCFS):** The FCFS algorithm is a fundamental method employed for determining the placement of applications in FC [11].

      Edge-ward utilizes a FCFS approach, which ensures that data is positioned close to the edge of the network, specifically on Fog nodes [12]. This algorithm has been chosen because it specifically emphasizes placement strategies that prioritize the Edge. If a particular Fog node fails to meet the demands of an application, Edge ward chooses supplementary Fog devices. The technique generates tuples of devices that reflect the paths through which "application modules" (services) are run. Application requests are processed in the order they are received until there are sufficient computing resources at each hierarchical level. If the chosen Fog device lacks computing capacity, the algorithm will seek a Fog device with sufficient capacity at the highest level of the network topology hierarchy. The algorithm adheres to a straightforward principle of allocating tasks to Fog nodes by their arrival sequence.

   b. **Greedy Algorithm:** The greedy algorithm is a fundamental method for placing applications in FC. It employs a strategy of making decisions that are locally optimal at each phase, to attain the most favorable immediate result. The author proposed an optimization methodology for efficiently organizing the timing of microservices in FC [13]. The proposed approach is a greedy method termed Fog Application Placement (FAP) that utilizes a gradient strategy. The FAP algorithm chooses a collection of categorized apps that are relevant to a particular area. It then deploys these applications on Fog nodes iteratively, ultimately achieving the most efficient deployment. Another research in this area suggests an adaptable method for distributing IoT data with numerous copies in the Fog infrastructure, while also evaluating and optimizing the total latency [14]. To address the changes in IoT applications, we have developed L backtracking greedy algorithm known as adaptive multiple data replicas placement of IoT apps with external event changes in the Fog infrastructure (iFogDPE). The algorithm is a consistent many-to-many matching algorithm that reduces write, read, and migration time by utilizing a limited number of data copies. Subsequently, it allocates each data consumer with its corresponding data replica.

c. **Branch and Bound Algorithm:** It optimizes FAP by dividing problems into smaller subproblems, exploring different branches, and bounding the search based on criteria [15]. Iterative branching and bounding narrow the search space for optimal solutions. Implementing this technique can optimize the efficiency and effectiveness of placing Fog applications, resulting in improved resource allocation, decreased latency, and greater QoS.

d. **Graph Coloring:** Graph coloring is a fundamental approach employed for application placement in FC. The algorithm assigns distinct colors to the nodes in a network, ensuring that no two neighboring nodes share the same hue. This approach facilitates the optimization by distributing the resources and load balancing in Fog settings. These basic algorithms offer uncomplicated and direct methods for determining the positioning of applications in a Fog environment. They can serve as first answers or as benchmarks for assessing the efficacy of more sophisticated placement algorithms. Zhang et al. [16] proposed an approach utilizing MEC to execute dispersed collaborative computations for optimizing the distribution of resources and offloading in distributed networks. The distributed potential game method was used to model and analyze the offloading strategy.

4. **Fuzzy-Based Techniques:** To implement a fuzzy inference system, one must first fuzzify the inputs, then apply the fuzzy rules, and finally defuzzify. By applying an input membership function to a set of discrete values, fuzzification converts them into membership grades in the set under consideration. Using appropriate language variables, explain the connection between specified input variables and outputs while using fuzzy rules. To defuzzify means to transform a set of fuzzy values into a single, unambiguous value. There are three fuzzy-based techniques [17, 18].

a. **The Fuzzy inference system:** Input fuzzification, fuzzy rule implementation, and output defuzzification are the distinct stages that constitute it. These are all performed in succession. Fuzzification is The process of converting exact figures into fuzzy membership ratings, applying fuzzy rules to describe input–output relationships, and defuzzification converts fuzzy values into precise numerical values. It involves linguistic variables.

b. **Fuzzy-based decision-making**: Fuzzy-based techniques are employed to elucidate the correlation between input variables and outputs, relying on linguistic variables. This enables decision-making by utilizing appropriate linguistic principles and considering the QoE and QoS.

c. **Fuzzy logic models**: These models are utilized for evaluating Fog instances based on their performance metrics, such as the app's expectation grade and the instances' capacity category rating. These models provide priority to application placement requests and assign them to computational instances. FAP utilizes fuzzy-based strategies that employ fuzzy logic and inference systems to manage uncertainty and imprecision throughout the decision-making process effectively. These solutions employ QoE, QoS, and performance measures to enhance the implementation of applications in FC environments.

5. **Mathematical-Based Techniques**: These are approaches and methods that use mathematical models and algorithms to solve issues and make choices. These strategies utilize mathematical equations, formulas, and optimization algorithms to assess and optimize different elements of a given situation. Mathematical techniques are employed in the context of FAP to model and reform the allocation of resources and placement of applications in FC environments. These strategies take into account criteria such as the availability of resources, the delay in network communication, the capabilities of devices, and the need for QoS to make informed decisions on the best location for applications in Fog nodes. The application placement problem is addressed using mathematical models, namely graph partitioning algorithms and optimization algorithms. These models are employed to formulate and solve the problem, to achieve effective resource usage, increased performance, and enhanced QoS [19, 20].

6.3 AI-BASED APPROACHES IN FOG APPLICATION PLACEMENT (FAP)

Artificial Intelligence is being used in broad ways, such as in the agriculture field, in smart cities, and many more, so it is important to take a very deeper look at AI to understand it in depth. Figure 6.3 shows that algorithms that

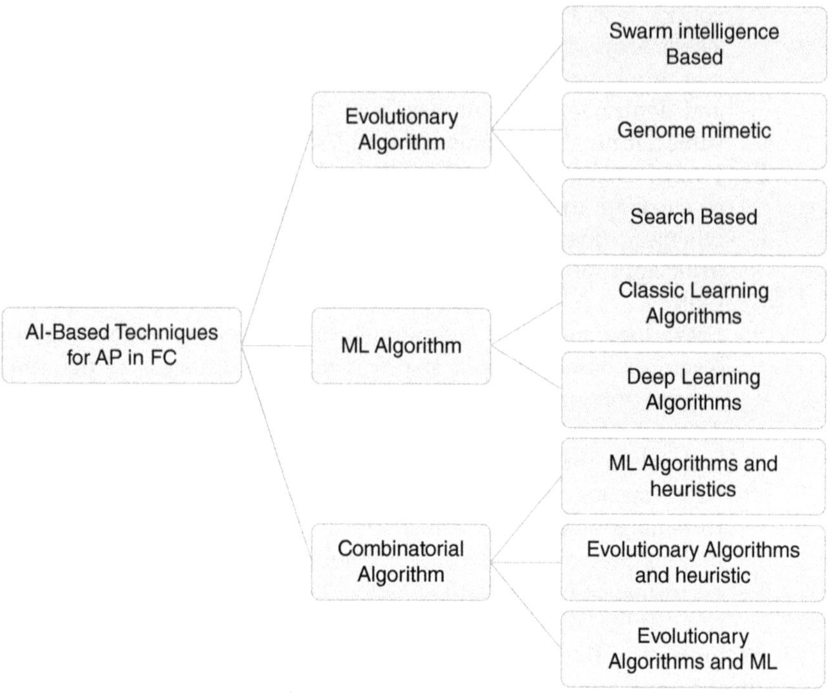

Figure 6.3 AI-based techniques for AP in FC.

use AI for FAP can be broadly classified into three types: ML [21], EV (Evolutionary algorithms) [22], and TC (combinatorial algorithms) [23].

6.3.1 Evolutionary algorithms

Evolutionary algorithms have been used in FAP to optimize resource management and load balancing. These algorithms mimic natural evolution processes, such as genetic recombination and mutation, to iteratively improve candidate solutions over generations. They operate on a population of potential solutions and use fitness functions to evaluate their performance. Evolutionary algorithms are categorized into subcategories, including swarm intelligence-based, genome mimetic, and search-based algorithms. For example, [24] integrated the Artificial Bee Colony (ABC) algorithm with Ant Colony Optimization (ACO) to overcome the issue of getting stuck in regional optima and locate the overall optimal solution. Combinatorial algorithms in FAP combine evolutionary algorithms with heuristics, machine learning algorithms, or both, to achieve better results [23]. These algorithms aim to optimize metrics for efficiency, such as makespan time or energy consumption, while considering constraints. Evolutionary algorithms have been proven to be efficient in optimizing resource allocation and enhancing service quality in FC environments [25].

6.3.1.1 Swarm intelligence algorithms

The collective behavior of social insects inspires these algorithms and aims to find optimal solutions through iterative optimization processes [26]. The swarm intelligence-based includes algorithms as shown in Figure 6.4.

- Particle Swarm Optimization (PSO): It is a widely used technique in swarm intelligence that emulates particle movement within a defined search area [27]. Every individual particle in the swarm represents a possible solution and shares information with other particles to update its position, taking into account its own experience and the best solution discovered so far [28]. PSO algorithm has been utilized in the placement of fog applications to effectively manage and optimize the overall and individual capabilities of particles. Every single particle is depicted as an itinerary determined by the number of workflow operations.
- Artificial Bee Colony (ABC): ABC is a metaheuristic algorithm that draws inspiration from honey bees' searching habits. The method employs a swarm of artificial bees to systematically navigate the search space and identify the most optimal solutions [29]. ABC has been employed in FAP to ascertain an appropriate equilibrium among cloud, fog, and IoT for application placement in the real topology of fog, taking into account aspects such as overall latency violations and system energy usage. Using few control parameters, ABC finds the

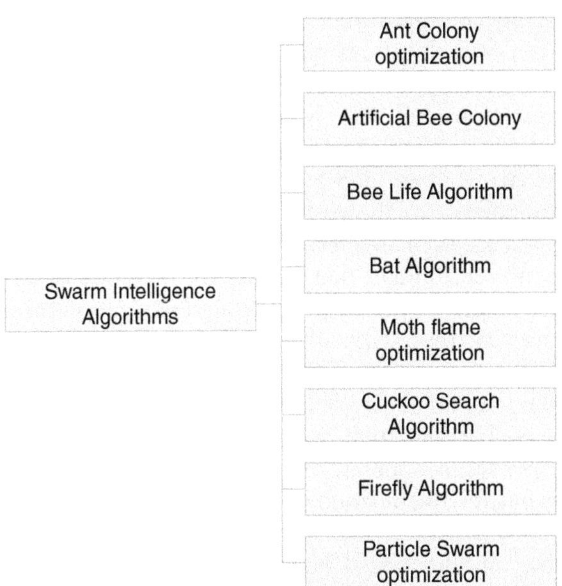

Figure 6.4 Swarm intelligence algorithms.

global best solution, which is very flexible, and has strong robustness. This algorithm's precision and sluggish convergence time are its drawbacks. To address the FC task scheduling issue, [29] suggested Bees Life Algorithm (BLA).

- The Bat Algorithm (BA) is a metaheuristic algorithm that is based on the echo behavior to locate things by bats. It can be utilized to enhance the positioning of programs in FC settings [30]. BA uses echolocation to navigate the search environment and identify the most optimal options for application placement, considering parameters like response time and resource utilization [31]. The benefits of BA include being quick to find the best solution in complex situations and being straightforward, adaptable, and efficient for nonlinear issues. BA's drawbacks include its rapid convergence and poor precision when evaluating low functions. Zafar et al. [30] has proposed the utilization of the BA to efficiently allocate computing resources on Fog for energy conservation purposes.

- The Cuckoo Search Algorithm (CSA) is a metaheuristic algorithm that draws inspiration from the brood parasitism behavior exhibited by cuckoo birds. It can be employed to optimize the allocation of applications in FC [32]. CSA emulates the behavior of cuckoo birds, which utilize the nests of numerous avian species to lay their eggs, as a means of seeking the most advantageous solutions. CSA employs randomization and local search techniques to effectively navigate the search area and identify the most favorable positions for fog applications, taking into account variables like latency and delay.

- Ant Colony optimization (ACO): Finding optimal pathways via graphs is a probabilistic computing problem that can be solved using the ACO. Artificial ants are a symbol for multiagent systems that draw inspiration from real ants' behavior. The most common paradigm is frequently that of biological ants, and their pheromone-based communication. Artificial ant and local search algorithm combinations have gained popularity as the go-to approach for many graph-based optimization tasks, such as internet and automobile routing. One class of optimization techniques that mimics the activities of an ant colony is called ant colony optimization. Through navigation through a parameter space that represents all possible solutions, artificial 'ants', such as simulation agents, find optimal solutions. As they explore their surroundings, real ants leave behind pheromones that guide one another to supplies. The simulated 'ants' save their locations and the caliber of their responses in a similar manner, allowing more ants to find better answers in subsequent simulation iterations. The bee algorithm is one version of this strategy that is more akin to the foraging habits of the honey bee, another social insect. As a result of its positive feedback accounts, ACO can cope with distributed problems effectively, and it quickly finds good solutions. Because of its versatility and ease of integration with other algorithms, ACO finds utility in dynamic applications. Even while ACO ensures convergence, the exact time it takes to get there is unclear; the probability distribution fluctuates with each iteration; the method often falls into local optimums, and large-scale issues are not fit for its lengthy search duration.

- Bee Life Algorithm (BLA): To pick cloud computing services that meet QoS requirements, the BLA is used. It is regarded as a swarm-based algorithm that mimics a bee colony's daily activities. It mimics the two key actions of foraging for food and reproduction, just like bees. The bee life approach can be used to solve both functional and combinatorial optimization issues.

- Firefly algorithms: The Firefly Algorithm is one of the greatest nature-inspired metaheuristic algorithms that encourage natural fireflies to flash. Natural fireflies entice other fireflies with their flashing abilities. The following behavior assumptions were made when the algorithm was put into practice [22, 33]. Because they are all unisexual, one firefly will always be drawn to another. The brighter of any two fireflies will attract the less brilliant one, which will cause it to travel toward it. However, if the distance between them rises, the brightness may decrease. Furthermore, a firefly will travel at random if none are brighter than the one it is.

- Moth flame optimization: Moths are beautiful insects that have a striking resemblance to the butterfly family. In general, this insect exists in approximately 160,000 different species in the natural world. Larvae and adults are the two major life stages for them. Through cocoons, the larvae become moths. The unique ways that moths navigate at night are one of the most captivating facets of their biology. They have evolved to use the light of the moon to fly at night. For navigation, they used a system known as transverse orientation. Using this technique, a moth flies by keeping a constant angle toward the moon, which is a particularly efficient way to fly great distances straight. The moon is far from the moth; therefore, this mechanism ensures a straight flight path. Humans are capable of using the same navigating technique. Assume that an individual wants to journey in the direction of the east and the moon is in the southern hemisphere of the sky. To maintain a straight trajectory toward the east, he should ensure that the moon remains positioned to his left while he walks. The low number of setup parameters, and quick early convergence during the shifting transition from the phase of research to that of exploitation, while maintaining a delicate equilibrium between the two, which is the advantage of MFO. The algorithm suffers when it diverts from the research phase by altering the number of flames. Furthermore, this method becomes stuck at local optima because of its slow convergence time. [31] suggested a task scheduling technique utilizing MFO to minimize the time it takes to execute jobs and to support QoS parameters.

6.3.1.2 Genome mimetic algorithms

A type of optimization algorithm grounded in the principles of genetics and evolution is known as a genome mimic algorithm as shown in Figure 6.5. Genome mimetic algorithms are widely employed in diverse domains, such as FC. [34] presented an FC architecture to solve the service placement issue in IoT applications using a genetic algorithm. The system aimed to optimize network communication delay and resource consumption. The techniques encompassed in this category are Memetic Algorithms (MA), Differential Evolution (DE), and Genetic Algorithms (GA).

- Genetic Algorithms (GA) imitate the process of natural selection to generate the subsequent generation. GA mimic natural selection to produce the next generation. The capacity of GAs to search a large region and provide a workable qualitative answer in a polynomial amount of time is one of its main advantages [34]. They involve several stages, including the starting population, fitness function, selection, mutation, and crossover.
- Memetic Algorithms (MA): A memetic algorithm is a type of evolutionary algorithm that combines local search heuristics with genetic operators to improve the quality of results. MA have been utilized in diverse optimization challenges, encompassing the placement of fog applications. MA facilitate FAP by systematically navigating the extensive and intricate search space of potential solutions, ultimately identifying solutions that are nearly optimal or optimal and align with the intended objectives and limitations. MA possess the ability to adjust to the everchanging and diverse characteristics of the FC environment, effectively managing the compromises that arise from conflicting goals [35].
- Differential Evolution (DE) is an optimization process that relies on a population of people to develop new solutions by calculating the difference between them. A heuristic method called Differential Evolution (DE) has successfully handled several optimization problems. The processes involved in this context are initiation, mutation, recombination, and selection [36].

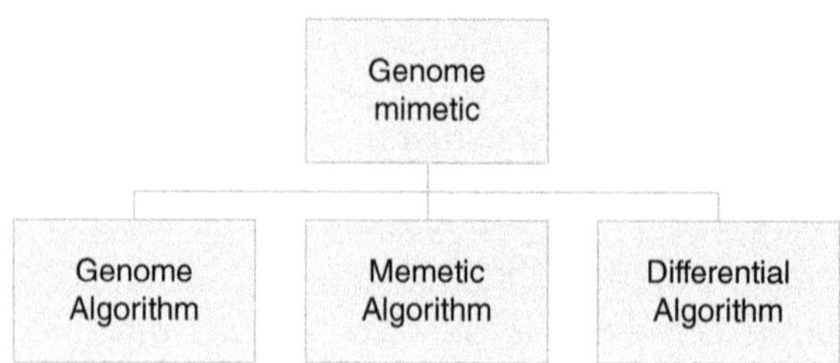

Figure 6.5 Genome mimetic algorithms.

6.3.1.3 Search-based algorithms

Search-based algorithms, use search techniques, such as genetic algorithms or PSO, that crawl through the solution space to identify the best places for applications in FC as shown in Figure 6.6.

Tabu Search: It is a metaheuristic technique made to effectively search the space for optimum solutions without getting caught in local optima. The system utilizes a "tabu list" to retain a temporary memory of previously explored solutions and limits the search from revisiting those solutions [37]. Tabu Search can be utilized for optimizing resource allocation and job scheduling in FAP, taking into account parameters like reaction time and resource utilization.

- The Gravitational Search Technique (GSA): It is an optimization technique that draws inspiration from the law of gravity and the behavior of celestial bodies. Every particle possesses a mass that serves as a measure of its performance [38]. The search process is conceptualized as the interaction between masses, with each mass representing a potential solution. The masses exhibit either attractive or repulsive forces toward each other, depending on their respective fitness values, resulting in the identification of optimal solutions. The GSA algorithm can be utilized to strategically assign fog applications to fog nodes, to optimize the allocation based on characteristics such as latency, energy consumption, and load balancing.
- Simulated Annealing (SA): To increase system performance, one of the main problems in FC is placing workloads among many fog nodes in an optimum manner. In this context, SA shows itself to be a potent optimization tool, offering a methodical way to efficiently distribute computing jobs over the network's edge. SA is a heuristic method used to solve NP-complete optimization problems [39]. The procedure begins with the creation of an objective function that includes variables such as task execution time, communication delays, and resource limitations and captures the desired performance metric. The state representation establishes the fog node configurations and workload distributions, generating an extensive state space that investigates various

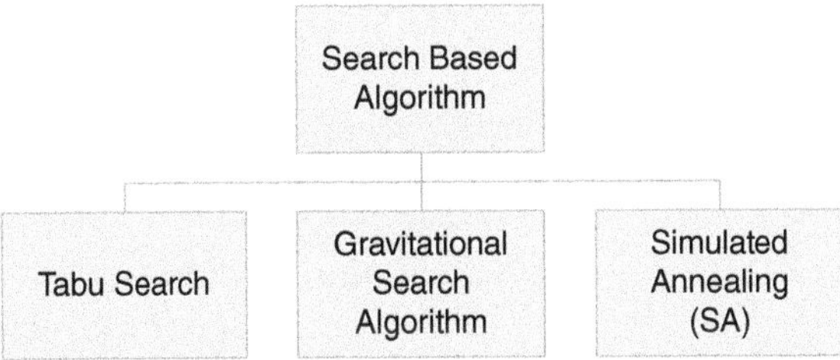

Figure 6.6 Search-based algorithms.

task placement combinations. Next, SA adds an adjacent state genera-tion mechanism, which enables the algorithm to investigate different workload distributions. By assessing each state's quality, the energy function directs the algorithm to find configurations that minimize energy usage, latency, or other pertinent factors.

6.3.2 Machine learning algorithms in FC

FC application placement on fog nodes involves strategically allocating com-puting resources to optimize system performance and efficiency via the use of machine learning techniques. To dynamically identify the best fog nodes for implementing particular applications, machine learning models analyze data traffic patterns, computing workloads, and other pertinent aspects. Data sen-sitivity, edge resource availability, and latency needs are just a few of the variables that are considered in this intelligent allocation. Machine learning algorithms can make real-time decisions on the deployment of applications through continuous learning and adaptation. This allows the processing of key workloads in closer proximity to the data source, which reduces latency and enhances response times. These models can also take into account the different computational capacities of fog nodes, assigning jobs to the most capable nodes. This method contributes to energy efficiency and cost-effectiveness in FC settings by improving overall system responsiveness and optimizing resource use. In conclusion, machine learning integration with fog node application placement allows for data-driven, dynamic decision-making that optimizes the distribution of computing activities around the edge of the network for improved efficiency and performance. The machine learning algorithms utilized in the FC are categorized into two sections: deep learn-ing algorithms and classical learning algorithms. The traditional learning Algorithms are creating algorithms that can learn from existing observations (or data sets) and predict new findings or input–output. A form of machine learning called deep learning uses many layers to train on high-dimensional input data and extract features, giving the data a more thorough description. To improve their heuristics and maximize performance metrics, a number of FAP's machine learning algorithms have integrated deep learning models. The sections that follow provide a brief summary of these algorithms.

6.3.3 Classic learning algorithm (CLA)

The CLA is divided into three major groups: supervised learning, unsuper-vised learning, and reinforcement learning as shown in Figure 6.7.

1. Supervised ML: It is the classification method employed to forecast the category of data samples from these specific groups. Supervised learning is classified into two parts: regression and classification as shown in Figure 6.8.

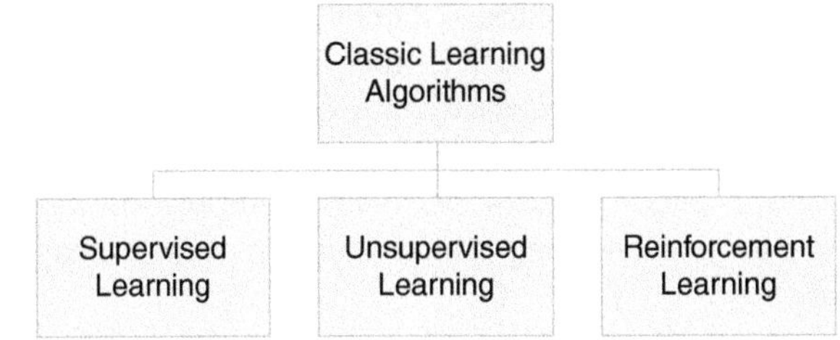

Figure 6.7 Classic learning algorithms.

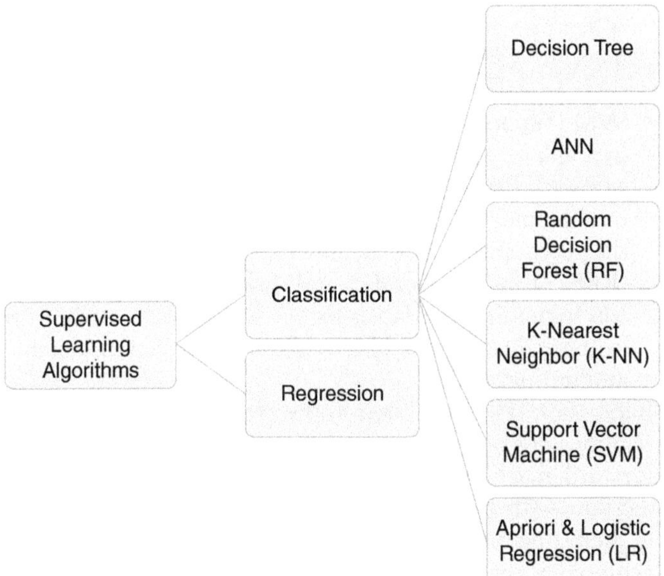

Figure 6.8 Supervised learning algorithms

- Classification Learning: FC involves developing intelligent systems capable of effectively categorizing data while also adapting seamlessly to the dynamic and unpredictable foggy environment. Classification learning aids in locating fog applications based on their requirements, enabling them to be allocated to optimal fog devices or cloud locations for enhanced performance and efficiency [40].
  - Decision Tree: Utilized in FC to analyze complex data by iteratively dividing it according to different criteria. Every split is comparable to selecting a direction in a forest obscured by fog, directing the data along several branches until it ultimately reaches a conclusive choice or categorization. The decision tree has significant benefits such as transparency, which minimizes uncertainty in decision-making, and enables a thorough analysis [21].

- ANNs: ANNs can be utilized in FC for a range of activities, including data processing, pattern identification, and decision-making. These adaptable algorithms are commonly used for supervised, unsupervised, and reinforcement learning [41]. Artificial neural network (ANN) nodes perform simultaneous data processing and analysis, enabling effective information management at the periphery.
- K-Nearest Neighbor (KNN): This algorithm is a simple and direct approach to regression and classification in lazy, instance-based learning [21]. KNN can be likened to a kind neighbor who provides guidance based on the experiences of its nearest associates. When presented with a novel data point, the algorithm examines its KNN (data points with comparable attributes) and utilizes their qualities to make determinations.
- SVM (Support Vector Machine): The SVM method finds the best hyperplane in the input space to divide two classes to solve classification issues. The hyperplane might be a two-dimensional line or a flat plane in many planes [21]. SVM kernel functions include the polynomial, sigmoid, linear kernel, and Radial Basic Function (RBF). Using training data, SVM classification trains and generalizes to predict class labels from test data. Two methods were used in the industrial machine work error detection system: training and testing.
- Regression: This is utilized in the area of FC to analyze and model relationships between different data points. Regression is a straightforward machine-learning approach that produces accurate predictions with few errors [21]. Regression algorithms are essential in analyzing and predicting patterns based on various sensor data acquired at the edge, such as temperature and humidity [42].

2. Unsupervised ML The fog filters employ unsupervised machine learning techniques to cluster the unlabeled data, categorizing it based on the value ranges observed in all tests. Unsupervised learning branches into three distinct segments: K-means clustering, Hierarchical clustering, and the Fuzzy Clustering Algorithm with Particle Swarm Optimization (FCAP) as shown in Figure 6.9 [21].

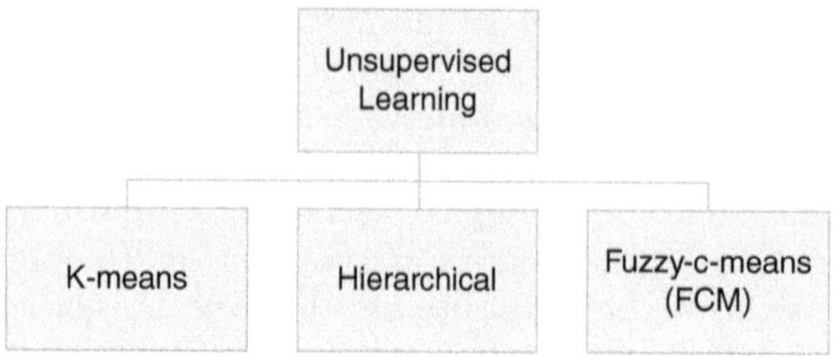

Figure 6.9 Unsupervised algorithms

- K-means Clustering: K-means clustering is an unsupervised learning method for analytical exploration of unlabeled data. K-means vector quantization is widely used in data mining[43]. This technique aims to identify the number of groups represented by the variable K in the data. The method iteratively assigns data points to K groups depending on specified attributes [21]. The algorithm requires features and K value input. Starting with K randomly picked centroids, the method iterates until convergence. This algorithm minimizes the squared error function.

- Hierarchical clustering: It is a widely recognized clustering method employed in unsupervised machine learning. K-means and hierarchical clustering are widely recognized as the main and most efficient clustering techniques. The backend employs a functional method that enables them to deliver exceptional performance. An inquiry was carried out to examine the utilization of the Hierarchical Clustering technique for the task of organizing job schedules and allocating resources in Fog environments. While considering fog application deployment, hierarchical clustering can be employed to categorize fog nodes according to their proximity and resource availability. This can aid in determining the optimal deployment site for an application module, taking into account the module's requirements and the resources available in the fog nodes [44].

- Fuzzy Clustering Algorithm with Particle Swarm Optimization (FCAP) Utilizing the conventional Fuzzy C-Means (FCM) clustering algorithm with particle swarm optimization (PSO) algorithm [45]. The FCAP method is suggested to accomplish resource scheduling in FC. The primary concept is to integrate the PSO algorithm into the FCM algorithm. The local optimization algorithm climbs hills to find the best solution.

6.3.3.1 *Deep learning algorithm*

A neural network-based representation of a learning subset of machine learning is called deep learning. Layered networks are called "deep" in deep learning as shown in Figure 6.10. It has three parts: Deep Reinforced Learning, Deep Neural Network, and Deep Robot Learning [46]. Deep learning is a subset of machine learning that utilizes multiple layers to extract features and train on high-dimensional input data, providing a more comprehensive description of the data. Deep learning algorithms, such as Deep Neural Networks (DNN), have been applied in FAP to overcome the limitations of traditional machine learning methods. Deep learning algorithms excel in handling large datasets and learning high-level features incrementally, making them suitable for FC environments with significant data sizes. Deep learning models in FAP aim to solve problems end-to-end, reducing the need for domain experts to identify specific features and improving overall

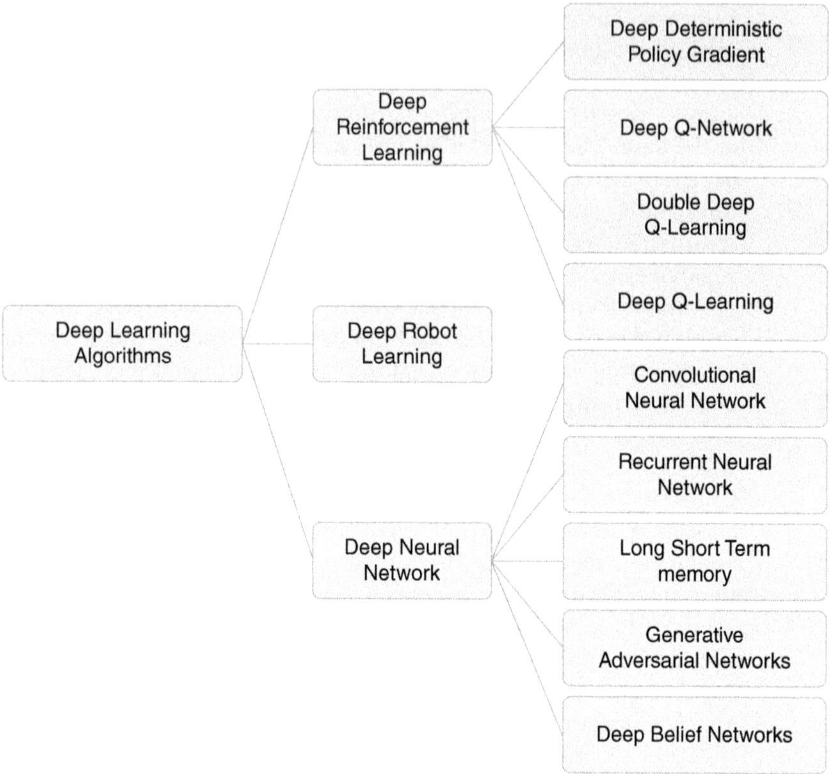

Figure 6.10 Deep learning algorithms.

efficiency. However, deep learning algorithms have longer training times due to a large number of parameters, and they may struggle with retaining previous learning and providing interpretable results. Despite these limitations, deep learning models offer improved accuracy and performance in FAP, enabling better resource allocation and optimization. Overall, deep learning models in FAP leverage their ability to handle large datasets and learn high-level features to optimize resource allocation and improve the efficiency of FC environments.

1. Deep Reinforced Learning: Deep Reinforcement Learning (DRL) is a state-of-the-art method that optimizes resource distribution and task allocation in FC environments by leveraging sophisticated neural networks for workload placement on fog nodes [47]. This approach uses a precise state representation to capture data on resource utilization, communication latencies, fog node statuses, and workload at any given time. The system can decide where to put workloads thanks to a well-designed action space, and its success is measured by a reward function that specifies goals like reducing latency or maximizing energy

usage [48]. Using this state information as input, the neural network architecture—which is frequently based on deep Q-networks or policy gradient methods—trains to provide the best possible behaviors by maximizing cumulative rewards. By finding a balance between exploration and exploitation, the DRL agent can adjust to the constantly changing FC environment [49]. The DRL agent, having been trained on a variety of scenarios, uses its acquired knowledge to make deft decisions on its own in the present while adjusting to changing network circumstances and workloads [50]. The method's capacity to continuously adjust and adjust to changes makes it a potent tool for maximizing workload placement in FC, which in turn enhances system responsiveness and efficiency at the edge.

2. Deep Neural Network (DNN): For the prediction model, deep learning was adapted. DNNs are a rapidly growing trend in intelligent data processing [51]. They provide a variety of machine learning (ML) methodologies that often exceed current models. For optimal application, deep models demand significant processing resources and numerous data examples. It is an advanced model that has been trained on large-scale labeled datasets [52]. This allows it to be used for a variety of tasks, including regression, and classification of different types of data. It basically uses several layers to extract complicated patterns and interpretations from input, which enables it to make sophisticated decisions.

   - Convolutional neural network(CNN): CNNs process the input information using convolutional layers, that are particularly useful for tasks including recognizing images. FC eliminates the necessity to send massive volumes of raw information to centralized cloud servers by using CNNs to process visual data locally at the edge. CNNs serve the purpose of object detection. For instance, [53] demonstrated the flexibility of CNNs beyond conventional image processing by proposing a four-layer CNN framework for resolving scheduling issues in FC.

   - Recurrent neural networks (RNN): RNNs have memory to capture dependencies across time and are made for processing data sequentially. RNNs are often used in FC to process data with time series, including sensor data from IoT devices. As a result, sequential data can be processed locally without depending on a remote cloud server. RNNs are used for applications like traffic volume prediction. Authors in [**CHEN20161**] showed how versatile RNN is by combining it with a hidden Markov model to forecast traffic volume.

   - Long Short-Term Memory (LSTM): An RNN type called LSTM is made especially to identify persistent dependencies in information that is sequential. LSTMs can be used in FC to forecast future outcomes in a time series or do other tasks that need the recollection of past occurrences. Authors in [51] demonstrated the effectiveness of LSTM in dynamic virtualized Edge nodes deployment for Edge cloud systems in a smartphone context by using it to anticipate requests for services and resource charges.

- Generative Adversarial Networks (GANs): GANs are aids to improving or creating new data. The discriminator and generator, which collectively make up a GAN work to produce novel instances. are GANs often utilized in FC to enhance the output of ML models by producing authentic data to enrich localized samples or for data augmentation purposes [54] focused on offloading tasks in smartphone Edge computing environments by employing GAN algorithms to forecast the upcoming trajectory of smartphone users.
- Deep Belief Networks (DBNs): DBNs are generating neural networks that can be effectively applied to problems related to classifying and feature extraction. DBNs can be used in FC to obtain useful attributes from local data before sending pertinent data to the cloud for additional processing. For example, [52] presented an effective ML model that uses logistic regression and DBN for the classification stage to facilitate compute offloading in Edge/Cloud contexts.

3. Deep Robot Learning: Fog Robotics allows for secure and distributed deep robot learning [55]. Secured processing and storage at the network edge offer many options for lower latency requirements and greater data control. However, standardizing robot communication with Edge resources is difficult for Fog Robotics adoption [56].

6.3.4 Combinatorial algorithms

Efficient utilization of resources and the implementation of dynamic task scheduling are crucial for optimizing performance and responsiveness in FC. Combinatorial algorithms are effective tools that can handle difficult optimization problems as they easily integrate with many aspects of FC.

6.3.4.1 Integration of machine learning algorithms and heuristics

The authors provide additional approaches or techniques in one of the categories of combinatorial algorithms for finding more effective responses than just basic ML techniques, which involves a mix of heuristics and ML techniques. A Module Placement technique using the Classification and Regression Tree Algorithm (MPCA) was presented by [42]. MPCA was employed for both the training and testing stages of ML algorithms, given the time-intensive characteristic of this step. MPMCP is an additional algorithm to enhance MPCA even further. The findings of MPMCP were used to determine which fog devices would work best. To schedule jobs in Fog-based IoT systems, [57] presented a dual deep Q-learning solution that includes experience replay techniques and a target network. They were able to successfully minimize the state and action space dimensions by using four different schedulers. To reduce latency, power consumption, and migration costs, [58] introduced a unique DQL (Deep Q-Learning) model for container migration in mobile activities. Using this heuristic, they maximized the random action selection throughout the exploration stage. Enhancements were also made to the Q-network's DNN training approach.

6.3.4.2 Integration of evolutionary algorithms with heuristic methods

Combining heuristics with evolutionary algorithms leverages the best features of both methodologies to provide a hybrid strategy that can solve difficult optimization issues by locating optimal solutions and traversing complicated problem domains with ease. This technique builds on the advantages of both approaches. Three evolutionary techniques were presented by [59] to solve the Fog service placement issue. The methods encompass a multiobjective adaptive algorithm that relies on the weighted sum genetic algorithm, the non-dominated sorting genetic algorithm II, and decomposition. Optimizing the latency of the network, service dispersion, and resource usage was the main goal. The optimization goals were to reduce network latency between service applications, distribute jobs among Fog nodes as efficiently as possible, and minimize free resources. A mixed heuristic approach was proposed by [60] for FC job scheduling. To reduce latency and energy consumption, this method combines improved Ant Colony Optimization (IACO) with improved Particle Swarm Optimization (IPSO). IPSO is used to collect the best answer, which is then used as the IACO algorithm's first input to find the best task-scheduling solution.

6.3.4.3 Integration of evolutionary algorithms with machine learning algorithms

The symbiotic link between evolutionary algorithms and ML algorithms aims to improve the overall efficacy of computing systems in activities that encompass optimizing for data-driven learning. A combination technique of GA and Reinforcement Learning (RL) is developed to reduce delay in resource allocation. The effective use of resources and load balancing in healthcare systems is the main goal of this approach. A Resource Allocator (RA) using RL algorithms for high load balancing and a Load Balancer Agent (LBA) for optimum request processing are both components of the Fog layer. An adaptive weight (AW), which is adjusted using a genetic algorithm to find the optimal AW value and maximize overall efficiency and minimize delay in resource allocation, determines the selection priority of a fog server [61]. PSO and FCM are used in a hybrid technique to avoid local minimum problems in fog resource scheduling.

Global optimization is ensured by the conjunction with PSO, which accelerates convergence as compared to FCM alone [62].

6.4 CONCLUSION

The survey explores challenges and solutions for sustainable development in application placement in the fog layer, a crucial aspect of FC. With increasing IoT devices, traditional cloud architectures face issues. Efficient placement requires AI and non-AI techniques to overcome secure issues. AI addresses local optima, resource limitations, heterogeneity, and security, while non-AI techniques tackle resource allocation, load balancing, and security. To progress, it is crucial for the research community to persist in investigating novel solutions, expanding the limits of both AI and non-AI methods to improve the effectiveness and safety of application deployment in FC. The survey findings enhance comprehension of the subject and offer a significant resource for academics, developers, and decision-makers dealing with the intricate landscape of FC and its problems related to application location. In order to fully harness the capabilities of FC, it is crucial to foster ongoing collaboration and interdisciplinary methods. This will enable us to effectively tackle the growing difficulties and unlock the complete potential of this sophisticated computing paradigm.

REFERENCES

1. Songhorabadi, M., Rahimi, M., MoghadamFarid, A. & Kashani, M. H. Fog computing approaches in IoT-enabled smart cities. *Journal of Network and Computer Applications* **211**, 103557 (2023).
2. Bhambri, P., Rani, S., Gupta, G. & Khang, A. *Cloud and fog computing platforms for internet of things* (CRC Press, 2022).
3. Zahmatkesh, H. & Al-Turjman, F. Fog computing for sustainable smart cities in the IoT era: Caching techniques and enabling technologies-an overview. *Sustainable cities and society* **59**, 102139 (2020).
4. Kai, K., Cong, W. & Tao, L. Fog computing for vehicular ad-hoc networks: paradigms, scenarios, and issues. *the journal of China Universities of Posts and Telecommunications* **23**, 56–96 (2016).
5. Chouikhi, S., Esseghir, M. & Merghem-Boulahia, L. Energy consumption scheduling as a fog computing service in smart grid. *IEEE Transactions on Services Computing* **16**, 1144–1157 (2022).
6. Yeruva, A. R. et al. *A smart healthcare monitoring system based on fog computing architecture* in 2022 2nd International Conference on Technological Advancements in Computational Sciences (ICTACS) (2022), 904–909.
7. Nayeri, Z. M., Ghafarian, T. & Javadi, B. Application placement in Fog computing with AI approach: Taxonomy and a state of the art survey. *Journal of Network and Computer Applications* **185**, 103078 (2021).

8. Mahmud, R., Ramamohanarao, K. & Buyya, R. Application management in fog computing environments: A taxonomy, review and future directions. *ACM Computing Surveys (CSUR)* **53**, 1–43 (2020).

9. Bhatia, M., Sood, S. K. & Kaur, S. Quantum-based predictive fog scheduler for IoT applications. *Computers in Industry* **111**, 51–67 (2019).

10. Zhu, H., Huang, C. & Zhou, J. *Edgechain: Blockchain-based multi-vendor mobile edge application placement* in 2018 4th IEEE Conference on Network Softwarization and Workshops (NetSoft), (2018), 222–226.

11. Bittencourt, L. F., Diaz-Montes, J., Buyya, R., Rana, O. F. & Parashar, M. Mobility-aware application scheduling in fog computing. *IEEE Cloud Computing* **4**, 26–35 (2017).

12. Gupta, H., Vahid Dastjerdi, A., Ghosh, S. K. & Buyya, R. iFogSim: A toolkit for modeling and simulation of resource management techniques in the Internet of Things, Edge and Fog computing environments. *Software: Practice and Experience* **47**, 1275–1296 (2017).

13. Faticanti, F., De Pellegrini, F., Siracusa, D., Santoro, D. & Cretti, S. *Cutting throughput with the edge: App-aware placement in fog computing* in 2019 6th IEEE International Conference on Cyber Security and Cloud Computing (CSCloud)/2019 5th IEEE International Conference on Edge Computing and Scalable Cloud (EdgeCom) (2019), 196–203.

14. Salah, N. B. & Saoud, N. B. B. Adaptive data placement in the Fog infrastructure of IoT applications with dynamic changes. *Simulation Modelling Practice and Theory* **119**, 102557 (2022).

15. Yu, Y., Bu, X., Yang, K., Wu, Z. & Han, Z. Green large-scale fog computing resource allocation using joint benders decomposition, Dinkelbach algorithm, ADMM, and branch-and-bound. *IEEE Internet of Things Journal* **6**, 4106–4117 (2018).

16. Zhang, J., Xia, W., Yan, F. & Shen, L. Joint computation offloading and resource allocation optimization in heterogeneous networks with mobile edge computing. *IEEE Access* **6**, 19324–19337 (2018).

17. Mahmud, R., Srirama, S. N., Ramamohanarao, K. & Buyya, R. Quality of Experience (QoE)aware placement of applications in Fog computing environments. *Journal of Parallel and Distributed Computing* **132**, 190–203 (2019).

18. Talaat, F. M., Ali, S. H., Saleh, A. I. & Ali, H. A. Effective load balancing strategy (ELBS) for realtime fog computing environment using fuzzy and probabilistic neural networks. *Journal of Network and Systems Management* **27**, 883–929 (2019).

19. Lera, I., Guerrero, C. & Juiz, C. Availability-aware service placement policy in fog computing based on graph partitions. *IEEE Internet of Things Journal* **6**, 3641–3651 (2018).

20. Kanika *et al.* *KELDEC: A recommendation system for extending classroom learning with visual environmental cues* in Proceedings of the 2019 3rd International Conference on Natural Language Processing and Information Retrieval (2019), 99–103.

21. Kumar, D. P., Amgoth, T. & Annavarapu, C. S. R. Machine learning algorithms for wireless sensor networks: A survey. *Information Fusion* **49**, 1–25 (2019).

22. Zedadra, O. et al. Swarm intelligence-based algorithms within IoT-based systems: A review. *Journal of Parallel and Distributed Computing* **122**, 173–187 (2018).

23. Talaat, F. M., Saraya, M. S., Saleh, A. I., Ali, H. A. & Ali, S. H. A load balancing and optimization strategy (LBOS) using reinforcement learning in fog computing environment. *Journal of Ambient Intelligence and Humanized Computing* **11**, 4951–4966 (2020).

24. Zahoor, S. et al. Cloud–fog–based smart grid model for efficient resource management. *Sustainability* **10**, 2079 (2018).

25. Rezazadeh, Z., Rahbari, D. & Nickray, M. *Optimized module placement in IoT applications based on fog computing in Electrical engineering (ICEE), Iranian conference on* (2018), 1553–1558.

26. Hussein, M. K. & Mousa, M. H. Efficient task offloading for IoT-based applications in fog computing using ant colony optimization. *IEEE Access* **8**, 37191–37201 (2020).

27. Nguyen, B. M., Thi Thanh Binh, H. H., The Anh, T. & Bao Son, D. Evolutionary algorithms to optimize task scheduling problem for the IoT based bag-of-tasks application in cloud–fog computing environment. *Applied Sciences* **9**, 1730 (2019).

28. Yadav, V., Natesha, B. & Guddeti, R. M. R. *Ga-pso: Service allocation in fog computing environment using hybrid bio-inspired algorithm* in TENCON 2019–2019 IEEE Region 10 Conference (TENCON) (2019), 1280–1285.

29. Bitam, S., Zeadally, S. & Mellouk, A. Fog computing job scheduling optimization based on bees swarm. *Enterprise Information Systems* **12**, 373–397 (2018).

30. Zafar, F. et al. *Resource allocation over cloud-fog framework using BA in Advances in Network-Based Information Systems: The 21st International Conference on Network-Based Information Systems (NBiS-2018)* (2019), 222–233.

31. Ghobaei-Arani, M., Souri, A., Safara, F. & Norouzi, M. An efficient task scheduling approach using moth-flame optimization algorithm for cyber-physical system applications in fog computing. *Transactions on Emerging Telecommunications Technologies* **31**, e3770 (2020).

32. Nazir, S. et al. *Cuckoo optimization algorithm based job scheduling using cloud and fog computing in smart grid in Advances in Intelligent Networking and Collaborative Systems: The 10th International Conference on Intelligent Networking and Collaborative Systems (INCoS-2018)* (2019), 34–46.

33. Hassan, K. et al. *A cloud fog based framework for efficient resource allocation using firefly algorithm in Advances on Broadband and Wireless Computing, Communication and Applications: Proceedings of the 13th International Conference on Broadband and Wireless Computing, Communication and Applications (BWCCA-2018)* (2019), 431–443.

34. Skarlat, O., Nardelli, M., Schulte, S., Borkowski, M. & Leitner, P. Optimized IoT service placement in the fog. *Service Oriented Computing and Applications* **11**, 427–443 (2017).

35. Goudarzi, M., Wu, H., Palaniswami, M. & Buyya, R. An application placement technique for concurrent IoT applications in edge and fog computing environments. *IEEE Transactions on Mobile Computing* **20**, 1298–1311 (2020).

36. Manasrah, A. M., Aldomi, A. & Gupta, B. B. An optimized service broker routing policy based on differential evolution algorithm in fog/cloud environment. *Cluster Computing* **22**, 1639–1653 (2019).

37. Mouradian, C. et al. Application component placement in NFV-based hybrid cloud/fog systems with mobile fog nodes. *IEEE Journal on Selected Areas in Communications* **37**, 1130–1143 (2019).

38. Karamoozian, A., Hafid, A. & Aboulhamid, E. M. *On the fog-cloud cooperation: How fog computing can address latency concerns of IoT applications* in *2019 Fourth International Conference on Fog and Mobile Edge Computing (FMEC)* (2019), 166–172.

39. Wu, Y., Tang, M. & Fraser, W. *A simulated annealing algorithm for energy efficient virtual machine placement* in *2012 IEEE international conference on systems, man, and cybernetics (SMC)* (2012), 1245–1250.

40. Aqib, M., Kumar, D. & Tripathi, S. Machine Learning for Fog Computing: Review, Opportunities and a Fog Application Classifier and Scheduler. *Wireless Personal Communications* **129**, 853–880 (2023).

41. Fei, X. et al. CPS data streams analytics based on machine learning for Cloud and Fog Computing: A survey. *Future Generation Computer Systems* **90**, 435–450 (2019).

42. Rahbari, D. & Nickray, M. Task offloading in mobile fog computing by classification and regression tree. *Peer-to-Peer Networking and Applications* **13**, 104–122 (2020).

43. Selimi, M. et al. A lightweight service placement approach for community network micro-clouds. *Journal of Grid Computing* **17**, 169–189 (2019).

44. Goudarzi, M., Palaniswami, M. & Buyya, R. *A distributed application placement and migration management techniques for edge and fog computing environments* in *2021 16th Conference on Computer Science and Intelligence Systems (FedCSIS)* (2021), 37–56.

45. Li, G., Liu, Y., Wu, J., Lin, D. & Zhao, S. Methods of resource scheduling based on optimized fuzzy clustering in fog computing. *Sensors* **19**, 2122 (2019).

46. Pouyanfar, S. et al. A survey on deep learning: Algorithms, techniques, and applications. *ACM Computing Surveys (CSUR)* **51**, 1–36 (2018).

47. Nguyen, D. C., Pathirana, P. N., Ding, M. & Seneviratne, A. Secure computation offloading in blockchain based IoT networks with deep reinforcement learning. *IEEE Transactions on Network Science and Engineering* **8**, 3192–3208 (2021).

48. Montague, P. R. Reinforcement learning: an introduction, by Sutton, RS and Barto, AG. *Trends in cognitive sciences* **3**, 360 (1999).

49. Farhat, P., Sami, H. & Mourad, A. Reinforcement R-learning model for time scheduling of ondemand fog placement. *The Journal of Supercomputing* **76**, 388–410 (2020).

50. Lu, H. et al. Edge QoE: Computation offloading with deep reinforcement learning for Internet of Things. *IEEE Internet of Things Journal* **7**, 9255–9265 (2020).

51. Yuan, X., Sun, M. & Lou, W. A dynamic deep-learning-based virtual edge node placement scheme for edge cloud systems in mobile environment. *IEEE Transactions on Cloud Computing* **10**, 1317– 1328 (2020).

52. Alelaiwi, A. An efficient method of computation offloading in an edge cloud platform. *Journal of Parallel and Distributed Computing* **127**, 58–64 (2019).

53. Li, H., Ota, K. & Dong, M. Deep reinforcement scheduling for mobile crowdsensing in fog computing. *ACM Transactions on Internet Technology (TOIT)* **19**, 1–18 (2019).

54. Wu, C., Peng, Q., Xia, Y. & Lee, J. *Mobility-Aware Tasks Offloading in Mobile Edge Computing Environment* in *2019 Seventh International Symposium on Computing and Networking (CANDAR)* (2019), 204–210.

55. Tanwani, A. K., Mor, N., Kubiatowicz, J., Gonzalez, J. E. & Goldberg, K. *A fog robotics approach to deep robot learning: Application to object recognition and grasp planning in surface decluttering* in *2019 international conference on robotics and automation (ICRA)* (2019), 4559–4566.

56. La, Q. D., Ngo, M. V., Dinh, T. Q., Quek, T. Q. & Shin, H. Enabling intelligence in fog computing to achieve energy and latency reduction. *Digital Communications and Networks* **5**, 3–9 (2019).

57. Gazori, P., Rahbari, D. & Nickray, M. Saving time and cost on the scheduling of fog-based IoT applications using deep reinforcement learning approach. *Future Generation Computer Systems* **110**, 1098–1115 (2020).

58. Tang, Z., Zhou, X., Zhang, F., Jia, W. & Zhao, W. Migration modeling and learning algorithms for containers in fog computing. *IEEE Transactions on Services Computing* **12**, 712–725 (2018).

59. Guerrero, C., Lera, I. & Juiz, C. Evaluation and efficiency comparison of evolutionary algorithms for service placement optimization in fog architectures. *Future Generation Computer Systems* **97**, 131–144 (2019).

60. Wang, J. & Li, D. Task scheduling based on a hybrid heuristic algorithm for smart production line with fog computing. *Sensors* **19**, 1023 (2019).

61. Talaat, F. M., Saraya, M. S., Saleh, A. I., Ali, H. A. & Ali, S. H. A load balancing and optimization strategy (LBOS) using reinforcement learning in fog computing environment. *Journal of Ambient Intelligence and Humanized Computing* **11**, 4951–4966 (2020).

62. Li, G., Liu, Y., Wu, J., Lin, D. & Zhao, S. Methods of resource scheduling based on optimized fuzzy clustering in fog computing. *Sensors* **19**, 2122 (2019).

Chapter 7

Fog computing for sustainable smart cities

Software deployment, data analytics and legal considerations

Swati Malik

Chitkara University Institute of Engineering and Technology, Chitkara University, Chandigarh, India

Swati Malik

Jindal Global Law School, OP Jindal Global University, Sonipat, India

Gaganpreet Kaur

Chitkara University Institute of Engineering and Technology, Chitkara University, Chandigarh, India

Shilpi Harnal

Chandigarh University, Mohali, India

Vinay Gautam

Chitkara University Institute of Engineering and Technology, Chitkara University, Chandigarh, India

7.1 INTRODUCTION

World fashion is shifting towards sustainable smart cities. Sustainable smart city concepts aim to improve inhabitants' quality of life while delivering services by leveraging cutting-edge technology. Many governments have organised large-scale initiatives to transform most cities into smart cities throughout the world. North America [1], Europe [2], Asia, Australia, China [3], and other regions can all be home to several smart city development initiatives. Smart cities are being made possible by a multitude of technologies, one of which is the Internet of Things (IoT). To gather data and recover data across the system and other devices, all of these are using the Internet [4]. In the foreseeable future, IoT will likely have many possible uses because it is always changing [5]. IoT provides a diversity of components that are either built-in or integrated into sensors, actuators, software, stimulating power, and connected appliances. This type of arrangement provides uninterrupted applications in smart cities, which may be set up with multiple IoT-implemented devices and structures, for instance, mobile phones in the

DOI: 10.1201/9781003494430-7

communication system, sensors in the transportation system, and cameras in the monitoring system, etc.

IoT combines multiple devices by using the Internet [6]. As a result, in order to monitor them from a distance, a smart city may include intelligent appliances and sensor networks such as controlling the lights, air conditioner, power usage, etc. Both quicker data processing and access are necessary for each of these applications. Coordinated action is required to address communication network delay since it can result in substandardised network performance. Fog computing can be highly important in supporting low latency access for IoT applications. IoT devices and sensors may be fully deployed at a number of places with the incorporation of fog computing, and data processing and analysis can be performed more quickly [7]. Using big data analytics in IoT applications can be facilitated by integrating the fog nodes.

In addition to smart medical services, smart signals, smart transportation, and smart houses, the smart city offers a host of other community services. A sustainable smart city uses a variety of connected services to enhance the quality of life for its people by integrating sensors, data, and algorithms [8]. When it comes to designing and executing sustainable smart city projects, most nations are still in the early phases of adopting cutting-edge technologies. In addition to the hazy parade, Deloitte estimated in 2018 that over 1000 smart city developments had been completed globally, with China accounting for half of these projects [9]. The market for smart cities is projected to be worth USD 83.9 billion in 2019 and is projected to increase at a compound annual growth rate (CAGR) of 24.7% between 2022 and 2032. Most importantly, the development of smart cities is anticipated to be aided by the technologies known as AI, IoT, and 5G (AIoT5G) [10]. Projected global revenue for smart cities from 2020 to 2032 is displayed in Figure 7.1, with smart infrastructure accounting for the largest portion of the market with an estimated revenue of over 120 billion USD in 2022 [7].

7.2 SMART CITY DEVELOPMENT THROUGH DATA AND SOFTWARE DEPLOYMENT

To improve the standard of services provided to citizens, numerous cities worldwide are contemplating the operation of sustainable smart city models. While maintaining a sustainable smart city requires a range of software, tools, and technologies. The main emphasis is on offering smart services to enhance healthcare, education, energy, transport, and various relevant sectors [11, 12]. Fog computing, IoT, cloud computing, robotics, big data analytics, clouds, wireless sensor networks (WSNs), cyber-physical systems (CPS), and unmanned aerial vehicles (UAVs) are a few examples of these software and technology [11, 13]. In smart cities, a combination of these state-of-the-art instruments and technologies offers different service levels.

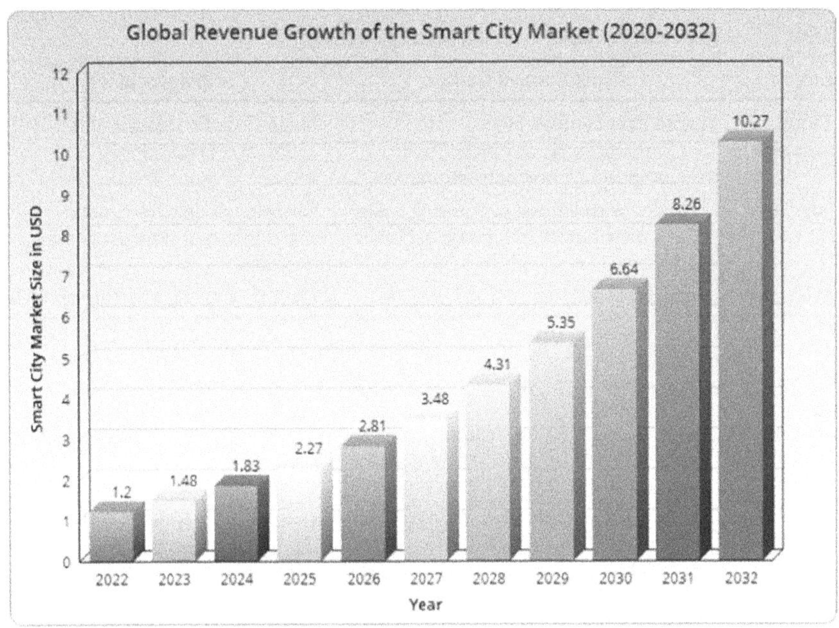

Figure 7.1 Global revenue growth for smart cities (2022–2032) [7].

IoT facilitates seamless integration among entities, into city networks [13]. In smart cities [14], cyber-physical methods are employed for facilitating secure communication between the physical and virtual realms. Functional and automation services are supported by smart cities by utilising robotics and UAVs [15]. While cloud computing offers an inexpensive, scalable computing environment and a data storage platform to support smart applications [10, 11]. Big data analytics offers intelligent and optimised data-driven assessments towards the smart solutions [16]. Fog computing offers low latency, improved mobility, real-time support, and location awareness for applications related to smart cities [17]. The most recent advancements in fog computing have made it feasible to support data storage and software deployment on the fog nodes with services like DaaS. As a result, through the application that enables data to be deployed on the fog node—reliance on fast data analysis becomes crucial. On the other hand, time-sensitive analysis uses the fog node's data and requires low latency. Table 7.1 shows the computational task of each layer in a fog environment. Figure 7.2 shows the layered architecture of the fog environment.

Simultaneously, a growing number of advancements in healthcare and transportation that are implemented through IoT-based solutions generate a massive amount of data [18]. Therefore, using contemporary transfer technologies is an efficient method for moving large amounts of data to fog servers. For example, the speed of IBM Aspera's immediate file transfer

Table 7.1 Computational tasks in multilayer architecture

Layer	Computational Devices	Computational Tasks
Cloud Layer	Large data centres, High Computational capability, Super Computers, Enormous storage.	Huge Data Processing and Storage, Data mining
Fog Layer	Devices with limited processing ability in a distributed architecture, Mobile devices.	Knowledge discovery, data analysis, real-time processing of data.
IoT Layer	Actuators, Sensors, and IoT devices.	Data Processing and Human GUI

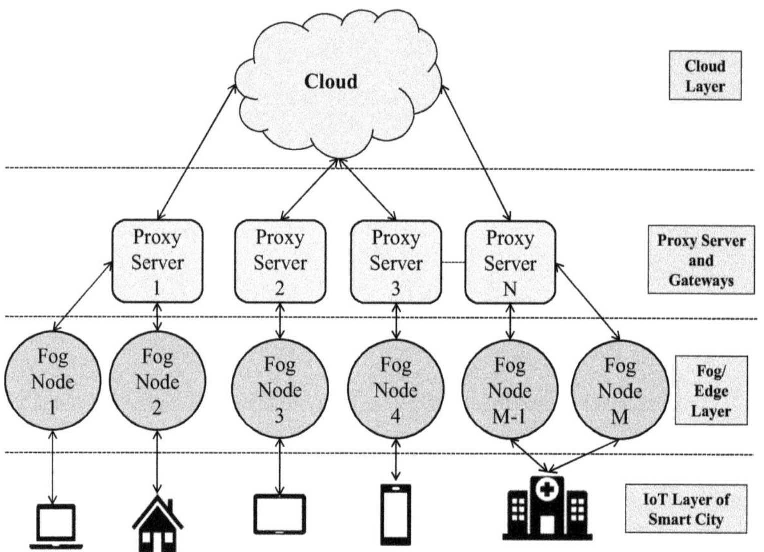

Figure 7.2 Layered architecture of fog computing for smart cities.

technology (Aspera FASP) is faster than that of traditional file transfer technologies. Many large corporations use this technology to speed up the transfer of massive volumes of data [19, 20]. Consequently, for workstation support, a multitude of software tools utilised by branches in various cities are considered. Therefore, cost-effective and efficient big data storage, sharing, and analysis are essential for smart cities. The public can access a vast amount of smart city data for use with a variety of software applications utilising fog-based services.

Figure 7.3 shows the deployment of software and data related to the IoT in fog nodes for applications in smart cities. Furthermore, smart cities leverage the design of smart services that utilise various technologies such as: (a) tracking patterns in device connections; (b) creative distributed smart control algorithms; (c) communication and location; (d) managing the computation of renewable power; (e) integrating power data into service security; and so

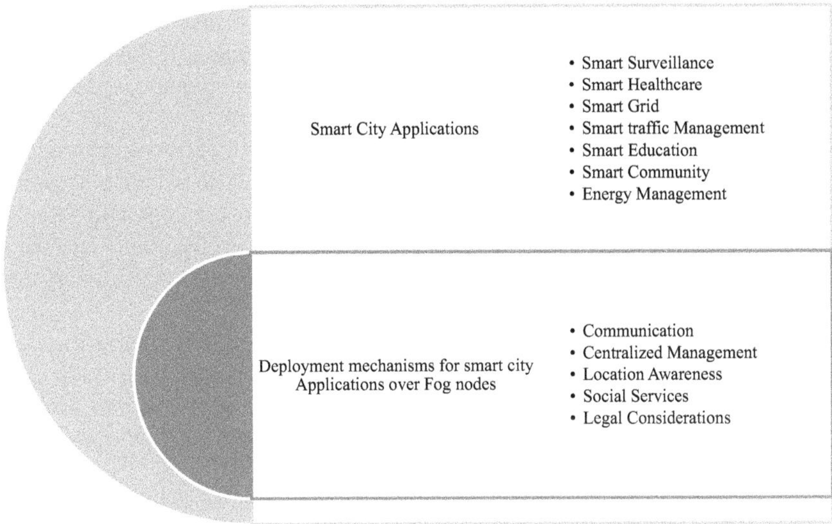

Figure 7.3 Deployment Mechanisms for smart city applications over fog nodes.

on. These technologies are used to minimise human activity through the collection of data through central management and power management [21]. It is possible to create smart cities with automation, security, and energy-saving features by putting software and data into fog nodes [6].

Fog and edge computing, which seeks to create a link between the widespread IoT deployment and the present cloud systems, is a promising paradigm for smart cities' development. Edge and fog computing features include scalability, localisation, security, low latency, and confidentiality [22]. Consequently, the idea of a "smart city" employs hardware for the deployment and support of complex algorithms, edge and fog node functionalities, and a range of devices and features. Fog nodes, IoT applications, and objects themselves are all interconnected in smart cities. These applications analyse and gather data to create new insights that can lead to improved performance and real added value [23, 24]. Value-added methods offer smart services and the capacity to create the next system's architecture. These factors mainly rely on interoperability, which is predicated on the potential to use data processing to provide intelligence, and heterogeneity, or managing various types of related objects [25]. The core concepts of edge and fog computing emerge as a paradigm for handling data for IoT applications and devices [21]. Fog environment also adheres to the security and quality of service (QoS) standards [26, 27]. The following steps are come across during the working of an application in fog environment [28]:

Analysis: In this stage, professionals with various backgrounds collaborate. This covers information and communications technology (ICT) technicians as well as managers from the security, power, and energy domains. Experts talk about and pinpoint the primary processes that need to be under control. ICT specialists participate in this process as an integrated link. This analysis turns into a very user-focused method for outlining the requirements for each subsystem.

Design: An edge, fog, and cloud-based three-level architectural design is employed. Layered planning combines techniques to provide increased smart services and subsystem interoperability at the edge and fog levels. The cloud layer connects with the Intranet through the integration of IoT devices, apps, and protocols. This allows the system to benefit from the edge and fog paradigms.

Execution and Data Analysis: Following installation and integration during the relevant phase, the subsystems are put into place. Rules in each subsystem give rise to services. Automated and sophisticated services are proposed by analysing the data generated by things or objects.

Establish: Directions for each subsystem are included in the expert rules. Rules with opinion or view processes are then installed. Intelligent tools and techniques are used to infer the automatic and modified rules. The user-centred method is used to develop a technique for designing, enhancing, and validating creative services that meet interoperability requirements.

The node requirements and specifications that are used for the management of things can be suggested by means of an extra design layer. Either fog nodes or edge nodes implement the service and procedure designs. Determine the services, internal operations, and communication of each node in detail. The data attains an extension of processing and intelligence capabilities. Nodes that are in close proximity to data sensors, controllers, and actuators are found in the edge and fog layers.

Edge Layer Functionality: This layer controls functionality and oversees the software that is created on inbuilt devices to link sensors and actuators. An edge node has several sophisticated algorithms installed on it. In order to facilitate integration into the local network, interfaces and communication devices are installed.

Functionality of the Fog Layer: The local area network observation's action, file configuration, communication, and storage are the foundations of the fog layer's functionality. The fog node uses the data to create an IoT device that can process, communicate, and store data, such as a server or gateway. All service integration is finished at this point only. The fog layer device is also capable of performing edge node functions while providing services. There are three more steps that need to be implemented at edge and fog nodes [11, 21].

Communication Services: Every device ought to be connected to a comparable network that has interoperable components. All of the actuators' and sensors' components ought to be able to use the services. For instance, obtaining environmental conditions, reading power parameters from far-off locations, and accessing open weather forecasts online. Additional features could be connection dependability, security, and interoperability. Consequently, providing seamless communication is the goal.

Data Processing and Control in Embedded Devices: Embedded devices can acquire new capabilities and implement essential control rules and data analysis services by means of data processing and control. This phase could be working with the meteorological data to compute, looking at energy consumption, finding patterns, identifying events, etc.

Developed Services on Gateway Nodes: This action focuses on IoT-based devices and applications with communication protocols, as well as advanced computing paradigms. By storing and filtering data, fog nodes intelligently analyse and communicate it. Additionally, it creates information about issues with cloud services or authorises control actions at lower levels. Counting water and power consumption, detecting clever patterns, and other analytical services are a few examples of the actions.

Created Services on Gateway Nodes: This step addresses IoT-based applications and devices that use communication protocols, as well as advanced computing paradigms. Fog nodes intelligently analyse data by communicating it through a store-filter process. Along with generating information about issues with cloud services, among the activities are the analysis of new patterns, intelligent detection, estimation of water and power consumption, and analytical applications.

7.3 ANALYTICS AND DATA MANAGEMENT USING FOG

A smart city continuously generates enormous amounts of data at ever-increasing rates. Digital platforms, social media, gadgets, sensors, services, and applications all generate data. Protecting momentum and dynamics as well as enhancing the quality of life in smart environments depend on effective data management [2, 21, 29]. The data generated by smart cities from an operational standpoint [28] is expected to have significant characteristics that support implementation of sustainable smart city technology. It is currently identified as a recently emerging field in the data economy.

Private Fog Node: A private fog node is usually intended for limited use by one company with several clients. The cluster could be owned, run, and managed by the organisation, or it could be entirely (or partially) managed by a for-profit entity.

Community Fog Node: A community fog node is typically intended for the limited use of a particular consumer public having numerous organisations with similar goals. It may be owned, run, and managed by one or more community organisations, or it may be managed by a different entity.

Public Fog Node: The general public is intended to have access to a public fog cluster. The cluster may be owned, run, and managed by a company, government organisation, university, or a mix of these three.

Hybrid Fog Node: A hybrid fog node is made up of two or more independent fog nodes, which could be private, community, or public. Proprietary or standardised technology ensures the transfer of data and applications among nodes.

Smart cities are implementing IoT projects to gather and analyse enormous amounts of continuously generated data. Figure 7.4 depicts data management in fog-cloud environment. This is done to enhance existing services, develop smart applications, and effectively respond to real-time emergencies. For achieving these goals, innovative methodologies and advanced data management techniques are required. While traditional databases and business intelligence architectures remain important for smart cities, IoT solutions necessitate specialised skills to effectively handle continuous data from diverse sources in varied forms.

Data management in the era of the IoT is evolving into a comprehensive field that involves various methods, tools, and platforms for storing, preparing, and processing data in both batch and real-time scenarios. Data management includes various additional disciplines linked to data, such as data integration (the process of combining and unifying data), data quality management (ensuring the accuracy and reliability of data), data provisioning (making data available to users), and data governance. Hence, the management of IoT data in organisations, including smart cities, is an intricate undertaking, particularly when leveraging extensive volumes of data from diverse and dissimilar sources for business intelligence and decision-making purposes. These organisations utilise algorithms to extract data from many sources, ensuring different levels of data quality before storing them in a centralised data repository. In addition, they employ automated data aggregation and classification technologies at the edge and in the fog to expedite the extraction of insights from data streams and safeguard data repositories from large data volumes and rapid data flow. Figure 7.4 illustrates the previously indicated data management tasks throughout the range from devices in the fog-cloud scenario to IoT.

Smart cities are implementing IoT efforts to gather and analyse enormous amounts of data in order to enhance existing services, develop new ones, and effectively respond to emergencies. This requires the adoption of innovative methodologies and data management techniques. While traditional databases and business intelligence architectures remain important for smart cities, IoT solutions necessitate specialised skills to effectively handle continuous data from diverse sources in varied forms.

Data management in the era of IoT is evolving into a comprehensive field that involves various methods, tools, and platforms for storing, preparing, and processing data in both batch and real-time scenarios. Data management

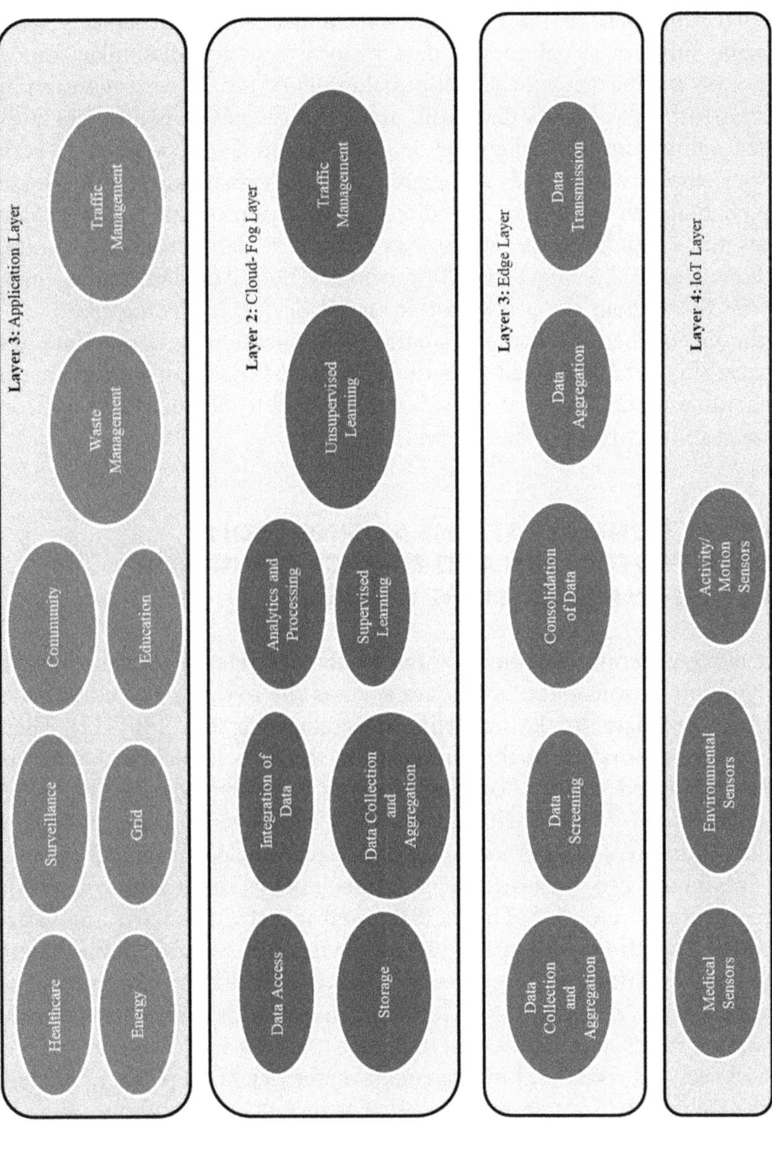

Figure 7.4 Data management in fog-cloud environments.

includes various other data-related fields such as data integration (the process of combining, consolidating, and connecting data), data quality management (ensuring the accuracy and reliability of data), data provisioning (supplying data to users or systems), and data governance (establishing rules and policies for data usage). Hence, the management of IoT data in organisations, particularly in smart cities, is an intricate undertaking, particularly when leveraging substantial volumes of data from diverse and dissimilar sources for business intelligence and decision-making purposes. These organisations utilise algorithms to extract data from many sources, ensuring different levels of data quality prior to storing them in a centralised data repository. In addition, they employ automated data aggregation and classification technologies at the edge and in the fog to expedite the extraction of insights from data streams and safeguard data repositories from large data volumes and rapid data flow. Figure 7.3 illustrates the previously indicated data management functions throughout the continuum from IoT devices to the fog/cloud.

In the rise of the technical era, smart city applications generate data at a fast pace; this data is utilised sometimes in shared mode and sometimes in non-sharable mode. So, legal considerations need to be addressed in order to protect the records and data from misuse.

7.4 LEGAL CONSIDERATIONS ARISING FROM DATA PROTECTION AND PRIVACY CONCERNS IN FOG-ENABLED SMART CITIES

Smart cities generate enormous volumes of data related to their citizens by relying on various data techniques such as the IoT, fog and cloud environments, big data analytics, artificial intelligence, etc. [30, 31] These data-driven systems amass this information through surveillance cameras, internet-connected sensors, location tracking, facial recognition, biometric data collection, video surveillance, etc [32]. Moreover, smart cities maintain digital archives related to demographic records, geographical records, municipal taxes, property records, legal documents about land ownership and registrations, etc [33]. This data is used to provide smart, innovative services and solutions like intelligent transportation systems, public safety, optimal conservation of energy, waste disposal, a healthy environment, and smart healthcare delivery [34]. Gathering massive quantities of personal information raises questions about likely misuse, concerns over data protection and security risks, and infringement of the "right to privacy," including "the right to be forgotten" [35, 36]. For instance, smartphones may be turned into tracking devices, as discussed by Justice Sotomayor in *United States v. Jones*, "G.P.S. monitoring can generate a precise, comprehensive record of a person's public movements that reflects a wealth of detail about a person's familial, political, professional, religious, and sexual associations like trips to the psychiatrist, the abortion clinic, the AIDS treatment centre, the strip club, the criminal defence attorney, the gay bar, etc. Moreover, the government can store such records and efficiently mine them for information years into the future. Because G.P.S. monitoring is cheaper than conventional surveillance techniques and proceeds surreptitiously by design, it evades the ordinary checks constraining abusive law enforcement practices" [37]. In light of these concerns, it becomes imperative that smart city designs aim to protect the right to privacy of an individual, including the "privacy of behaviour and habits, communication, data and images, ideas and feelings, location and space, and affiliation" [36, 38].

7.4.1 Understanding the legal position on data privacy and data protection in India

Considering these concerns, it becomes crucial to understand the statutory framework governing smart cities' gathering, handling, and storage of data [39, 40]. The Information Technology Act, 2000 ("I.T.A.") covers India's data protection and security laws. The Act imposes an *"obligation to pay compensation"* on corporate bodies like companies or firms that are careless in employing and supporting reasonable security measures related to any "sensitive personal data" possessed or handled by them "in a computer resource" [36, 41, 42]. This compensation is paid to the person suffering wrongful loss or gain because of this negligence. The Statute punishes *"breach of confidentiality and privacy"* [36, 41, 42]. The law provides that if an individual has obtained "access to any electronic record, information, document or other material without the consent of the person concerned and discloses such information or material to any other person, they shall be punished with imprisonment for a term which may extend to two years, or with fine which may extend to one lakh rupees, or with both" [36, 41, 42]. The divulgation of information "in breach of a lawful contract" is punishable under the law [36, 41, 42]. Any individual as well as an "intermediary who, while providing services under the terms of a lawful contract, has secured access to any material containing personal information about another person, with the intent to cause or knowing that he is likely to cause wrongful loss or wrongful gain discloses, without the consent of the person concerned, or in breach of a lawful contract, such material to any other person, shall be punished with imprisonment for a term which may extend to three years, or with fine which may extend to five lakh rupees, or with both" [36, 41, 42]. The Act provides an *exemption to intermediaries* "for any third-party information, data, or communication links provided or hosted by it, provided the intermediary's function is limited to granting access to a communication system where information provided by third parties is transmitted, temporarily stored, or hosted, or if the intermediary does not initiate the transmission, chooses the receiver of the transmission, select or modify the knowledge contained in the transmission. However, this exemption won't apply if the intermediary has conspired, abetted, aided, or induced, whether by threats, promise, or otherwise, in committing the unlawful act" [36, 41, 42]. The government can prevent the public from accessing internet URLs, websites, or applications on mobile phones if they disrupt public order [36, 41, 42]. However, these laws do not address all the legal issues arising from smart cities, and such limitations can be addressed by drafting privacy policies or entering into contracts or settlements integrating the requisites to safeguard the interests of IoT service providers as well as the users [41, 42].

7.4.2 Apprehensions about the right to privacy and safety in smart cities

With the evolution of smart cities as a key to urban problems, concerns over data protection and privacy emerge as real challenges [34, 43–45]. Achieving sustainable development requires balancing innovation and safeguarding individual privacy rights [39]. The sensitive data collected by the smart cities should be accessible only to the proper parties for a relevant purpose and must be safeguarded [33, 46]. The IoT service providers hired by the smart cities must have a detailed privacy policy to ensure transparency [47]. This policy should cover the restrictions on using such information and safety measures to secure data and prevent hacking or unauthorised use [29, 31]. Smart cities should adopt policies on data minimisation and purpose limitation [34, 42, 44]. For instance, the "European Parliament and Council of European Council" implemented "Regulation (EU) 2016/679" in 2018, which provides for a "hefty penalty upon the violators of the standards relating to privacy and security. The regulation is premised upon the principles of transparency, fairness and lawfulness, purpose limitation and limiting data collection; the accuracy of data; temporal limitations on storage; integrity and confidentiality and accountability" [48, 49]. The IoT service providers can also enter non-disclosure agreements with IoT users to cover such policies [47]. The issues concerning citizens' rights to access personal data, its control, and portability must be covered under such policy [36, 39, 40]. Citizens should give informed consent before data collection [34]. For instance, laws like the California Consumer Privacy Act of 2018 prevent the sharing and collection of consumer data without the knowledge of the concerned person [50]. Also, European Union Convention 108 obligates the nation-state parties to outlaw the collection and handling of information regarding "race, political or religious belief, health, and sexual life" without the individual's knowledge and proper legal safeguards [31, 42]. Additional guidelines are provided under this Convention, which provides for the rectification or erasure of such data obtained or processed in violation of domestic legislation [51, 52]. The IoT service providers should also anonymise and de-identify data to reduce risks related to the identification of citizens' personal information [34, 44]. For example, painting over or censoring licence plates, etc. Secure Digital infrastructure like Data Safes and Data Trusts can also significantly ensure security and privacy.

7.4.3 Concerns arising from third-party data handling in smart cities

Smart city projects often involve multiple stakeholders and depend on government and private players' collaboration. This involvement of third parties disseminates data to multiple sources [53]. Additionally, the IoT service provider may outsource the responsibility of data accumulation, processing, or safekeeping to third-party "specialist data brokers or vendors" [47]. For instance, Street Surveillance equipment may include sensors from one company, cameras from another, and an internet server from a third company. Moreover, independent contractors may be hired to analyse this data. In such situations, privacy protection becomes paramount by including data protection clauses in the non-disclosure agreements with these third parties [47]. Additionally, e-contracts like "End User Licensing Agreements (EULA)" may be entered into between IoT service providers and users to address data ownership, data protection, security, and privacy issues, especially considering the multiple stakeholders involved in an IoT environment [47]. This will also resolve issues dealing with data handling and ownership and establish clear liability for any damage caused to IoT users due to unauthorised data usage, cyber breaches, or attacks [32]. For instance, if the data generated by a smart city is transferred to another jurisdiction or country, the legality of such a transfer may become an issue as different countries will have different privacy and data protection rules [34, 36, 44]. Smart cities should also occasionally conduct privacy impact assessments (PIAs) to reduce privacy risks and to secure data [36]. This will ensure that smart cities integrate privacy by design [53, 54]. In smart cities, enormous amounts of data may be generated due to IoT environments. This may lead to data ownership and intellectual property rights (IPR) concerns and cause legal disputes between the IoT service providers and third parties. Entering into legal agreements regarding the title and claim to intellectual property rights will reduce such issues and potential concerns over the misuse of such data.

Smart cities must focus on adapting technologies compliant with the basic privacy laws [42]. While developing a smart city project, the emphasis should be on creating sustainable urban spaces that balance innovation and citizen-centric approaches [55]. Ignorance of privacy and data protection concerns resulting from a lack of compliance with the legal framework would only prevent the smart city vision from becoming a reality.

7.5 CONCLUSION

This chapter examined important aspects of fog computing in smart cities, including deploying software and data on fog nodes, analytics conducted on fog platforms, and legal considerations about data privacy, security, and safety. The chapter underscored the significance of smooth integration and showcased the game-changing possibilities of real-time insights derived from fog-based analytics. The statement emphasised the importance of having strong legislative frameworks to address privacy issues and ensure the ethical implementation of fog computing in sustainable smart cities. It is crucial to consider the technological, analytical, and legal aspects while developing sustainable smart cities in order to ensure responsible and efficient growth in urban environments that are interconnected ecosystems. To fully leverage the advantages of fog computing in building intelligent and adaptable urban environments, it is crucial to adopt a comprehensive strategy that includes innovation, privacy protection measures, and ethical guidelines.

REFERENCES

1. Chen, C., Wang, Z., & Guo, B. (2016). The road to the Chinese smart city: progress, challenges, and future directions. *IT Professional*, *18*(1), 14–17.
2. European (2019). European smart cities project. http://www.smart-cities.eu/. Accessed 05 Aug 2019.
3. Guo, M., Liu, Y., Yu, H., Hu, B., & Sang, Z. (2016). An overview of smart city in China. *China Communications*, *13*(5), 203–211.
4. Yang, Y., Wu, L., Yin, G., Li, L., & Zhao, H. (2017). A survey on security and privacy issues in Internet-of-Things. *IEEE Internet of things Journal*, *4*(5), 1250–1258.
5. Yousuf, O., & Mir, R. N. (2019). A survey on the Internet of Things security: State-of-art, architecture, issues and countermeasures. *Information & Computer Security*, *27*(2), 292–323.
6. Talari, S., Shafie-Khah, M., Siano, P., Loia, V., Tommasetti, A., & Catalão, J. P. (2017). A review of smart cities based on the internet of things concept. *Energies*, *10*(4), 421.
7. Smart Cities Market Size by Component (Hardware, Services, and Software), Application (Smart Infrastructure, Smart Energy, Smart Healthcare, Smart Education, Smart Governance, Smart Buildings, Smart Mobility Management, Smart Security, and Others), Regions, Global Industry Analysis, Share, Growth, Trends, and Forecast 2023 to 2032 https://www.thebrainyinsights.com/report/smart-cities-market-13548#:~:text=The%20global%20smart%20cities%20market,the%20global%20smart%20cities%20market. (Accessed on 12 Dec 2023)
8. Lata, M., & Kumar, V. (2021). Internet of energy IoE applications for smart cities. In *Internet of Energy for Smart Cities* (pp. 127–144). CRC Press.
9. Chou, W., Ma, C., & Chung, R. (2018). Supercharging the smart city-smarter people and better governance. *Deloitte Perspective*, *7*, 87–93.
10. Dublin (2020) Smart cities market by strategy, technology, and outlook for solutions, applications and services 2020–2025.
11. Jawhar, I., Mohamed, N., & Al-Jaroodi, J. (2018). Networking architectures and protocols for smart city systems. *Journal of Internet Services and Applications*, *9*, 1–16.

12. Jawhar, I., Mohamed, N., & Al-Jaroodi, J. (2018). Networking architectures and protocols for smart city systems. *Journal of Internet Services and Applications, 9*, 1–16.
13. Zanella, A., Bui, N., Castellani, A., Vangelista, L., & Zorzi, M. (2014). Internet of things for smart cities. *IEEE Internet of Things Journal, 1*(1), 22–32.
14. Gurgen, L., Gunalp, O., Benazzouz, Y., & Gallissot, M. (2013, March). Self-aware cyber-physical systems and applications in smart buildings and cities. In *2013 Design, Automation & Test in Europe Conference & Exhibition (DATE)* (pp. 1149–1154). IEEE.
15. Ermacora, G., Rosa, S., & Bona, B. (2015, March). Sliding autonomy in cloud robotics services for smart city applications. In *Proceedings of the Tenth Annual ACM/IEEE International Conference on Human-Robot Interaction Extended Abstracts* (pp. 155–156).
16. Clohessy, T., Acton, T., & Morgan, L. (2014, December). Smart city as a service (SCaaS): A future roadmap for e-government smart city cloud computing initiatives. In *2014 IEEE/ACM 7th International Conference on Utility and Cloud Computing* (pp. 836–841). IEEE.
17. Singh, J., & Kumar, V. (2014). Multi-disciplinary research issues in cloud computing. *Journal of Information Technology Research (JITR), 7*(3), 32–53.
18. Anthopoulos, L., & Attour, A. (2018, April). smart transportation applications' business models: a comparison. In *Companion Proceedings of the The Web Conference 2018* (pp. 927–928).
19. Lata, M., & Kumar, V. (2021). Standards and regulatory compliances for IoT security. *International Journal of Service Science, Management, Engineering, and Technology (IJSSMET), 12*(5), 133–147.
20. López, P., Fernández, D., Jara, A. J., & Skarmeta, A. F. (2013, March). Survey of internet of things technologies for clinical environments. In *2013 27th International Conference on Advanced Information Networking and Applications Workshops* (pp. 1349–1354). IEEE.
21. Malik, S., & Gupta, K. (2021). Smart City: A new phase of sustainable development using fog computing and IoT. In *IOP Conference Series: Materials Science and Engineering* (Vol. 1022, No. 1, p. 012093). IOP Publishing.
22. Ahmad, S., & Afzal, M. M. (2020). Deployment of fog and edge computing in IoT for cyber-physical infrastructures in the 5G era. In *Sustainable Communication Networks and Application: ICSCN 2019* (pp. 351–359). Springer International Publishing.
23. Intel Building Management Platform (2018) Smart building with internet of things technologies. https://www.intel.com/content/www/us/en/smartbuildings/overview.html. Accessed 16 Dec 2023.
24. Ferrández-Pastor, F. J., Mora, H., Jimeno-Morenilla, A., & Volckaert, B. (2018). Deployment of IoT edge and fog computing technologies to develop smart building services. *Sustainability, 10*(11), 3832.
25. Ferrández-Pastor, F. J., Mora, H., Jimeno-Morenilla, A., & Volckaert, B. (2018). Deployment of IoT edge and fog computing technologies to develop smart building services. *Sustainability, 10*(11), 3832.
26. Malik, S., Gupta, K., & Singh, M. (2020). Resource management in fog computing using clustering techniques: a systematic study. *Annals of the Romanian Society for Cell Biology, 24*(2), 77–92.

27. Malik, S., & Gupta, K. (2019). Resource scheduling in fog: Taxonomy and related aspects. *Journal of Computational and Theoretical Nanoscience*, *16*(10), 4313–4319.

28. Kanika, Chakraverty, Chakraborty, P., Aggarwal, A., Madan, M., & Gupta, G. (2022). Enriching WordNet with subject specific out of vocabulary terms using existing ontology. In *Data Engineering for Smart Systems: Proceedings of SSIC 2021* (pp. 205–212). Springer Singapore.

29. Moustaka, V., Vakali, A., & Anthopoulos, L. G. (2017, April). CityDNA: smart city dimensions' correlations for identifying urban profile. In *Proceedings of the 26th international conference on world wide web companion* (pp. 1167–1172).

30. Tikhaleva, E. Y. (2023). "Smart cities": Legal regulation and potential of development. *Journal of Digital Technologies and Law*, *1*(3), 803–824.

31. Chakraborty, C., Lin, J. C. W., & Alazab, M. (2021). *Data-Driven Mining, Learning and Analytics for Secured Smart Cities*. Springer International Publishing.

32. Elmaghraby, A. S., & Losavio, M. M. (2014). Cyber security challenges in Smart Cities: Safety, security and privacy. *Journal of advanced research*, *5*(4), 491–497.

33. Sharma, Dr Abhay, Kumar, Hari Dilip. Security and Privacy in Indian Smart Cities (December 15, 2021). https://www.digitalfirstmagazine.com/security-and-privacy-in-indian-smart-cities/

34. Fabrègue, B. F., & Bogoni, A. (2023). Privacy and security concerns in the smart city. *Smart Cities*, *6*(1), 586–613.

35. "Justice K.S. Puttaswamy (Retd.) v. Union of India", (2017) 10 SCC 1. https://main.sci.gov.in/supremecourt/2012/35071/35071_2012_Judgement_26-Sep-2018.pdf

36. Walters, R., Trakman, L., & Zeller, B. (2019). *Data protection law*. Springer Nature.

37. United States v. Jones, 615 F. 3d 544. https://www.law.cornell.edu/supremecourt/text/10-1259

38. Finn, R. L., Wright, D., & Friedewald, M. (2013). Seven types of privacy. *European Data Protection: Coming of Age*, 3–32.

39. Eshita Deb, Assessing the Legal Framework for Regulating Data Collection and Privacy in the Context of Smart Cities, (July 26, 2023). https://legalvidhiya.com/assessing-the-legal-framework-for-regulating-data-collection-and-privacy-in-the-context-of-smart-cities/

40. Wexler, R. (2022). Life, Liberty, and Date Privacy: The Global CLOUD, the Criminally Accused, and Executive versus Judicial Compulsory Process Powers. *Texas Law Review*, *101*, 1341.

41. Ss. 43A, 72, 72A, 79, 69A, The Information Technology Act, 2000, Chapter IX, XI, XII. www.meity.gov.in

42. Chakrabarty, S. P., Ghosh, J., & Mukherjee, S. (2021). Privacy issues of smart cities: Legal outlook. In *Data-Driven Mining, Learning and Analytics for Secured Smart Cities: Trends and Advances* (pp. 295–311). Cham: Springer International Publishing.

43. Habibzadeh, H., Nussbaum, B. H., Anjomshoa, F., Kantarci, B., & Soyata, T. (2019). A survey on cybersecurity, data privacy, and policy issues in cyber-physical system deployments in smart cities. *Sustainable Cities and Society*, *50*, 101660.

44. What are the Legal Considerations for Privacy and Security in Smart Cities. (June 23, 2023. https://onlinelegaladvisor.in/legal/what-are-the-legal-considerations-for-privacy-and-security-in-smart-cities
45. Khan, Z., Pervez, Z., & Abbasi, A. G. (2017). Towards a secure service provisioning framework in a smart city environment. *Future Generation Computer Systems, 77*, 112–135.
46. Ismagilova, E., Hughes, L., Rana, N. P., & Dwivedi, Y. K. (2020). Security, privacy and risks within smart cities: Literature review and development of a smart city interaction framework. *Information Systems Frontiers*, 1–22.
47. Sarin, Anirudh. "India: Legal Issues Pertaining To Internet Of Things (IOT)", (12 April 2018). https://www.mondaq.com/india/privacy-protection/691560/legal-issues-pertaining-to-internet-of-things-iot
48. Regulation (EU) 2016/679. https://eur-lex.europa.eu/legal-content/EN/TXT/PDF/?uri=CELEX:32016R0679#:~:text=This%20Regulation%20is%20intended%20to,well-being%20of%20natural%20persons
49. Varkonyi, G. G., Varadi, S., & Kertesz, A. (2019). Legal aspects of operating IoT applications in the Fog. *Fog and Edge Computing: Principles and Paradigms.*
50. California Consumer Privacy Act of 2018 (CCPA). https://oag.ca.gov/privacy/ccpa
51. European Union Convention 108, Convention for the Protection of Individuals with regard to Automatic Processing of Personal Data (ETS No. 108). https://www.coe.int/en/web/conventions/full-list?module=treaty-detail&treatynum=108
52. Edwards, L. (2016). Privacy, security and data protection in smart cities: A critical EU law perspective. *The European Data Protection Law Review, 2*, 28.
53. Garg, R., Varadi, S., & Kertész, A. (2019, February). Legal considerations of IoT applications in fog and cloud environments. In *2019 27th Euromicro International Conference on Parallel, Distributed and Network-Based Processing (PDP)* (pp. 193–198). IEEE.
54. Rawat, D. B., & Ghafoor, K. Z. (Eds.). (2018). *Smart cities cybersecurity and privacy.* Elsevier.
55. Verma, A., Khanna, A., Agrawal, A., Darwish, A., & Hassanien, A. E. (2019). Security and privacy in smart city applications and services: Opportunities and challenges. *Cybersecurity and Secure Information Systems: Challenges and Solutions in Smart Environments*, 1–15.

Chapter 8

Sustainable data-driven irrigation and fertilization based on artificial intelligence

Vinay Gautam, Gaganpreet Kaur, and Swati Malik
Chitkara University Institute of Engineering and Technology,
Chitkara University, Chandigarh, India

Shuchi Upadhyay
SOHST, UPES, Dehradun, India

8.1 INTRODUCTION

Devising smart solutions for the agriculture domain is essential to resolve issues related to productivity. In a country like India, where 70% of the population depends on agriculture for their daily needs [1]. Therefore, an appropriate solution is needed to resolve agriculture issues. Hence, the quality and quantity of agricultural produce can be improved by introducing smart solutions based on artificial intelligence. Precision agriculture involves technologies such as the Internet of Things (IoT) and artificial intelligence technology to resolve agricultural issues. Here, IoT is used to retrieve and disseminate appropriate information that can be used further to identify the quantity of fertilizer and water for the field. Irrigation and fertilization are major issues that majorly impact the productivity of agricultural produce. Recent research, has identified that the right amount of fertilization and irrigation can enhance the productivity of crops. Efficiency can be attained after figuring out the amount of fertilizer and water needed for the field. In existing research, it has been observed that the water and fertilizer requirements are identified with the help of sensors. The objective of this research is to develop a data-driven smart system to support irrigation and fertilization. Both can be chosen based on the types of crops. Therefore, the system sensor is used to collect crop information before recommending the amount of fertilizer and water.

As per reports, in agriculture, the resources are limited and need to be utilized properly, and cannot waste valuable resources. Therefore, it is our prime duty to provide effective solutions to utilize resources appropriately. Various solutions for agriculture improvement already exist, but most agricultural people are not aware of them. The government has introduced various ICT-based [2] policies to improve agriculture produce quality and quantity. But most farmers are not aware of it. These policies provide useful information to farmers regarding crop cultivation. With the availability of these policies, farmers can independently make decisions on crop cultivation,

DOI: 10.1201/9781003494430-8

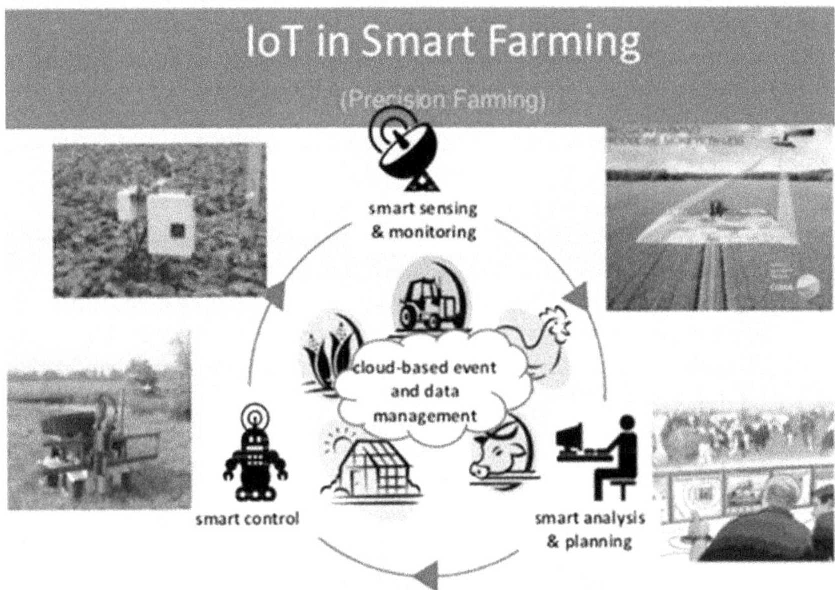

Figure 8.1 IoT-based smart agriculture.

irrigation, and fertilization. Afterward, the agriculture industries benefited from various smart solutions based on the IoT, as depicted in Figure 8.1.

An agriculture solution was introduced in [3], that uses IoT and machine learning methods to provide effective crop management. The association of two advanced technologies [4] has completely replaced legacy agriculture practices. The association of these recent and advanced technologies helped to utilize resources optimally. Information dissemination is quick due to advanced technology that reduces the gaps [5]. IoT infrastructure quickly provides valuable input on climate conditions and other factors. Different types of sensors have been employed in IoT-based systems to garb real-time data and utilized for further decision-making. The data is kept in the cloud storage area and can be accessed from any remote location with smart devices. IoT-based solutions provide valuable input to users and solve all issues related to farms [6–7]. IoT-based solutions can help farmers regularly monitor water content and soil nutrient value. A smart system was developed to analyze the moisture and nutrient contents in the soil [8–10].

IoT-based system power can be enhanced with the latest and advanced AI techniques [11]. Research has used AI in IoT systems to provide valuable insight, as given below:

An IoT-based recommendation system was developed in [1] to provide valuable inputs for soil. This system has been employed to inform users about moisture content and nutrient value in the soil. The data-driven IoT solutions are useful for users and keep them updated. These systems guide users to optimize their valuable resources, such as water and fertilizer. This system improves agriculture productivity by providing valuable insights. A few benefits of data-driven IoT systems are listed below:

a. Soil Management: Data-driven systems are used to provide valuable insight into soil. It can be employed for analyzing soil for moisture and nutrient values.

b. Irrigation Management: Data-driven system can be used to properly manage irrigation based on data generated through sensors.

c. Pest Management: Data-driven systems can be used to identify the appropriate amount of pests needed for the field.

d. Data-driven systems provide information about a sufficient amount of fertilization in the field.

8.2 RELATED WORK

The state-of-the-art paper is further described under two sub-sections as listed below:

8.2.1 Irrigation management

Irrigation is managed with different methods, as discussed and depicted in Figure 8.2:

a. Surface irrigation is a classical irrigation method. Here, the soil surface is utilized to apply water in the field that needs an experienced person.

Surface Irrigation

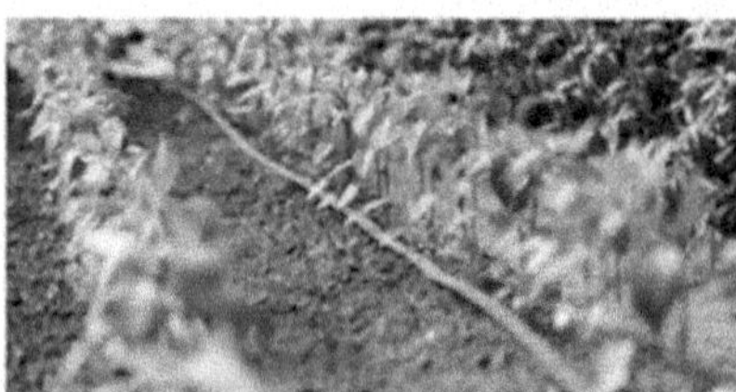

Drip Irrigation

Sprinkle Irrigation

Figure 8.2 Types of irrigation.

b. The drip or trickle irrigation method is micro-irrigation used to save water and nutrient value of a field. This method employs a drip of water for field irrigation.

c. Sprinkle Irrigation: In this method, pipes are used to distribute water just like rainfall. Here, the distribution of water will be the same for each part of the field.

Various issues need to be taken care of while selecting an appropriate irrigation method, as listed below:

a. Moisture content in the soil [12, 13, 15, 16–27, 28–35, 38–42, 45–51, 53–63]
b. Soil temperature [13, 15, 25, 28, 39, 46]
c. pH value of soil [20, 23, 35, 46, 51, 56, 58]
d. Nutrient Contents in Soil(NPK) [23, 46, 56]
e. Air temperature and moisture [13–15, 16–20, 22–25, 28–30, 35–37, 40–42, 45–47, 49–51, 53–55, 57, 59–63]
f. Light intensity in the soil [12, 15, 16, 29, 30, 34, 43, 44, 52, 60].

The problem related to irrigation was resolved with smart irrigation methods based on artificial intelligence described in the next section:

8.2.2 Smart irrigation

A Raspberry-based smart system was developed for field irrigation based on the data collected from sensors [64]. Initially, the system was programmed to collect water contents in the soil and take appropriate action based on the data received. The system was directly operated with smartphones. An intelligent system was developed that collects data from soil and takes appropriate action on irrigation. Here, a microcontroller was employed to transmit signals to the pump for irrigation [65, 66]. A wireless sensor-based smart system was proposed for irrigation that was remotely operated through a smartphone [67, 68]. Table 8.1 depicts the details of the smart irrigation method.

The smart systems elaborated above were better solutions, eliminated manual irrigation, and eased the lives of users. However, the state-of-the-art

Table 8.1 Smart irrigation techniques

Ref.	Processing	Aim
[69]	Simplify data received from sensors	Moisture level identification
[70]	Remove erroneous data	Estimation of water required
[71]	Avoid inappropriate data entries	Reduce irrigation efforts and cost
[72]	Avoid erroneous values	Appropriate action on irrigation
[73]	Remove erroneous data	Irrigation for the specific zone
[74–78]	Remove erroneous data	Irrigation management

solutions were implemented without artificial intelligence techniques. The use of AI can enhance the power of these systems. Therefore, this research implemented a smart irrigation system empowered with AI techniques.

8.2.2.1 Fertilizer management

Fertilizers are prime to enhance the fertility of land. This can be a reason for huge quality produce. Therefore, an effective and efficient fertilization method is needed to distribute fertilizers in the field. Fertilizers can improve the state of nutrients in the field, which is a prime reason for quality production. Mineral fertilization is best for crops and fields as well. This can be a reason for huge growth in crops. Fertilization can impact positive and negative as well. Therefore, an appropriate amount of fertilizer needs to be applied to the field. Initially, manual methods were employed by farmers to apply fertilizer, which might be the reason for the inappropriate distribution of fertilizers. Hence, it must employ an artificial intelligence-based smart system to predict fertilizer. The state-of-the-art methods for smart fertilization are described in the next section.

8.2.3 Smart fertilization

Fertilization is an aspect that is prime for crop growth. An appropriate amount of fertilization is required to be applied otherwise, it harms soil fertility. Therefore, researchers have proposed various methods to address said problem. A few methods are described below:

A smart fertilization system was proposed in [79] that recommends an appropriate amount of fertilizer based on climate factor analysis. In [80–81] the researchers have proposed a method that uses big data to further estimate the appropriate amount of fertilization.

A smart system was developed in [82] that captures valuable insight into soil nutrients and analyzes them to take appropriate action. Mostly the color of the soil is used to estimate the nutrient values. Here, the smart system analyzes the soil color and estimates the fertilizer amount.

Table 8.2 depicts various fertilization methods, and Figure 8.3 depicts the fertilizer management system:

Table 8.2 Smart fertilization details

Ref.	Objective	Techniques
[83–94]	Identify Nitrogen Status	PLT Model
[95]	Nitrogen and Phosphorous prediction	No
[96]	Identify Nitrogen Status	LNC-indices (prediction model)
[97]	NPK Prediction	PLSR (prediction model)
[98–106]	Identify Nitrogen Status N-content	PLSR (prediction model)

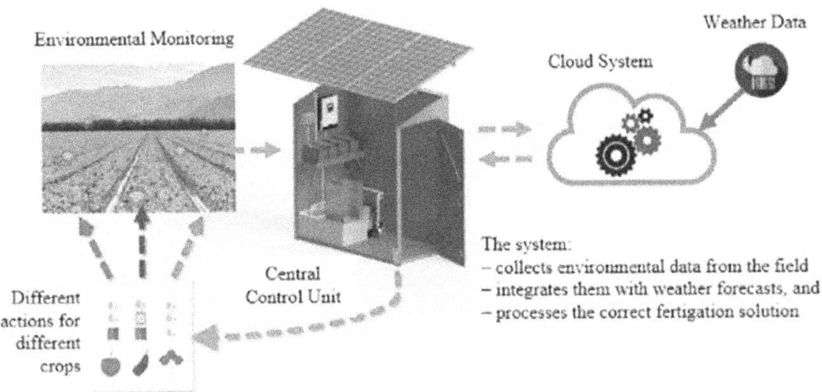

Figure 8.3 Smart fertilization management.

Various studies have been employed in the state-of-the-art for fertilizer status identification. Therefore, this paper describes a model that resolves issues related to agriculture with a precise solution based on IoT.

8.3 SUSTAINABLE SMART SOLUTION BASED ON AIoT

The proposed system architecture is completely defined with four stages as described below and shown in Figure 8.4.

A. Sensor layer
B. Network Layer
C. Service Layer
D. Application Layer

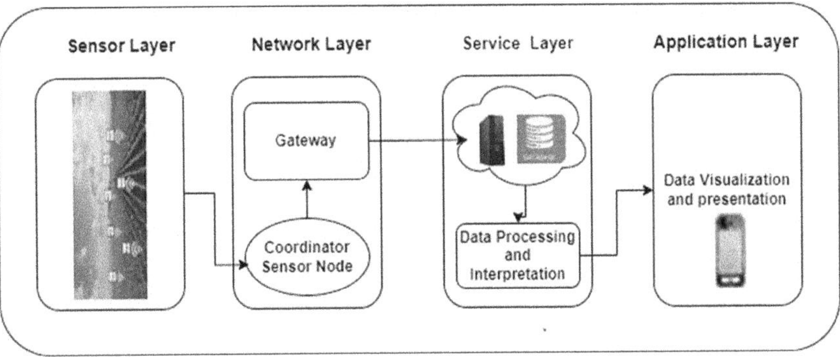

Figure 8.4 System architecture.

8.3.1 Sensor layer

In the system, various types of sensors were deployed to collect data from the farm. Sensors were deployed within the field to collect data related to soil moisture and nutrient values. Sensors collect signals, convert them into appropriate data, and send it to the coordinator node. The details of sensors deployed in the field are given below

1. Optical sensors are used to measure the temperature of the soil. The sensor receives a single reflection from the soil, and the same is used to predict the moisture content in the soil.
2. Volumetric soil moisture sensors are deployed to measure the water content in the soil.
3. ISE (Ion Selective Electrode) sensors are deployed in the field to collect data for nutrient value in the soil. The same is used to predict fertilizer requirements in the field.

The data collected in the sensor layer is further transmitted to the network layer for further access.

8.3.2 Network layer

The primary goal of the network layer is to connect the sensor layer to the service layer. The data from the sensor layer is transferred to the service layer data storage through the network access point. This layer makes sure to create a communication connection between the two layers. The gateway in the network layer is the main point of connection between data and storage. A list of devices is given below that are used to establish a connection.

a. Wireless Routers are used in the network to establish a connection between two layers.
b. GPRS and ETSI are used to establish communication between layers.

8.3.3 Service layer

The major task of the service layer is data storage and interpretation.

a. Data Storage
b. Data Interpretation
a. Data Storage
   The data generated through sensors are raw and need to be filtered to get appropriate information. The service layer applies different operations before it in the database. At the service level, the data is cleansed by various cleansing operations, and afterward, the data is checked for duplicacy.

b. Data Interpretation

After the cleaning operation, the data is kept in the cloud storage, and the same is used to apply an appropriate algorithm to generate prediction results. The results are further used to make decisions.

8.3.4 Application layer

The application layers contain different ways to present data to users. Smart farming systems provide ease to users with the help of mobile apps. The farming system provides visual data and information regarding fertilization and irrigation on users' or farmers' smart mobiles with the help of the mobile app. Based on the real-time information, the farmer will get the real-time information about irrigation and fertilization situations.

8.4 PROPOSED ALGORITHM

Artificial intelligence-based smart systems employ deep learning algorithms here in this service layer and produce analysis results. In research, we have proposed a composite algorithm that comprises the power of two deep learning algorithms, such as ANN and Bi-LSTM. Various state-of-the-art models have been discussed in related work to tackle issues related to irrigation and fertilization, but the proposed model produced a better result with real-time data. The flow of the proposed methodology is depicted in Figure 8.5.

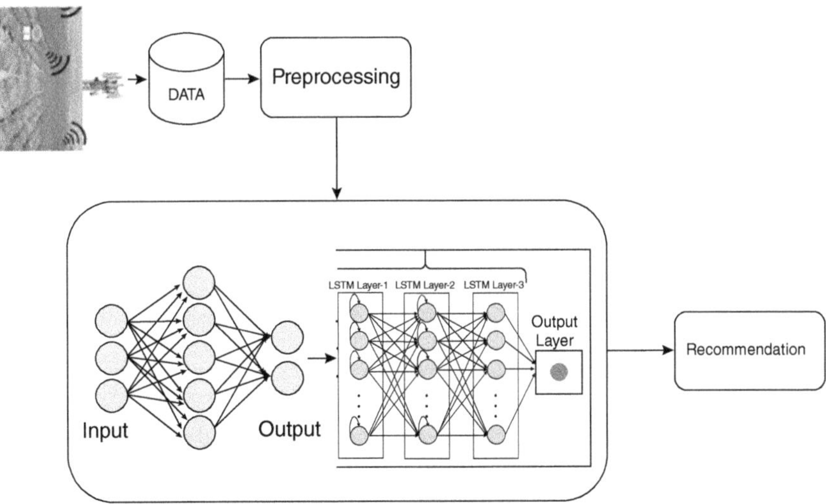

Figure 8.5 Flow of methodology.

The proposed methodology generates the desired result in different steps, as discussed below:

1. Data Preprocessing and augmentation
2. Feature extraction and Prediction
3. Recommendation

The model has been elaborated with the following equations:

$$I_i = w_{ij} * x_{ij} + u_i \qquad (8.1)$$

Where:

I_i : Input weighted with i weight
w_{ij} : weight for neuron i to j
x_{ij} : output from i to j neuron
u_i : *baseline input*

$$Y_i = \sum_{\substack{i=1 \\ j=1}}^{n} w_{ij} * x_{ij} \qquad (8.2)$$

Bi-LSTM representation is as follows:

$$v_i = B\left(Y_{1:n} : i\right) \qquad (8.3)$$

8.5 RESULT AND DISCUSSION

Artificial intelligence-based IoT systems use various machine learning and deep learning algorithms. Here, in the proposed system, we have employed a composite model comprised of ANN and Bi-LSTM. The results of the proposed method are compared with those of other deep learning models, and the proposed model outperformed with a 98.25 accuracy rate.

Finally, the observation is that the hybrid model outperformed and showed a prediction accuracy of 98.75%.

8.6 CONCLUSION

In research, a sustainable agriculture system was proposed to take care of two major issues of irrigation and fertilization. The agriculture system was based on data-driven that was generated in real-time from the field. Generated data from the field was utilized to produce a better outcome. Here, the outcome was produced with a composite model composed of ANN and LSTM.

The proposed system was compared with other state-of-the-art deep learning models to show comparative results. Findings show that the proposed composite model outperformed with 98.75%.

REFERENCES

[1] MINAGRI, "Minagri annual report 2019-2020," *Minagri Annu. Rep. 2019-2020*, no. decrease in small animal population, pp. 52–52, 2020.

[2] J. C. Aker, I. Ghosh, and J. Burrell, 2016. "The promise (and pitfalls) of ICT for agriculture initiatives," *Agricultural Economics*, 47(S1): 35–48.

[3] K. Kapitanova and S. H. Son, 2012. "Machine learning basics," *Intell. Sens. Networks Integr. Sens. Networks, Signal Process. Mach. Learn.*, no. Ml, pp. 3–29.

[4] D. Mishra, A. Abbas, T. Pande, A. K. Pandey, K. K. Agrawal, and R. S. Yadav, 2019. "Smart agriculture system using IoT," *ACM Int. Conf. Proceeding Ser.*

[5] D. Orn, L. Duan, Y. Liang, H. Siy, and M. Subramaniam, 2020. "Agro-AI education: Artificial intelligence for future farmers," *SIGITE 2020 - Proc. 21st Annu. Conf. Inf. Technol. Educ.*, pp. 54–57.

[6] S. Jain and D. Ramesh, 2020. "Machine learning convergence for weather based crop selection," *2020 IEEE Int. Students' Conf. Electr. Electron. Comput. Sci. SCEECS 2020*, no. February.

[7] A. Rehman, T. Saba, M. Kashif, S. M. Fati, S. A. Bahaj, and H. Chaudhry, 2022. "A revisit of internet of things technologies for monitoring and control strategies in smart agriculture," *Agronomy*, 12(1): 1–21.

[8] Syaza Norfilsha Binti Ishak, 2008. "Smart home garden irrigation system with raspberry Pi," *IEEE*, 16(June): 24.

[9] B. Swaminathan, S. Palani, K. Kotecha, V. Kumar, "IoT driven artificial intelligence technique for fertilizer recommendation model," *IEEE Consum. Electron. Mag.*, no. February, 2022.

[10] A. F. Suhaimi, N. Yaakob, S. A. Saad, and K. Azami, 2021. "IoT based smart agriculture monitoring, automation and intrusion detection system IoT based smart agriculture monitoring, automation and intrusion detection system,".

[11] D. Wang, W. Cao, F. Zhang, Z. Li, S. Xu, and X. Wu, 2022. "A Review of Deep Learning in Multiscale Agricultural Sensing," *Remote Sensing*, 14(3).

[12] A. Imteaj et al., 2016. "Iot based autonomous percipient irrigation system using raspberry pi," in *19th International Conference on Computer and Information Technology (ICCIT)*, pp. 563–568, IEEE.

[13] M. N. Rajkumar et al., 2017. "Intelligent irrigation systemâC"an iot based approach," in *International Conference on Innovations in Green Energy and Healthcare Technologies (IGEHT)*, pp. 1–5, IEEE.

[14] W. Zhao et al., 2017. "Design and implementation of smart irrigation system based on lora," in *IEEE Globecom Workshops (GC Wkshps)*, pp. 1–6, IEEE.

[15] W. Difallah et al., 2018 "Design of a solar powered smart irrigation system (spsis) using wsn as an iot device," in *Proceedings of the 2018 International Conference on Software Engineering and Information Management*, pp. 124–128.

[16] I. N. R. Hendrawan et al., 2019. "Fuzzy based internet of things irrigation system," in *1st International Conference on Cybernetics and Intelligent System (ICORIS)*, vol. 1, pp. 146–150, IEEE.

[17] N. A. M. Leh et al., 2019. "Smart irrigation system using internet of things," in *9th International Conference on System Engineering and Technology (ICSET)*, pp. 96–101, IEEE.

[18] H. Laksiri et al., 2019. "Design and optimization of iot based smart irrigation system in sri lanka," in *14th Conference on Industrial and Information Systems (ICIIS)*, pp. 198–202, IEEE.

[19] M. A. Sayed et al., 2019. "An iot based robotic system for irrigation notifier," in *International Conference on Robotics, Automation, Artificial-intelligence and Internet-of-Things (RAAICON)*, pp. 77–80, IEEE.

[20] A. Dasgupta et al., 2019. "Smart irrigation: Iot-based irrigation monitoring system," in *Proceedings of international ethical hacking conference 2018*, pp. 395–403, Springer.

[21] A. Shufian et al., 2019. "Smart irrigation system with solar power and gsm technology," in *5th International Conference on Advances in Electrical Engineering (ICAEE)*, pp. 301–305, IEEE.

[22] J. Karpagam et al., 2020. "Smart irrigation system using iot," in *6th International Conference on Advanced Computing and Communication Systems (ICACCS)*, pp. 1292–1295, IEEE.

[23] B. R. Prakash et al., 2020. "Super smart irrigation system using internet of things," in *2020 7th International Conference on Smart Structures and Systems (ICSSS)*, pp. 1–5, IEEE.

[24] M. G. B. Palconit et al., 2020. "Iot-based precision irrigation system for eggplant and tomato," in *9th Int. Symp. Comput. Intell. Ind. Appl.*, no. November, pp. 0–6.

[25] D. S. J. Muneeswari, 2017. "Smart irrigation system using iot approach," *International Journal of Engineering Research & Science (IJOER)*, March.

[26] S. Rawal, 2017. "Iot based smart irrigation system," *International Journal of Computer Applications*, 159(8): 7–11.

[27] I. Ahmed et al., 2020. "Iot based smart irrigation system at university of chittagong, bangladesh," *International Journal of Computers and Applications*, 176: 39–45.

[28] W. Difallah et al., 2018. "Intelligent irrigation management system," *Energy*, 9(9): 429–433.

[29] Z. Ahmad et al., "Client server smart irrigation system".

[30] P. Naik et al., 2018. "Automation of irrigation system using iot," *International Journal of Engineering and Manufacturing Science*, 8(1): 77–88.

[31] S. Kathuria et al., 2018. "E-krishi: An iot based smart irrigation system," *International Journal of Electrical, Electronics and Data Communication (IJEEDC)*, 6(6): 8–13.

[32] D. Thakur et al., 2020. "Smart irrigation and intrusions detection in agricultural fields using iot," *Procedia Computer Science*, 167: 154–162.

[33] N. Sudharshan et al., 2019. "Renewable energy based smart irrigation system," *Procedia Computer Science*, 165: 615–623.

[34] A. Pathak et al., 2019. "Iot based smart system to support agricultural parameters: A case study," *Procedia Computer Science*, 155: 648–653.

[35] V. Balaji, 2019. "Smart irrigation system using iot and image processing," *International Journal of Engineering and Advanced Technology*, 8(6S): 115–120.

[36] V. Manimegalai et al., "Smart irrigation system with monitoring and controlling using iot," *The International Journal of Engineering and Advanced Technology*, 9: 1373–1376, 2020.

[37] S. Akter et al., 2018. "Developing a smart irrigation system using arduino," *International Journal of Research Studies in Science, Engineering and Technology*, 6(1): 31–39.

[38] P. Sachin et al., 2019. "Iot based smart irrigation system," *International Journal of Advanced Research in Electrical, Electronics and Instrumentation Engineering*, 8(5): 1604–1608.

[39] K. Singh et al., 2019. "Iot based approach for smart irrigation system suited to multiple crop cultivation," *International Journal of Research in Engineering and Technology*, 12: 357–363.

[40] N. Ismail et al., 2019. "Smart irrigation system based on internet of things (iot)," *Journal of Physics: Conference Series*, 1339: 012012, IOP Publishing.

[41] D. Joshi, 2019. "Smart irrigation system using iot for farmers," *International Journal of Emerging Technologies and Innovative Research*, 6(1): 207–209.

[42] D. Baravade et al., 2019. "Study paper on smart irrigation system," *International Journal of Engineering Sciences Research Technology*, 8(4): 99–102.

[43] M. Ayaz et al., 2019. "Internet-of-things (iot)-based smart agriculture: Toward making the fields talk," *IEEE Access*, 7: 129551–129583.

[44] A. Abdelmoamen Ahmed et al., 2021. "A distributed system for supporting smart irrigation using internet of things technology," *Engineering Reports*, 3(7): e12352.

[45] J. Jayadevan et al., 2020. "A novel architecture for internet of things in precision agriculture," *International Journal of Applied Engineering Research*, 15(3): 204–211.

[46] C. Subramani et al., 2020. "Iot-based smart irrigation system," in *Cognitive Informatics and Soft Computing*, pp. 357–363, Springer.

[47] S. E. Babaa et al., 2020. "Smart irrigation system using arduino with solar power," *International Journal of Engineering Research & Technology (IJERT)*, 9(05): 7.

[48] S. Velmurugan, 2020. "An iot based smart irrigation system using soil moisture and weather prediction".

[49] K. Kannan et al., 2020. "An internet of things based smart irrigation using solenoid valve," *International Journal of Recent Technology and Engineering (IJRTE)*, 9(1): 2018–2023.

[50] M. M. Islam et al., "Iot based smart irrigation monitoring & controlling system in agriculture,"

[51] W. Li et al., 2020. "Review of sensor network-based irrigation systems using iot and remote sensing," *Advances in Meteorology*, 2020.

[52] K. Singh et al., 2021. "Design of a low-cost sensor-based iot system for smart irrigation," in *Applications in Ubiquitous Computing*, pp. 59–79, Springer.

[53] M. S. R. Al Nahian et al., "Iot based soil monitoring and automatic irrigation system," 2021.

[54] M. E. Karar et al., 2021. "A pilot study of smart agricultural irrigation using unmanned aerial vehicles and iot-based cloud system," arXiv preprint arXiv:2101.01851.

[55] A. Hassan et al., 2021. "A wirelessly controlled robot-based smart irrigation system by exploiting arduino," *Journal of Robotics and Control (JRC)*, 2(1): 29–34.

[56] M. Jimenez-Buend et al., 2021. "High-density wi-fi based sensor network for efficient irrigation management in precision agriculture," *Applied Sciences*, 11(4): 1628.

[57] D. K. Suvra and T. Sen, "Iot based automated irrigation and smart agriculture monitoring system,"

[58] S. Khriji et al., 2021. "Precision irrigation: an iot-enabled wireless sensor network for smart irrigation systems," in *Women in precision agriculture*, pp. 107–129, Springer.

[59] M. S. Munir et al., 2021. "Intelligent and smart irrigation system using edge computing and iot," *Complexity*, 2021.

[60] R. Ullah et al., 2021. "Eewmp: an iot-based energy-efficient water management platform for smart irrigation," *Scientific Programming*, 2021.

[61] K. Pradeep, A. Balasundaram, et al., 2021. "Iot based smart irrigation for agricultural fields," *Annals of the Romanian Society for Cell Biology*, pp. 2000–2009.

[62] B. Supriya et al., 2021. "Iot based remote smart irrigation system," *International Journal of Modern Agriculture*, 10(2), 3805–3811.

[63] A. K. Podder et al., 2021. "Iot based smart agrotech system for verification of urban farming parameters," *Microprocessors and Microsystems*, 82: 104025. "Wireless iot protocol table.glow labs," 2017.

[64] S. Vaishali, S. Suraj, G. Vignesh, S. Dhivya, and S. Udhayakumar, "Mobile integrated smart irrigation management and monitoring system using IOT," *Proc. 2017 IEEE Int. Conf. Commun. Signal Process. ICCSP 2017*, vol. 2018-Janua, pp. 2164–2167, 2018.

[65] J. Karpagam, "2021 7th international conference on advanced computing and communication systems, ICACCS 2021," *2021 7th Int. Conf. Adv. Comput. Commun. Syst. ICACCS 2021*, pp. 1–4, 2021.

[66] A. Triantafyllou, P. Sarigiannidis, and S. Bibi, "Precision agriculture: A remote sensing monitoring system architecture," *Infection*, vol. 10, no. 11, 2019.

[67] S. Hwang. 2017. "Monitoring and controlling system for an iot based smart home," *International Journal of Control, Automation*, 10(2), 339–348.

[68] M. Z. M. Noor and R. A. Ramlee. 2021. "Performances analysis of iot based smart greenhouse system," *International Journal of Electrical and Electronics Engineering Applied Sciences*, 4(2): 1–8.

[69] P. Padalalu, S. Mahajan, K. Dabir, S. Mitkar, D. Javale, 2017. Smart water dripping system for agriculture/farming. In *2017 2nd International Conference for Convergence in Technology (I2CT)*, pp. 659–662. https://doi.org/10.1109/I2CT.2017.8226212

[70] R.G. Perea, E.C. Poyato, P. Montesinos, J.A.R. Díaz, 2019. Prediction of irrigation event occurrence at farm level using optimal decision trees. *Computers and Electronics in Agriculture*

[71] T. Xie, Z. Huang, Z. Chi, T. Zhu, 2017. Minimizing amortized cost of the on-demand irrigation system in smart farms. In *3rd International Workshop on Cyber-Physical Systems for Smart Water Networks*, pp. 43–46. https://doi.org/10.1145/ 3055366.3055370

[72] G. Kokkonis, S. Kontogiannis, D. Tomtsis, 2017. A smart IoT fuzzy irrigation system. *Power (mW)*, 100(63): 25.

[73] C. Goumopoulos, B. O'Flynn and A. Kameas, 2014. Automated zone-specific irrigation with wireless sensor/actuator network and adaptable decision support. *Computers and Electronics in Agriculture*, 105: 20–33.

[74] A. Goldstein, L. Fink, A. Meitin, S. Bohadana, O. Lutenberg and G. Ravid, 2018. Applying machine learning on sensor data for irrigation recommendations: revealing the agronomist's tacit knowledge. *Precision Agriculture*, 19: 421–444.

[75] S. Kang, Z. Liang, Y. Pan, P. Shi and J. Zhang, 2000. Alternate furrow irrigation for maize production in an arid area. *Agricultural Water Management*, 45(3): 267–274.

[76] H. Hou, S. Peng, J. Xu, S. Yang and Z. Mao, 2012. Seasonal variations of CH4 and N2O emissions in response to water management of paddy fields located in Southeast China.

[77] C. Perinbam, Automatic Irrigation System using Fuzzy Logic* Jose Anand. *Hemosphere*, 89(7): 884–892.

[78] A.K. Mousa, M.S. Croock and M.N. Abdullah, 2014. Fuzzy based decision support model for irrigation system management. *International Journal of Computer Applications*, 104(9).

[79] R. Maheswari, H. Azath, P. Sharmila, and S. Sheeba Rani Gnanamalar, "Smart village: Solar based smart agriculture with iot enabled for climatic change and fertilization of soil," *2019 IEEE 5th Int. Conf. Mechatronics Syst. Robot. ICMSR 2019*, pp. 102–105, 2019.

[80] R. Prabha, E. Sinitambirivoutin, F. Passelaigue, and M. V. Ramesh, "Design and Development of an IoT based smart irrigation and fertilization system for chilli farming," *2018 Int. Conf. Wirel. Commun. Signal Process. Networking, WiSPNET 2018*, pp. 1–7, 2018.

[81] S. L. Ullo and G. R. Sinha. 2020. "Advances in smart environment monitoring systems using iot and sensors," *Sensors (Switzerland)*, 20(11): 1–18.

[82] D. L. Mary and M. Ramakrishnan. 2021. "A novel approach to optimize water and fertilizers in agriculture using IoT," *International Journal on Cybernetics & Informatics*, 10(2): 57–64.

[83] Á. Maresma, J. Lloveras, J.A. Martínez-Casasnovas, 2018. Use of multispectral airborne images to improve in-season nitrogen management, predict grain yield and estimate economic return of maize in irrigated high yielding environments. *Remote Sensing*, 10(4): 543.

[84] L. P. Osco, A. P. M. Ramos, D. R. Pereira, M. Éas, N. N. Imai, E. T. Matsubara, N. Estrabis, et al. 2019b. "Predicting canopy nitrogen content in citrus-trees using random forest algorithm associated to spectral vegetation indices from UAV-imagery." *Remote Sensing*, 11 (24): 2925. doi:10.3390/rs11242925

[85] H. Zheng, T. Cheng, D. Li, X. Yao, Y. Tian, W. Cao, and Y. Zhu. 2018. "Combining Unmanned Aerial Vehicle (Uav)-based multispectral imagery and ground-based hyperspectral data for plant nitrogen concentration estimation in rice." *Frontiers in Plant Science* 9: 936. doi:10.3389/ fpls.2018.00936

[86] H. Herrero, P. Waylen, J. Southworth, R. Khatami, D. Yang, and B. Child. 2020. "A healthy park needs healthy vegetation: The story of gorongosa national park in the 21st century." *Remote Sensing*, 12 (3): 476. doi:10.3390/rs12030476

[87] Y. Shendryk, J. Sofonia, R. Garrard, Y. Rist, D. Skocaj, and P. Thorburn. 2020. "Fine-scale prediction of biomass and leaf nitrogen content in sugarcane using uav lidar and multispectral imaging." *International Journal of Applied Earth Observation and Geoinformation* 92: 102177. doi:10.1016/j.jag.2020. 102177

[88] S. Liu, L. Li, W. Gao, Y. Zhang, Y. Liu, S. Wang, and J. Lu. 2018. "diagnosis of nitrogen status in winter oilseed rape (Brassica Napus L.) Using in-situ hyperspectral data and Unmanned Aerial Vehicle (UAV) multispectral images." *Computers and Electronics in Agriculture*, 151: 185–195. doi:10.1016/j.compag.2018.05.026

[89] N. Lu, W. Wang, Q. Zhang, D. Li, X. Yao, Y. Tian, Y. Zhu, et al. 2019. "Estimation of nitrogen nutrition status in winter wheat from unmanned aerial vehicle based multi-angular multispectral imagery." *Frontiers in Plant Science*, 10: 1601. doi:10.3389/fpls.2019.01601

[90] J. Geipel, J. Link, J. A. Wirwahn, and W. Claupein. 2016. "A programmable aerial multispectral camera system for in-season crop biomass and nitrogen content estimation." *Agriculture*, 6 (1): 4. doi:10.3390/ agriculture6010004

[91] M. Maimaitijiang, A. Ghulam, P. Sidike, S. Hartling, M. Maimaitiyiming, K. Peterson, E. Shavers, et al. 2017. "Unmanned Aerial System (Uas)-based phenotyping of soybean using multi-sensor data fusion and extreme learning machine." *ISPRS Journal of Photogrammetry and Remote Sensing*, 134: 43–58. doi:10.1016/j.isprsjprs.2017.10.011

[92] S. Ji-Yong, Z. Xiao-Bo, Z. Jie-Wen, W. Kai-Liang, C. Zheng-Wei, H. Xiao-Wei, Z. De-Tao, and M. Holmes. 2012. "nondestructive diagnostics of nitrogen deficiency by cucumber leaf chlorophyll distribution map based on near infrared hyperspectral imaging." *Scientia Horticulturae*, 138: 190–197. doi:10.1016/j.scienta.2012.02.024

[93] S. Ji-Yong, Z. Xiao-Bo, Z. Jie-Wen, W. Kai-Liang, C. Zheng-Wei, H. Xiao-Wei, Z. De-Tao, and M. Holmes. 2012. "Nondestructive diagnostics of nitrogen deficiency by cucumber leaf chlorophyll distribution map based on near infrared hyperspectral imaging." *Scientia Horticulturae*, 138: 190–197. doi:10.1016/j.scienta.2012.02.024

[94] D. Zhang, H. Song, H. Cheng, D. Hao, H. Wang, G. Kan, H. Jin, and D. Yu. 2014. "The acid phosphatase-encoding gene GmACP1 contributes to soybean tolerance to low-phosphorus stress." *PLoS Genetics*, 10(1): 1. doi:10.1371/journal.pgen.1004061

[95] Y. Liu, Q. Lyu, S. He, S. Yi, X. Liu, R. Xie, Y. Zheng, and L. Deng. 2015. "Prediction of nitrogen and phosphorus contents in citrus leaves based on hyperspectral imaging." *International Journal of Agricultural and Biological Engineering*, 8: 80–88. doi:10.3965/j.ijabe.20150802.1464

[96] S. Moharana, S. Dutta. 2016. "Spatial variability of chlorophyll and nitrogen content of rice from hyperspectral imagery." *ISPRS Journal of Photogrammetry and Remote Sensing*, 122: 17–29. doi:10.1016/j. isprsjprs.2016.09.002

[97] P. Pandey, Y. Ge, V. Stoerger, and J. C. Schnable. 2017. "High throughput in vivo analysis of plant leaf chemical properties using hyperspectral imaging." *Frontiers in Plant Science*, 8: 1348. doi:10.3389/fpls.2017.01348

[98] M. Corti, P. M. Gallina, D. Cavalli, and G. Cabassi. 2017. "hyperspectral imaging of spinach canopy under combined water and nitrogen stress to estimate biomass, water, and nitrogen content." *Biosystems Engineering*, 158: 38–50. doi:10.1016/j.biosystemseng.2017.03.006

[99] X. Yu, H. Lu, and Q. Liu. 2018. "Deep-learning-based regression model and hyperspectral imaging for rapid detection of nitrogen concentration in oilseed rape (Brassica Napus L.) Leaf." *Chemometrics and Intelligent Laboratory Systems*, 172: 188–193. doi:10.1016/j.chemolab.2017.12.010

[100] B. Bruning, H. Liu, C. Brien, B. Berger, M. Lewis, and T. Garnett. 2019. "The development of hyperspectral distribution maps to predict the content and distribution of nitrogen and water in wheat (triticum aestivum)." *Frontiers in Plant Science*, 10. doi:10.3389/fpls.2019.01380

[101] B. Banerjee, S. Joshi, E. Thoday-Kennedy, R. K. Pasam, J. Tibbits, M. Hayden, G. Spangenberg, and S. Kant. 2020. "High-throughput phenotyping using digital and hyperspectral imaging-derived biomarkers for genotypic nitrogen response." *Journal of Experimental Botany*, 71(15): 4604–4615. doi:10.1093/jxb/eraa143

[102] X. Ye, S. Abe, and S. Zhang. 2020. "Estimation and mapping of nitrogen content in apple trees at leaf and canopy levels using hyperspectral imaging." *Precision Agriculture*, 21 (1): 198–225. doi:10.1007/s11119-019-09661-x

[103] M. Pagola, R. Ortiz, I. Irigoyen, H. Bustince, E. Barrenechea, P. Aparicio-Tejo, C. Lamsfus, and B. Lasa. 2009. "New method to assess barley nitrogen nutrition status based on image colour analysis: Comparison with SPAD-502." *Computers and Electronics in Agriculture*, 65(2): 213–218. doi:10.1016/j. compag.2008. 10.003

[104] Y. Li, D. Chen, C. N. Walker, and J. F. Angus. 2010. "Estimating the Nitrogen Status of Crops Using a Digital Camera." *Field Crops Research* 118 (3): 221–227. doi:10.1016/j.fcr.2010.05.011

[105] A. Mercado-Luna, E. Rico-García, A. Lara-Herrera, G. Soto-Zarazúa, R. Ocampo-Velázquez, R. Guevara-González, G. Herrera-Ruiz, and I. Torres-Pacheco. 2010. "Nitrogen determination on tomato (Lycopersicon Esculentum Mill.) seedlings by color image analysis (RGB)." *African Journal of Biotechnology*, 9: 5326–5332.

[106] N. Lu, W. Wang, Q. Zhang, D. Li, X. Yao, Y. Tian, Y. Zhu, et al. 2019. "Estimation of nitrogen nutrition status in winter wheat from unmanned aerial vehicle based multi-angular multispectral imagery." *Frontiers in Plant Science*, 10: 1601. doi:10.3389/fpls.2019.01601

Security challenges and solutions in cloud, fog, and edge computing for sustainable development

Sachin Lalar
Kurukshetra University, Kurukshetra, India

Tajinder Kumar and Sonam Kamboj
JMIETI, Radaur, India

Rajender Kumar
Punjabi University, Patiala, India

9.1 INTRODUCTION

In the field of distributed computing, cloud, edge, and fog computing are three different paradigms that have all had a significant impact on the development of the contemporary digital environment. Fundamentally, cloud computing is an internet-based centralized paradigm that provides computer resources and services. This paradigm allows organizations to grow their infrastructure flexibly by facilitating on-demand access to a shared pool of programmable resources. Conversely, edge computing reduces latency and improves real-time processing capabilities by moving computation and data storage closer to data production sources. This decentralized approach is precious when rapid decision-making and low-latency communication are critical, such as in Internet of Things (IoT) applications. Fog computing, on the other hand, extends the computing paradigm to the network's edge and enables localized data processing and analysis. It is situated between the cloud and the edge. Fog computing provides a middle ground between cloud and edge computing benefits, making it particularly useful in settings with limited bandwidth or sporadic access [1]. These computing paradigms create a flexible ecosystem that enables enterprises to tailor their computing infrastructure to specific requirements, maximizing cloud-based centralized resource management, reducing edge latency, or striking a balance between fog computing and diverse and dynamic environments.

9.1.1 Explanation of cloud, fog, and edge computing

  a. *Cloud Computing*:
    The delivery and consumption of computer services are being completely transformed by cloud computing. It entails giving users online

DOI: 10.1201/9781003494430-9

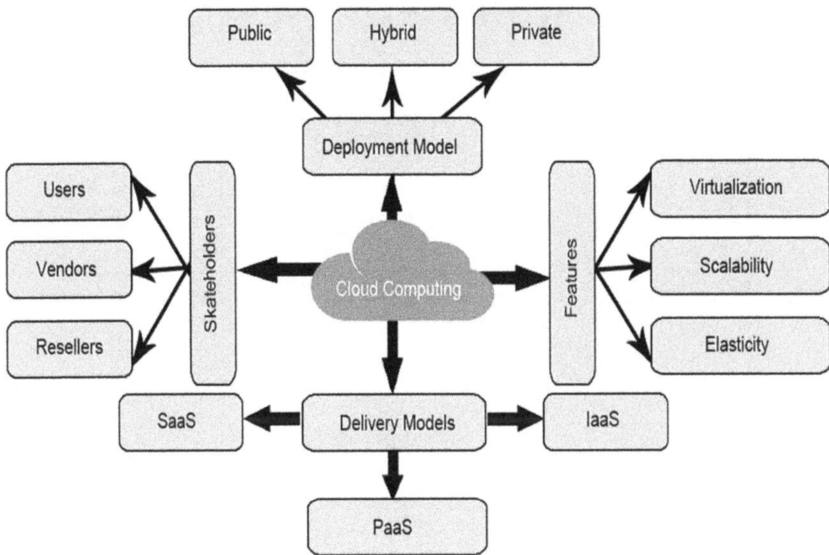

Figure 9.1 Overview of cloud computing.

access to a shared pool of computer resources whenever needed as shown in Figure 9.1. With this approach, businesses may use virtualized resources like servers, storage, and apps without spending money on or maintaining physical infrastructure. Modern computer designs require cloud computing because it offers scalability, flexibility, and cost-effectiveness. Its responsibilities include large-scale data processing, centralized data storage, and support for various services and apps available to everyone with an internet connection. The main characteristics of cloud computing are [2]:

- **Centralized Resource Management:** Cloud computing enables online access to a shared pool of reconfigurable resources by centralizing computer services and resources. Organizations benefit from this centralized approach's cost-effectiveness, scalability, and efficient resource allocation.
- **Scalability and Flexibility:** Cloud services give businesses unmatched flexibility by enabling them to scale resources up or down in response to demand. This flexibility is beneficial for companies whose workloads change often.
- **Remote Data Storage and Processing:** Cloud computing allows users to access apps and data from almost anywhere with an internet connection. It is based on distant data centers for processing and storage. For accessibility and cooperation, centralized data management is essential.

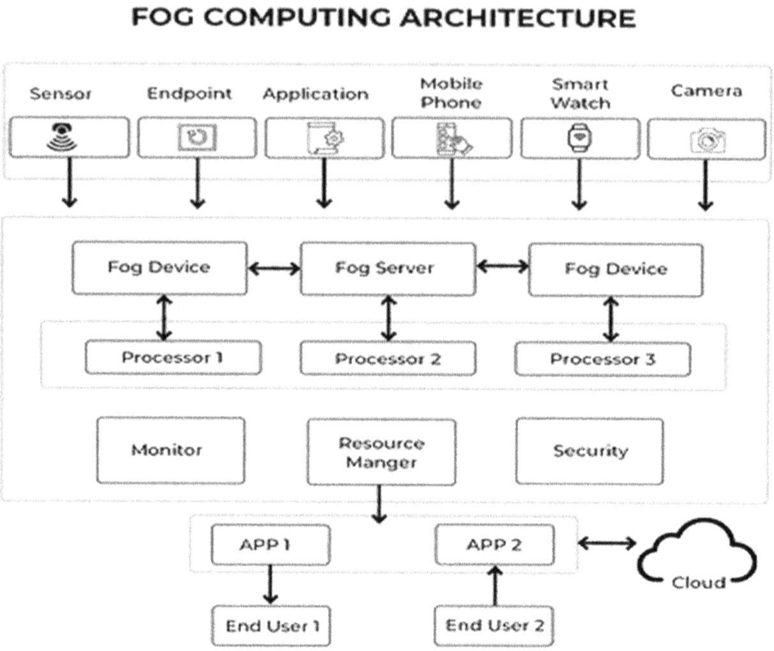

Figure 9.2 Overview of fog computing.

b. *Fog Computing*

Fog computing increases the possibilities of cloud computing by decreasing latency and boosting efficiency by moving processing resources closer to the network's edge as shown in Figure 9.2. It functions closer to the data source or data-generating equipment at the network's edge. In situations like the IoT, smart cities, and industrial automation, where real-time processing and low-latency communication are essential, fog computing is especially pertinent. Its functions include effective data filtering, real-time analytics, and local data processing, which relieves pressure on centralized cloud servers. The main characteristics of fog computing are [3]:

- **Decentralized Processing:** By bringing processing closer to the data source, edge computing lowers latency and improves real-time processing capabilities. This is particularly crucial for applications like the IoT, where quick decisions are crucial.
- **Bandwidth Optimization:** Edge computing reduces the need to transport massive amounts of data to centralized cloud servers by processing data locally at the edge. In situations when latency requirements are strict or bandwidth is restricted, this optimization is essential.
- **Privacy and Security:** Edge computing reduces the need to transfer sensitive data across networks by processing it locally, improving privacy and security. This is especially important for situations where data security is the main priority.

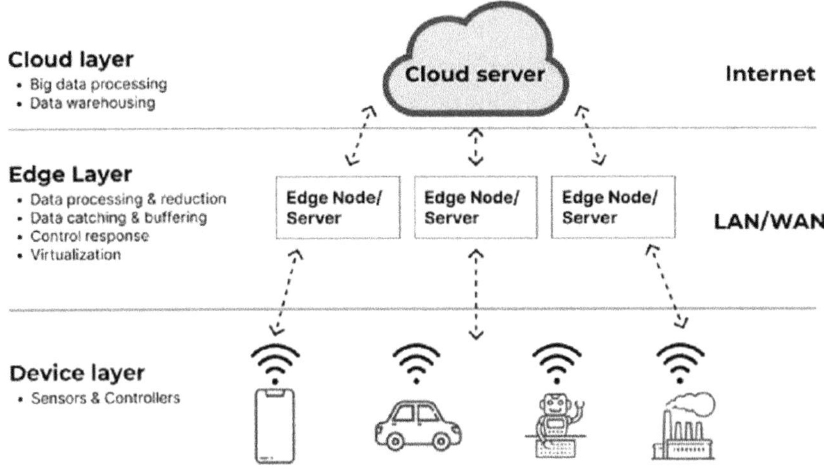

Figure 9.3 Overview of edge computing.

c. *Edge Computing*

By shifting computational duties to edge devices, edge computing expands on decentralized processing. Reducing the need to send data to centralized cloud servers entails processing data locally on IoT devices or gateways. Applications like robots, augmented reality, and driverless cars that require instantaneous processing and decision-making require edge computing. Its functions include enabling immediate data processing, lowering latency, and improving computer systems' overall efficiency. The main characteristics of edge computing are [4]:

- **Distributed Processing at the Network Edge:** By extending the cloud computing paradigm to the network's edge, fog computing enables dispersed processing and analysis. When a balance between centralized and edge processing is necessary, this improves the effectiveness of data processing.
- **Intermediary between Cloud and Edge:** As a bridge between the cloud and edge, fog computing keeps links to the more significant cloud architecture while offering localized processing capabilities. This is useful when the requirements for computation are varied and dynamic.
- **Enhanced Reliability:** Fog computing improves dependability by decreasing reliance on a single centralized point of failure through its dispersed processing capabilities. This is especially useful in situations where uninterrupted functioning is essential.

With various alternatives, these computing paradigms enable organizations to customize their approach to suit unique business requirements and technological constraints. These possibilities include basing their computing infrastructure design on scalability, latency, privacy, and dependability.

9.2 ROLES IN MODERN COMPUTING ARCHITECTURES

The combination of cloud, fog, and edge computing in contemporary computer architectures allows a wide range of applications, from real-time, low-latency processing at the edge to data-intensive operations in the cloud [5].

- **Cloud Computing**: Centralized data storage, scalable processing, and universal accessibility.
- **Fog Computing**: Low-latency processing, real-time analytics, and efficient data handling at the network edge.
- **Edge Computing**: Immediate data processing, reduced dependence on centralized cloud servers, and enhanced efficiency for time-sensitive applications.

This decentralized method maximizes resource use, improves responsiveness, and facilitates various applications vital to the changing digital environment. Table 9.1 shows how cloud, edge, and fog computing may play complementary roles in satisfying the unique needs of different use cases and sectors, boosting efficiency, responsiveness, and creativity in the digital world.

Table 9.1 Applications of cloud, edge, and fog computing [6–7]

Cloud Computing Applications	Edge Computing Applications	Fog Computing Applications
Enterprise Resource Planning (ERP): Cloud-based ERP solutions facilitate real-time collaboration and data access for enterprises by streamlining business operations.	**Internet of Things (IoT)**: In applications like smart cities and linked devices, edge computing reduces latency and improves responsiveness by enabling real-time processing of data provided by the Internet of Things.	**Smart Manufacturing**: In the industrial sector, fog computing improves real-time data analysis for process optimization, predictive maintenance, and quality control.
Data Storage and Backup: Cloud storage solutions enable companies to shift their storage requirements to distant servers by providing scalable and secure data storage.	**Autonomous Vehicles**: In autonomous cars, edge computing handles data locally to speed up decision-making for safety and navigation.	**Retail Analytics**: Fog computing helps with inventory management and personalized consumer experiences by supporting real-time analytics in retail settings.
Software as a Service (SaaS): Cloud-based applications that provide services, like Microsoft 365 and Google Workspace, enhance collaboration and accessibility.	**Smart Grids**: Edge computing facilitates the effective control of energy distribution of smart grids, allowing for a quicker reaction to variations in supply and demand.	**Augmented Reality (AR) and Virtual Reality (VR)**: By processing data closer to the end-user, fog computing lowers latency in AR and VR applications, improving the user experience.

(Continued)

Table 9.1 (Continued) Applications of cloud, edge, and fog computing [6–7]

Cloud Computing Applications	Edge Computing Applications	Fog Computing Applications
Development and Testing: Scalable infrastructure for application creation, testing, and deployment is made available to developers by cloud platforms.	**Healthcare Monitoring**: Edge devices handle health monitoring data locally, minimizing the requirement for continuous data transmission and guaranteeing quick insights.	**Smart Cities**: Through localized data processing, fog computing in smart cities facilitates effective traffic control, surveillance, and public service delivery.

Combined Applications:

- **Healthcare Data Processing**: While edge and fog computing allow for quick analysis of medical equipment and real-time patient data monitoring, cloud computing securely saves electronic health records.
- **Supply Chain Management**: Cloud-based solutions oversee the whole supply chain, while edge and fog computing enhance warehouse inventory control and local logistics.

Emerging Applications:

- **Blockchain Networks**: While edge and fog computing provide distributed processing for decentralized apps (DApps) and smart contracts, cloud computing provides the backbone of blockchain networks.
- **5G Networks**: In 5G networks, edge and fog computing are essential because they process data closer to the source, lower latency, and improve network performance.

9.3 IMPORTANCE OF CLOUD, FOG, AND EDGE COMPUTING IN DRIVING SUSTAINABLE DEVELOPMENT

The combination of edge, cloud, and fog computing is a powerful force behind sustainable growth, providing creative answers to pressing problems of the day. Resource optimization is made possible by cloud computing's centralized and scalable architecture, which lowers the demand for substantial physical hardware and energy use. By allocating processing jobs closer to the data source, fog computing increases efficiency by reducing latency and data transfer, saving energy and bandwidth on the network. At the cutting edge of data creation, edge computing improves real-time processing, allowing IoT devices and smart cities to run more effectively and with less environmental impact. When taken as a whole, these computing paradigms support sustainability by maximizing energy efficiency, reducing resource consumption, and developing technology solutions that support social advancement and environmental responsibility [8].

The integration of cloud, fog, and edge computing transforms the digital landscape and emerges as a catalyst for building a more sustainable and resilient future. Cloud, fog, and edge computing technologies are critical for sustainable development because they encourage resource efficiency, energy conservation, and environmentally conscious behavior. Because of their ability to support sustainable initiatives, enhance accessibility, and permit real-time decision-making, they will be crucial in building a more technologically advanced and sustainable future as [9]:

a. **Resource Efficiency:**
   - *Cloud Computing*: The effective use of resources made possible by centralized cloud servers eliminates the need for individual businesses to maintain a sizable physical infrastructure.
   - *Fog and Edge Computing*: These technologies enhance resource efficiency by processing data closer to the source, saving energy and bandwidth needed for data transmission to centralized facilities.

b. **Energy Conservation:**
   - *Cloud Computing*: By managing data centers and virtualizing servers, cloud providers may optimize energy utilization, which lowers energy consumption overall.
   - *Fog and Edge Computing*: Localized processing improves these systems' sustainability and energy efficiency by reducing the energy required for data transmission.

c. **Reduced Environmental Impact:**
   - *Cloud Computing*: Compared to dispersed computing methods, centralized data centers may operate more sustainably thanks to economies of scale.
   - *Fog and Edge Computing*: Green computing principles are supported by minimizing data transmission to centralized places, lowering data transit's environmental effect.

d. **Real-time Decision-Making:**
   - *Fog and Edge Computing*: Real-time decision-making is made possible by local processing capabilities, and this is critical for applications like autonomous cars, smart grids, and healthcare. In certain situations, this prompt action improves safety and efficiency, supporting sustainable development objectives.

e. **Accessibility and Inclusivity:**
   - *Cloud Computing*: Cloud-based services facilitate worldwide access to computer resources, promoting inclusion and closing the digital gap among different groups.
   - *Fog and Edge Computing*: These technologies improve accessibility by bringing computer power closer to end users, facilitating growth in many areas and communities.

f. **Optimized Infrastructure Utilization:**
   - *Cloud Computing*: Resource scalability enables businesses to modify their computing capacity in response to demand, avoiding overspending and maximizing the use of their infrastructure.
   - *Fog and Edge Computing*: By maximizing the utilization of edge devices, centralized cloud servers are well-rested, and computing resources are utilized efficiently, thanks to localized processing.

g. **Support for Sustainable Initiatives:**
  - *All Technologies*: Accomplishing more general sustainability objectives is facilitated by the capacity to handle and analyze data effectively. These initiatives include smart city projects, precision agriculture, and environmental monitoring.

9.4 SECURITY CHALLENGES

Cloud computing must overcome obstacles to ensuring data privacy and compliance, controlling multi-tenancy hazards, and resolving network security issues. Due to device deployment in easily accessible areas, edge computing faces challenges such as resource-constrained situations, decentralized security complexity, and physical solid protection requirements. Maintaining consistent security architecture between the cloud and edge, maximizing security in distributed processing, and reducing risks connected to its intermediate position are among the issues fog computing faces. For any paradigm to be implemented securely and efficiently, these issues must be considered carefully [10–12].

9.4.1 Security challenges in cloud computing

Because cloud computing relies on shared, centralized, and remotely accessible computer resources, it presents various security issues. Organizations' significant obstacle is ensuring data privacy and compliance when entrusting critical information to third-party cloud service providers. Data breaches, unauthorized access, and noncompliance are always possible. Additionally, maintaining data security and isolation among several users using the same infrastructure is made more difficult by cloud environments' multi-tenancy models. Table 9.2 addresses the security challenges in cloud computing and combines technological solutions, effective policies, and ongoing awareness to ensure a robust and resilient security posture [12].

9.4.2 Security Challenges in Fog Computing

Although fog computing offers dispersed and decentralized processing capabilities, it poses unique security problems that should be carefully considered. Sensitive data exposure to possible attacks at the network edge is one of the leading causes of concern. Because fog computing is spread, it might be challenging to implement uniform security measures on a wide range of edge devices. Furthermore, fog computing is vulnerable to illegal access, data breaches, and cyber-attacks because it depends on networked equipment. It becomes essential to guarantee the confidentiality and integrity of data handled at the network edge. Table 9.3 explains the security challenges in fog computing [16].

Table 9.2 Security challenges in cloud computing [13–15]

	Challenge	Solution
Data Privacy and Compliance	Cloud computing involves storing and processing data on external servers, raising concerns about the privacy and security of sensitive information.	Robust encryption, strict access controls, and adherence to compliance standards (such as GDPR and HIPAA) are essential to mitigate data privacy risks.
Multi-Tenancy Risks	Multiple users sharing the same infrastructure may pose risks like unauthorized access or data leakage between tenants.	Implement strong isolation measures, conduct regular security audits, and employ advanced authentication mechanisms to enhance multi-tenancy security.
Network Security Concerns	Relying on networks for data transfer creates vulnerabilities, including interception, eavesdropping, and man-in-the-middle attacks.	Utilize encryption protocols (e.g., TLS/SSL), implement robust firewalls, and deploy intrusion detection and prevention systems to safeguard network communications.
Insecure APIs	Application Programming Interfaces (APIs) in the cloud can be vulnerable, leading to unauthorized access and data breaches.	Regularly audit and secure APIs, use strong authentication mechanisms, and keep API software up-to-date to prevent exploitation.
Insufficient Identity and Access Management (IAM)	Weak identity and access controls may result in unauthorized access, privilege escalation, or misuse of user credentials.	Implement robust IAM policies, conduct regular access reviews, and enforce the least privilege principle to control user access effectively.
Emerging Threat Landscape	The dynamic nature of the threat landscape requires continuous adaptation to new attack vectors, malware, and vulnerabilities.	Stay informed about emerging threats, regularly update security measures, and proactively share threat intelligence to address new security challenges.
Data Loss and Leakage	The potential for accidental data loss or intentional leakage poses a significant threat to data integrity and confidentiality.	Employ data loss prevention (DLP) tools, implement encryption for sensitive data, and establish clear policies on data handling and sharing.
Inadequate Incident Response	A lack of preparedness to respond effectively to security incidents may lead to prolonged system vulnerabilities and increased damage.	Develop and regularly test an incident response plan, establish clear communication channels, and train personnel to promptly address security incidents.

(Continued)

Table 9.2 (Continued) Security challenges in cloud computing [13–15]

	Challenge	Solution
Shared Responsibility Model Misunderstandings	Organizations may need to understand their responsibilities in a shared responsibility model, leading to security gaps.	Clearly define and understand the shared responsibilities between the cloud service provider and the customer, and actively address the customer's role in securing their data and applications.
Vendor Lock-In	Dependence on a single cloud service provider can create challenges regarding migration, service continuity, and negotiating favorable terms.	Adopt interoperable standards, diversify cloud service providers where feasible, and plan for a potential transition to mitigate vendor lock-in risks.

Table 9.3 Security challenges in fog computing [17]

	Challenge	Solution
Data Privacy and Compliance	Processing sensitive data by fog computing raises questions about data integrity and privacy. A major risk is posed by unauthorized access to or modification of data.	Put robust encryption mechanisms in place for data at rest and in transit. Ensure that only authorized parties can access and alter data by implementing access restrictions and authentication methods. Maintain and audit data integrity regularly.
Device Security	Fog computing depends on a wide range of edge devices, such as actuators and sensors, which could have weak security. Devices that have been compromised might serve as attack vectors.	Use secure boot procedures, frequent software upgrades, device authentication, and other security precautions on edge devices. Use device management solutions to monitor and regulate edge device security.
Network Security Concerns	Because fog computing is spread, there are more points of vulnerability in the network, which raises the possibility of man-in-the-middle attacks, illegal access, and other network-level threats.	Put strong firewalls and intrusion detection systems in place at the network edge and use secure communication protocols. To prevent eavesdropping, encrypt all communications between fog nodes.

(Continued)

Table 9.3 (Continued) Security challenges in fog computing [17]

	Challenge	Solution
Resource Constraints	Because fog computing edge devices frequently have limited processing and storage power, applying conventional security measures without affecting performance might be challenging.	Enhance security procedures for environments with limited resources as a solution. Employ low-complexity encryption techniques and security measures that reduce processing overhead. Sort essential security features according to the particular use case.
Interoperability and Standardization	Standardized security measures might be challenging to apply smoothly in fog computing settings due to the vast diversity of devices and technologies that can cause interoperability issues.	Encourage the development of industry-wide fog computing security standards. Promote the use of compatible security frameworks and protocols. Establish best practices for safeguarding diverse fog environments by working with stakeholders.
Data Redundancy and Replication	The challenge of fog computing is durability through data redundancy and replication across numerous nodes. This makes it difficult to guarantee consistent data security across all copies.	Apply uniform access rules and encryption to every copy of the copied data. To have a consistent security posture, audit and synchronize security measures regularly. Make use of safe data replication techniques.
Dynamic Environment	Environments for fog computing are dynamic, with devices often joining and departing the network—access management and authentication present difficulties when managing security in such a dynamic environment.	Put access limits and dynamic authentication systems that can adjust to environmental changes. Use blockchain technology to maintain identities in a safe, decentralized manner.
Lack of Security Awareness	Users and managers might need to be made aware of fog computing's best practices and security implications, which might leave them vulnerable.	Run frequent user and administrator awareness and training campaigns. Encourage a culture of security awareness and ensure all relevant parties know the most recent security risks and countermeasures.

9.4.3 Security challenges in edge computing

While edge computing has revolutionary potential, certain security risks must be carefully considered. Because edge environments are decentralized, there is a more excellent attack surface, making it difficult to monitor and safeguard every point of access adequately. The installation of strong security measures may be hampered by limited resources on edge devices, which might lead to vulnerabilities. Logistical issues arise regarding ensuring edge devices operating in remote regions receive regular updates and fixes. The resolution of these security issues necessitates an all-encompassing strategy drawn in Table 9.4 that includes steps for monitoring, encryption, authentication, and industry-wide cooperation to create uniform security procedures for the changing edge computing environment [17].

Table 9.4 Security challenges in edge computing [18–20]

	Challenge	*Solution*
Limited Resources	Because edge devices sometimes have limited resources, putting strong security measures in place can take time and effort.	Use lightweight encryption techniques, prioritize security measures according to the importance of the data, and optimize security protocols for resource-constrained contexts.
Decentralized Nature	Because edge computing is decentralized, there is a more excellent attack surface, which makes it more challenging to keep an eye on and safeguard every point of access.	Using network segmentation and distributed security techniques like secure boot procedures to protect essential components from such attacks.
Physical Vulnerability	Edge devices are often deployed in physically accessible locations, making them susceptible to physical tampering or theft.	To identify and respond to unauthorized access, use device attestation and physical security measures like tamper-resistant hardware and secure enclosures.
Data Transmission Security	Sending confidential information between edge devices and the cloud poses security issues, mainly when bandwidth is at a premium.	Adopt edge-to-cloud security integration for end-to-end protection, use encryption techniques for safe data transfer, and implement secure communication routes like VPNs.

(Continued)

Table 9.4 (Continued) Security challenges in edge computing [18–20]

	Challenge	*Solution*
Authentication and Authorization	As edge computing is spread, it might be challenging to verify the identification of people and devices at the edge.	To guarantee that only authorized entities may access sensitive information, use role-based access restrictions and robust authentication techniques like multi-factor authentication.
Updates and Patch Management	It is challenging to routinely update and repair security flaws in edge devices because they frequently operate in isolated or difficult-to-reach places.	Provide automatic update systems, use over-the-air (OTA) updates where it is practical, and give preference to devices with integrated secure firmware update methods.
Lack of Standardization	The lack of uniform security standards among diverse edge devices and platforms challenges the creation of coherent security solutions.	Encourage secure development techniques, push for industry-wide security standards, and work with manufacturers to ensure security-by-design guidelines are followed.
Insufficient Monitoring and Logging	Due to resource limitations, comprehensive monitoring and logging features may not be implemented on edge devices.	For a comprehensive understanding of edge security, prioritize critical logging for security events, use anomaly detection algorithms, and put in place centralized monitoring systems.
Vendor-Specific Security Risks	Vendors of edge devices differ in their security postures, which leads to inconsistencies and possible vulnerabilities.	Before deploying edge devices, thoroughly evaluate their security, select reliable suppliers prioritizing security, and push for industry-wide security best practices.
Integration with Cloud Security	Ensuring end-to-end protection requires a smooth integration of edge and cloud security mechanisms.	Use secure APIs for communication, apply uniform security policies to cloud and edge settings, and conduct routine security audits to spot and fix any integration problems.

9.5 DATA PRIVACY AND COMPLIANCE ISSUES IN CLOUD, FOG, AND EDGE COMPUTING

Data privacy and compliance are critical challenges since cloud, fog, and edge computing systems handle and store sensitive data across traditional borders. Ensuring adherence to regional, national, and sector-specific regulations, especially in decentralized edge environments, becomes challenging. To preserve user privacy and meet ever-changing compliance regulations, finding a balance between innovation and stringent data protection measures is imperative [21].

1. **Cloud Computing:**
   - *Data Privacy Concerns*:
     - **Third-Party Data Handling:** Cloud providers manage enormous volumes of user data, which raises questions regarding the security, processing, and access to this data.
     - **Data Residency:** Geographical data locations in the cloud may affect adherence to local data protection regulations and raise concerns about data sovereignty.
   - *Compliance Challenges*:
     - **Regulatory Variability:** Ensuring consistent compliance with various international and industry-specific standards, including GDPR, HIPAA, or PCI DSS, presents issues.
     - **Auditability:** It cannot be easy to prove compliance through audits, particularly when working with shared resources and dynamic cloud settings.

2. **Fog Computing:**
   - *Data Privacy Concerns*:
     - **Localized Processing of Sensitive Data:** Since fog computing processes data closer to the source, there may be questions about localized processing of sensitive data.
     - **Data Ownership and Control:** Identifying the ownership and management of data handled at the edge can be complex, which raises privacy concerns.
   - *Compliance Challenges*:
     - **Regulatory Alignment:** Thorough thought and adaptation may be necessary to guarantee that fog computing installations comply with local, state, and federal rules.
     - **Edge Device Compliance:** Edge devices might need help managing and applying comprehensive compliance procedures due to resource constraints.

3. **Edge Computing:**
   - *Data Privacy Concerns*:
     - **Decentralized Data Processing:** The handling of data on peripheral devices gives rise to apprehensions over the security of that equipment and the possible disclosure of confidential data.
     - **Data Transmission Security:** Unauthorized access or interception may occur between edge devices and central systems during data transport.

- *Compliance Challenges*:
  - **Dynamic Edge Environments**: When devices enter and exit the network regularly, compliance attempts may encounter difficulties.
  - **Interoperability and Standards**: If edge computing compliance measures are not standardized, interoperability problems and disparities in security levels may arise throughout edge implementations.

9.6 MULTI-TENANCY RISKS IN CLOUD, FOG, AND EDGE COMPUTING

In cloud, fog, and edge computing settings, multi-tenancy—the practice of several users or organizations using the same computer resources—introduces unique security issues. It is essential to recognize and manage these risks to guarantee the availability, confidentiality, and integrity of data and services [22–23]:

a) **Unauthorized Access**:
   a. *Cloud Computing*: In a shared cloud setting, insufficient access restrictions might allow one tenant from another to get illegal access to confidential information or apps.
   b. *Fog and Edge Computing*: The possibility of unwanted access to devices and data increases because fog and edge devices are more physically accessible.
b) **Data Leakage Between Tenants**:
   a. *Cloud Computing*: When there is insufficient separation between tenants, data from one tenant may inadvertently be viewed or made public by another.
   b. *Fog and Edge Computing*: Data may travel over shared networks in distributed edge settings, raising the possibility of unintentional data exposure between tenants.
c) **Resource Contention**:
   a. *Cloud Computing*: Resource sharing in the cloud can result in contention problems, affecting how one tenant's apps run because other tenants use those resources.
   b. *Fog and Edge Computing*: The quality of service may be impacted by competition among tenants for processing power, memory, or bandwidth due to limited resources at the edge.
d) **Inadequate Isolation**:
   a. *Cloud Computing*: Inadequate virtualization or containerization can undermine tenant isolation, allowing one tenant to affect the security or performance of other tenants.
   b. *Fog and Edge Computing*: Strong isolation methods are required since the proximity of edge devices may raise the possibility of interference between tenants.

e) **Cross-Tenant Data Interception:**
  a. *Cloud Computing*: In shared networks, improper network segmentation might result in data being intercepted while transferred between tenants.
  b. *Fog and Edge Computing*: If tenants don't employ appropriate encryption, using common communication channels at the edge may make data interception more likely.
f) **Insufficient Authentication Mechanisms:**
  a. *Cloud Computing*: Inadequate authentication protocols can compromise user credentials, permitting unapproved access outside tenant boundaries.
  b. *Fog and Edge Computing*: Inadequate authentication on edge devices makes them vulnerable to illegal access, endangering the information and services of several tenants.
g) **Data Residuals:**
  a. *Cloud Computing*: Incomplete or inefficient data removal procedures might leave leftover data on shared resources, endangering the security and privacy of several renters.
  b. *Fog and Edge Computing*: Sensitive information may be exposed to incoming tenants if data remains on edge devices.
h) **Inadequate Monitoring and Auditing:**
  a. *Cloud Computing*: Inadequate auditing and monitoring tools may make it difficult to identify illicit activity or gauge how security events affect different tenants.
  b. *Fog and Edge Computing*: Inadequate insight into edge device operations may make monitoring and auditing more complex and might result in the missed detection of security events.

Robust access controls, encryption, isolation techniques, and ongoing monitoring are necessary to manage multi-tenancy cloud, fog, and edge computing risks. This allows for the safe coexistence of multiple tenants while preserving the integrity and confidentiality of their data and applications.

9.7 NETWORK SECURITY CONCERNS OF CLOUD, FOG, AND EDGE COMPUTING

a. **Data in Transit Vulnerabilities:**
  - *Cloud Computing*: Data transmission between the cloud server and the client might be subject to man-in-the-middle attacks, eavesdropping, or interception.
  - *Fog and Edge Computing*: The decentralized data processing at the network edge exposes it to comparable hazards associated with fog and edge computing.

b. **Insufficient Encryption:**
   - *Cloud Computing*: Cloud Computing: Sensitive data may be exposed during transmission or storage in the cloud due to inadequate or poorly designed encryption techniques.
   - *Fog and Edge Computing*: Encryption is essential at the edge to safeguard data as it travels between devices and local processing centers.
c. **Insecure APIs and Interfaces:**
   - *Cloud Computing*: Vulnerabilities in APIs and interfaces linking cloud services may allow for data tampering or unwanted access.
   - Fog and Edge Computing: Strong security measures are necessary since localized fog and edge settings APIs might be exploited.
d. **Distributed Denial of Service (DDoS) Attacks:**
   - *All*: DDoS attacks can overload network resources in cloud, fog, and edge settings, disrupting services.
   - Solution: Use DDoS mitigation techniques, including content delivery networks (CDNs), rate limitation, and traffic filtering.
e. **Network Segmentation Challenges:**
   - *Fog and Edge Computing*: Network segmentation is complicated by the scattered nature of fog and edge devices, which might provide attackers with lateral mobility.
   - Solution: The solution is to prohibit illegal access by using firewall rules, access restrictions, and effective network segmentation mechanisms.
f. **Interoperability and Integration Risks:**
   - *Fog and Edge Computing*: Edge Computing and Fog: If several platforms and devices are integrated at the edge, improper security measures may result in vulnerabilities.
   - *Solution*: Ensure safe integration by implementing robust authentication, standardizing protocols, and conducting frequent security audits.
g. **Man-in-the-Middle Attacks:**
   - *All*: Man-in-the-middle attacks can jeopardize data integrity and confidentiality in cloud, fog, and edge environments.
   - *Solution*: One possible solution is to utilize encryption, such as TLS/SSL, establish secure communication channels, and implement measures to identify and stop man-in-the-middle attacks.
h. **Authentication and Access Control Issues:**
   - *All*: Unauthorized access to cloud, fog, or edge resources might be caused by weak authentication procedures or insufficient access restrictions.
   - *Solution*: Adopt robust authentication procedures, uphold the least-privileged principle, and routinely examine and modify access constraints.
i. **Routing and Switching Insecurities:**
   - *All*: Unauthorized access or data interception may result from unsafe routing protocols and switching settings.

- *Solution*: Use network monitoring secure routing protocols and routinely audit and update switching setups.
j. **Data Packet Sniffing:**
    - *All*: Data exposure and possible security breaches may result from unauthorized data packet interception in transit.
    - *Solution*: To identify and stop packet sniffing efforts, use intrusion detection systems, encrypt your network, and monitor traffic.
k. **Lack of Visibility and Monitoring:**
    - *Fog and Edge Computing*: Due to limited visibility, it might be challenging to detect and respond to network security problems in decentralized fog and edge systems.
    - *Solution*: To improve visibility and proactively uncover security vulnerabilities, use log analysis, anomaly detection, and comprehensive network monitoring technologies.

Strong security policies, encryption procedures, access controls, and continual monitoring are needed to address these network security issues in cloud, fog, and edge computing environments and protect the integrity and confidentiality of data moving through these distributed architectures.

9.8 DECENTRALIZED SECURITY CHALLENGES IN CLOUD, FOG, AND EDGE COMPUTING

Compared to the more centralized cloud computing architecture, the decentralized nature of fog and edge computing creates significant security concerns that call for particular attention. Maintaining the availability, confidentiality, and integrity of data and services in decentralized computing systems requires addressing these issues [26]:

a. **Distributed Attack Surface:**
    - *Cloud Computing*: Because centralized cloud servers have a smaller attack surface, implementing thorough security measures is simpler.
    - *Fog and Edge Computing*: In fog and edge settings, the scattered nature of devices and nodes increases the attack surface, necessitating careful security measures across various endpoints.
b. **Physical Security Risks:**
    - *Cloud Computing*: Cloud servers are generally kept in safe data centers with limited physical access.
    - *Fog and Edge Computing*: Because Edge devices may be physically accessible, there is a direct physical security risk due to the possibility of theft, tampering, or illegal access.
c. **Limited Computational Resources:**
    - *Cloud Computing*: Cloud servers generally have enough processing power to put strong security measures in place.

- *Fog and Edge Computing*: More resources on edge devices must be needed to implement advanced security measures, which might lead to vulnerabilities.

d. **Network Vulnerabilities:**
  - *Cloud Computing*: Network traffic in the cloud is frequently restricted to secure and regulated data center facilities.
  - *Fog and Edge Computing*: Using open or shared networks at the edge raises the possibility of network-based threats like eavesdropping and man-in-the-middle assaults.

e. **Dynamic Topologies:**
  - *Cloud Computing*: Topologies in cloud settings are mostly steady and dependable.
  - *Fog and Edge Computing*: Establishing and maintaining secure communication routes and settings can be challenging due to edge environments' dynamic and ever-changing nature.

f. **Interoperability Challenges:**
  - *Cloud Computing*: Most of the time, cloud interoperability issues are limited to cloud service integration.
  - *Fog and Edge Computing*: It may take a lot of work to achieve smooth interoperability with diverse edge devices due to their differing designs and capabilities, which might lead to security vulnerabilities.

g. **Decentralized Identity Management:**
  - *Cloud Computing*: Cloud-based centralized identity management solutions are frequently utilized.
  - *Fog and Edge Computing*: Ensuring safe authentication and authorization processes becomes more complicated when devices and users at the periphery are managed through decentralized identity management.

h. **Edge Device Lifecycle Management:**
  - *Cloud Computing*: Cloud servers undergo centralized lifecycle management.
  - *Fog and Edge Computing*: It becomes a decentralized task to manage the lifespan of various edge devices, including security setups, fixes, and upgrades.

i. **Data Integrity and Consistency:**
  - *Cloud Computing*: Centralized databases are frequently used to guarantee data consistency and integrity.
  - *Fog and Edge Computing*: Dispersed data processing at the edge may make it more challenging to maintain data integrity and consistency across several locations.

j. **Lack of Centralized Security Oversight:**
  - *Cloud Computing*: Centralized cloud environments allow for unified security oversight.
  - *Fog and Edge Computing*: Implementing uniform security guidelines and reacting quickly to security events without centralized monitoring might be challenging.

A customized strategy is needed to address these decentralized security issues. This strategy includes developing dynamic security protocols that can adjust to the changing needs of edge environments, implementing decentralized identity and access management, and deploying lightweight security measures on edge devices.

9.9 SECURITY AS A PILLAR OF SUSTAINABLE DEVELOPMENT

As a fundamental tenet of sustainable development, security is crucial in cloud, fog, and edge computing. Understanding how security and sustainability objectives are related as these technologies become increasingly integrated into different industries is essential. The following points highlight the relevance of security in promoting sustainable development [22–25]:

a. **Data Protection and Privacy:**
   - *Cloud Computing*: Sophisticated security protocols safeguard private information in the cloud, guaranteeing legal compliance and privacy.
   - *Fog and Edge Computing*: By reducing data transit, localized data processing at the edge improves privacy and adheres to sustainable data protection rules.
b. **Resilience Against Cyber Threats:**
   - *Cloud Computing*: Resilient against cyber-attacks, a secure cloud architecture protects vital services and avoids interruptions that might harm sustainability initiatives.
   - *Fog and Edge Computing*: To ensure the durability of distributed systems, essential for sustained projects, security measures at the edge guard against localized attacks.
c. **Prevention of Environmental Impact:**
   - *Cloud Computing*: By reducing the environmental effect of cloud services, energy-efficient data centers support sustainability.
   - *Fog and Edge Computing*: Localized processing optimizes energy use and conforms to green computing principles by minimizing the need for large-scale data transmission.
d. **Reliable and Secure Connectivity:**
   - *Cloud Computing*: Secure and dependable cloud connectivity is necessary for international cooperation and promotes sustainable development initiatives.
   - *Fog and Edge Computing*: Edge devices with secure connections make real-time data transmission possible, making dependable communication essential for sustainable applications like environmental monitoring and smart cities.

9.10 CONCLUSION

The mutually beneficial link between security and sustainable growth becomes apparent as a pivotal point in the evolving field of contemporary computing. This chapter has shed light on the complex issues that underpin the cloud, fog, and edge computing domains and has offered a path forward for strengthening these technical frontiers. The examination of cloud computing security issues highlighted the importance of strong policies for network security, data privacy, and multi-tenancy. Case studies from real-life situations acted as a warning, highlighting the severe consequences of security breaches. The further exploration of the decentralized environments of fog and edge computing revealed particular difficulties brought about by their dispersed nature and resource limitations, calling for a paradigm change in the direction of convincing security integration. Security is a symbiotic partner for sustainable development, which is a light directing the course of our global growth. Through successful examples of how security measures have harmonized with sustainable initiatives, this chapter has knitted security practices into the larger fabric of sustainability. As our investigation ends, the combination of problems and answers moves us closer to a safe and sustainable future. The story captures the complexities of protecting our digital borders and the necessity of coordinating these endeavors with the more general objectives of sustainable development. With the help of these realizations, we set out on a path where technology not only helps us advance but also does so in an ethical manner, guaranteeing a robust and long-lasting digital environment for future generations.

REFERENCES

[1] Puthal, Deepak, Mohammad S. Obaidat, Priyadarsi Nanda, Mukesh Prasad, Saraju P. Mohanty, and Albert Y. Zomaya. "Secure and sustainable load balancing of edge data centers in fog computing." *IEEE Communications Magazine* 56, no. 5 (2018): 60–65.

[2] Rezapour, Ronita, et al. "Security in fog computing: A systematic review on issues, challenges and solutions." *Computer Science Review* 41 (2021): 100421.

[3] Hazra, Abhishek, et al. "Fog computing for next-generation internet of things: fundamental, state-of-the-art and research challenges." *Computer Science Review* 48 (2023): 100549.

[4] Farooqi, Abdul Majid, Syed Imtiyaz Hassan, and M. Afshar Alam. "Sustainability and fog computing: applications, advantages and challenges." *2019 3rd International Conference on Computing and Communications Technologies (ICCCT)*. IEEE, 2019.

[5] Mishra, Narendra, R. K. Singh, and Sumit Kumar Yadav. "Analysis and Vulnerability Assessment of Various Models and Frameworks in Cloud Computing." In *Advances in Data Sciences, Security and Applications: Proceedings of ICDSSA 2019*, pp. 407–417. Springer Singapore, 2020.

[6] Tahirkheli, Abeer Iftikhar, Muhammad Shiraz, Bashir Hayat, Muhammad Idrees, Ahthasham Sajid, Rahat Ullah, Nasir Ayub, and Ki-Il Kim. "A survey on modern cloud computing security over smart city networks: Threats, vulnerabilities, consequences, countermeasures, and challenges." *Electronics* 10, no. 15 (2021): 1811.

[7] Alwakeel, Ahmed M. "An overview of fog computing and edge computing security and privacy issues." *Sensors* 21, no. 24 (2021): 8226.

[8] Mann, Zoltán Ádám. "Optimization problems in fog and edge computing." *Fog and Edge Computing: Principles and Paradigms* 103 (2019).

[9] Swamy, Sowmya Nagasimha, Dipti Jadhav, and Nikita Kulkarni. "Security threats in the application layer in IOT applications." In *2017 International conference on i-SMAC (iot in social, mobile, analytics and cloud)(i-SMAC)*, pp. 477–480. IEEE, 2017.

[10] Gormuş, Sedat, Hakan Aydin, and Guzin Ulutaş. "Security for the internet of things: a survey of existing mechanisms, protocols, and open research issues." *Journal of the Faculty of Engineering and Architecture of Gazi University* 33, no. 4 (2018): 1247–1272.

[11] Qiu, Meikang, Sun-Yuan Kung, and Keke Gai. "Intelligent security and optimization in Edge/Fog Computing." *Future Generation Computer Systems* 107 (2020): 1140–1142.

[12] Rezapour, Ronita, et al. "Security in fog computing: A systematic review on issues, challenges and solutions." *Computer Science Review* 41 (2021): 100421.

[13] Friedman, V. "On the edge: Solving the challenges of edge computing in the era of IoT." URL: https://data-economy.com/on-the-edge-solving-the-challenges-of-edge-computing-in-the-eraof-iot (2018).

[14] Angel, Nancy A., et al. "Recent advances in evolving computing paradigms: Cloud, edge, and fog technologies." *Sensors* 22, no. 1 (2021): 196.

[15] Malik, Swati, and Kamali Gupta. "Smart City: A new phase of sustainable development using fog computing and IoT." *IOP Conference Series: Materials Science and Engineering*, Vol. 1022. No. 1. IOP Publishing, 2021.

[16] Jain, Siddhant, et al. "Fog computing in enabling 5G-driven emerging technologies for development of sustainable smart city infrastructures." *Cluster Computing*, 25, no. 2 (2022): 1–44.

[17] Das, Resul, and Muhammad Muhammad Inuwa. "A review on fog computing: issues, characteristics, challenges, and potential applications." *Telematics and Informatics Reports*, 10 (2023): 100049.

[18] Hamm, Andrea, Alexander Willner, and Ina Schieferdecker. "Edge computing: a comprehensive survey of current initiatives and a roadmap for a sustainable edge computing development." arXiv preprint arXiv:1912.08530 (2019).

[19] Hurbungs, V., V. Bassoo, and T. P. Fowdur. "Fog and edge computing: concepts, tools and focus areas." *International Journal of Information Technology* 13 (2021): 511–522.

[20] Henze, Martin, Roman Matzutt, Jens Hiller, Erik Mühmer, Jan Henrik Ziegeldorf, Johannes van der Giet, and Klaus Wehrle. "Complying with data handling requirements in cloud storage systems." *IEEE Transactions on Cloud Computing* 10, no. 3 (2020): 1661–1674.

[21] Khanh, Quy Vu, et al. "An efficient edge computing management mechanism for sustainable smart cities." *Sustainable Computing: Informatics and Systems* 38 (2023): 100867.

[22] Sha, Kewei, T. Andrew Yang, Wei Wei, and Sadegh Davari. "A survey of edge computing-based designs for IoT security." *Digital Communications and Networks* 6, no. 2 (2020): 195–202.

[23] Alwarafy, Abdulmalik, Khaled A. Al-Thelaya, Mohamed Abdallah, Jens Schneider, and Mounir Hamdi. "A survey on security and privacy issues in edge-computing-assisted internet of things." *IEEE Internet of Things Journal* 8, no. 6 (2020): 4004–4022.

[24] Bourechak, Amira, et al. "At the Confluence of Artificial Intelligence and Edge Computing in IoT-Based Applications: A Review and New Perspectives." *Sensors* 23, no. 3 (2023): 1639.

[25] Harnal, S., & Chauhan, R. K. (2016, January). Multimedia support from cloud computing: A review. In *2016 International Conference on Microelectronics, Computing and Communications (MicroCom)* (pp. 1–6). IEEE.

[26] Burhan, Muhammad, et al. "A comprehensive survey on the cooperation of fog computing paradigm-based iot applications: layered architecture, real-time security issues, and solutions." *IEEE Access* (2023).

Chapter 10

Empowering fog computing for achieving sustainability

Gaganpreet Kaur, Vinay Gautam, Swati Malik, Jatin Arora, and Kanika

Chitkara University Institute of Engineering and Technology, Chitkara University, Chandigarh, India

A V Senthil Kumar

Hindustan College of Arts and Science, Coimbatore, India

10.1 INTRODUCTION

Fog computing, a new computing paradigm, brings cloud computing closer to the network's edge. This facilitates data processing near the source [1]. This paradigm tackles challenges in real-time processing, low latency, data transmission, and analysis in typical cloud computing.

Fog computing moves computer resources and services closer to data generation and consumption along the network edge. Computing is done by edge devices like routers, switches, gateways, cell phones, and Internet of Things (IoT) sensors.

The following are fog computing essentials:

1. Fog computing near data sources minimises latency [2] and speeds up real-time applications.
2. A heterogeneous environment integrates multiple systems and components into a single computing environment.
3. Fog computing allows resource adaptability, maximising computing capacity while meeting changing demands.
4. Data Processing and Analysis: This technology provides edge data processing, analysis, and storage, removing the need to send data to a cloud server [3].
5. By dispersing computing duties, fog computing improves network performance, reliability, and efficiency.

Fog computing enhances cloud computing by bringing computational resources closer to users and their devices, improving response times [4], bandwidth usage, and support for diverse applications in manufacturing, smart cities, healthcare, and agriculture. A basic Fog computing design is shown in Figure 10.1. It consists of three distinct layers: the edge/sensor layer, the fog layer, and the cloud layer. The edge layer contains user devices that serve as the primary consumers or generators of data [5]. For example,

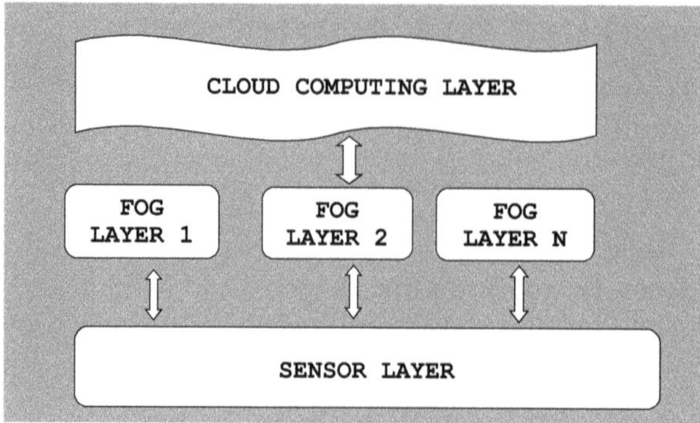

Figure 10.1 Fog computing basic architecture.

sensor nodes produce data, whereas actuators execute certain tasks based on conveyed commands and portable devices [6].

In further sections, the importance of fog computing in achieving SDGs is presented. It discusses how fog computing plays a vital role in different disciplines. Further different case studies along with examples are presented in the chapter. Moreover, fog in achieving SDG 2, 3, 11, 13, and 4 is discussed. Finally, ethical considerations and challenges are presented.

10.1.1 Sustainable Development Goals (SDGs)

In 2015, the United Nations introduced the 2030 Agenda for Sustainable Development, encompassing a set of 17 interconnected global objectives referred to as the Sustainable Development Goals (SDGs). These goals were established in response to many worldwide social, economic, and environmental challenges, with the aim of creating a more sustainable and equitable future for everyone. All available SDGs are presented in Figures 10.2 and 10.3.

In order to create a more prosperous and sustainable world for the present as well as the future, governments, businesses, civil society organisations,

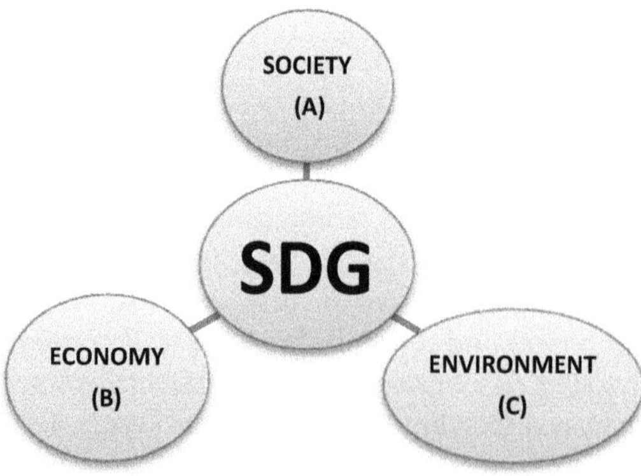

Figure 10.2 Classification of various SDGs.

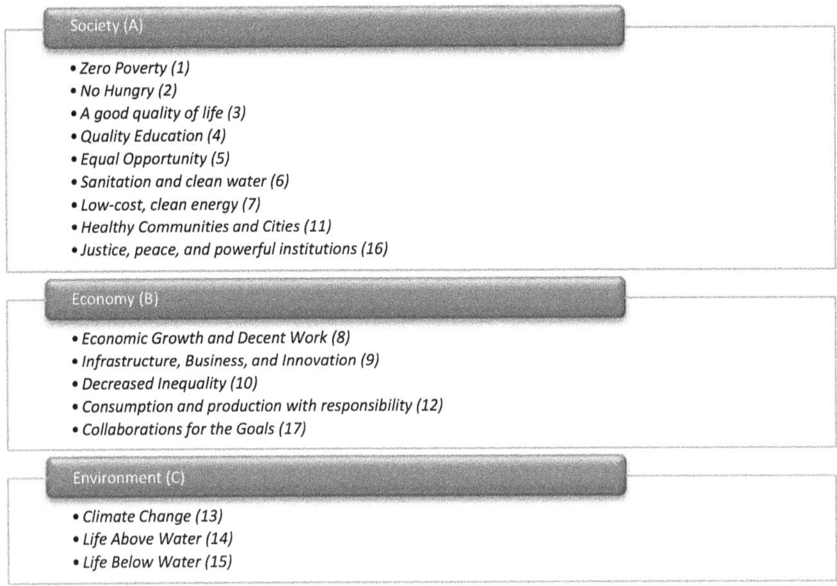

Figure 10.3 Available SDGs.

and individuals must collaborate, as shown by the SDGs, which highlight the interconnectedness of many global challenges.

10.1.2 Importance of fog computing in achieving SDGs

Fog computing greatly enhances the achievement of the SDGs by offering various technological solutions so as to meet critical challenges in industries [7]. It is essential for achieving the SDGs in several significant ways.

1. Fog computing supports seamless connectivity, in remote and under-privileged areas. This is significant as it will guarantee broader accessibility to technical resources [8], thus contributing to the attainment of Sustainable Development Goals 4 and 10. Additionally, it will aid in diminishing inequality and facilitating access to healthcare and top-notch education [9].

2. Engaging in real-time data processing for decision-making demands the ability to process data quickly and in real-time, a capability facilitated by fog computing. Prompt information and action are particularly crucial for SDGs such as the management of natural resources (SDG 15), disaster response (SDG 11), and climate change mitigation (SDG 13), since they can help mitigate adverse impacts and promote sustainable practices.

3. Fog computing contributes to the attainment of SDGs 11 and 12, which focus on sustainable cities and communities, by facilitating the development of smart cities and infrastructure. It facilitates the enhancement of waste management, energy management, optimisation of transportation systems, and public safety.

4. Fog computing assists in the remote monitoring of patients, facilitates telemedicine, and enhances the delivery of efficient healthcare services. These capacities expedite the achievement of the objectives outlined in SDG 3, which aim to improve health and well-being.

5. Fog computing facilitates precision agriculture by enhancing water efficiency, increasing agricultural production, and bringing real-time crop surveillance. This contributes to the achievement of SDG 2, which aims to eliminate hunger.

6. Fog computing empowers communities by offering technological solutions and tools. This initiative supports the SDGs of promoting inclusive technological advancement (SDG 9) and ensuring gender equality (SDG 5), ensuring that technology benefits all aspects of society. Figure 10.4 presents various SDGs focussing on fog computing utilisation. Figure 10.4 gives the pictorial representation for fog computing working for different SDG goals.

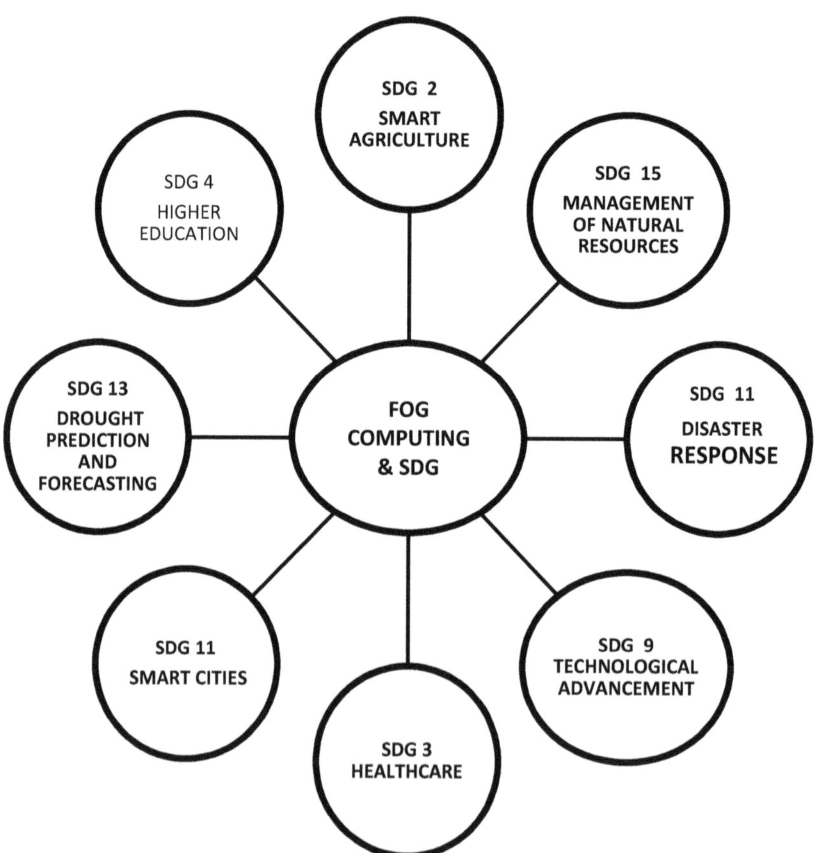

Figure 10.4 Fog computing & SDG's.

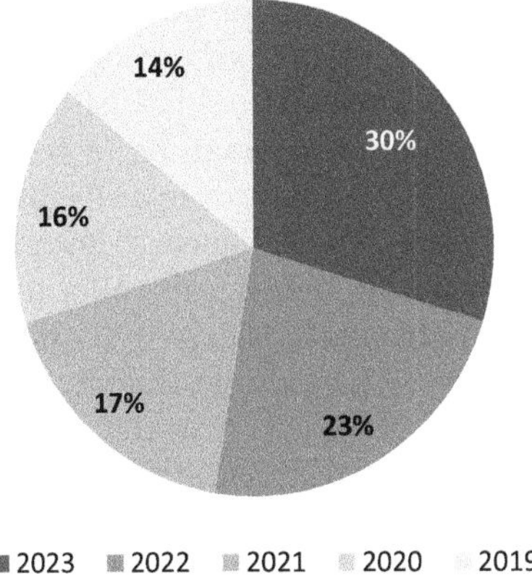

Figure 10.5 Fog computing & SDG's trend paper.

10.1.3 Fog computing and SDG: Playing major role in different disciplines

Fog has gained astonishing achievement in recent years in a variety of fields, including the healthcare sector, computer vision, intelligent sensing, higher education, and condition monitoring systems. It plays a significant role in enhancing productivity and efficiency [10]. From Figure 10.5, it is clear that fog computing is becoming more prevalent across a variety of application fields, which include telecommunications [11, 12], environmental science [13], education [14, 15], health care [16–19], and for sustainability, in addition to its traditional home in computer science.

10.2 FOG COMPUTING AND ENVIRONMENTAL SUSTAINABILITY

10.2.1 Resource optimisation and energy efficiency

Fog computing employs many approaches to dynamically allocate resources [20] and greatly enhance energy efficiency.

1. The processing of data can be performed in adjacent proximity to the edge devices that are producing the data by engaging fog computing. The need for large-scale data transmission is eliminated, by doing data processing on-site, resulting in energy savings that would otherwise be necessary for data transport. In contrast to cloud computing, which commonly involves transmitting data to remote data centres.
2. Fog computing minimises data traffic and network congestion by efficiently processing and storing essential data at the edge. This optimisation reduces the amount of unnecessary data transfers, which, particularly when handling huge amounts of data, decreases the energy usage during data transmission.

3. By employing local resources for computing tasks [21], fog nodes, which are located in close proximity to end devices, optimise the available processing capability. This distributed technique enables improved resource allocation, resulting in reduced total energy consumption and preventing overloads in centralised data centres.

4. Fog computing systems are designed to promptly adjust to varying resource requirements. By dynamically allocating resources based on task demands, they maximise energy utilisation, preventing unnecessary power usage during periods of inactivity and reducing overall carbon emissions.

5. Energy efficiency is a prevailing factor in the development of fog computing infrastructure and devices. Utilising energy-efficient components, improved hardware, and low-power processors can lead to a decrease in overall energy use.

6. The amalgamation of renewable energy sources such as solar, wind, or other sustainable sources can be achieved by integrating fog computing systems. This assimilation reduces carbon emissions by reducing reliance on fossil fuels and developing environmentally friendly and sustainable computing infrastructures.

The capability of fog computing to process data in close proximity to the edge, optimise bandwidth utilisation, dynamically assign resources, and provide effective decision-making greatly enhances resource optimisation and conservation. These practices are in accordance with the goals of sustainable development by minimising waste, preserving resources, and encouraging the optimal use of computing and network infrastructure.

10.2.2 Various available examples and case studies

Several existing case studies exemplify how fog computing contributes to environmental sustainability. These include:

1. Energy management and intelligent grids: Case Study: Pacific Gas and Electric Company (PG&E) in California created a smart grid system based on fog computing. This system works on edge computing to study real-time data from weather sensors, smart metres, and power distribution networks. By complete examination of this data at a regional level, PG&E enhanced the efficiency of energy distribution, reduced unnecessary energy consumption, and removed the cases of power outages.

2. Agriculture: Case Study: John Deere incorporated fog computing expertise into their range of products. These devices gathered and examined data on crop conditions, soil moisture levels, and temperature using sensors and edge computing. By utilising real-time data analysis, farmers can improve the utilisation of resources such as water, insecticides, and fertilisers. This allows them to make accurate

decisions on irrigation, fertilisation, and pest control, leading to improved crop productivity.

3. Conservation and monitoring of the environment: Case Study: The Amazon Conservation Association (ACA) deployed sensor networks with fog computing capabilities in the Amazon rainforest as part of a case study. The sensors gathered data pertaining to biodiversity, humidity, and temperature. The fog computing infrastructure locally processed this data, facilitating decision-making. The up-to-date information supported the adoption of upgraded forest management strategies and helped in the preservation of biodiversity.

4. Management of waste and recycling: Case Study: Waste Management Inc., a garbage management organisation, implemented fog computing expertise in their rubbish collection systems. Intelligent waste bins equipped with sensors gathered data regarding the varieties and quantities of trash. The implementation of fog computing technologies helped the optimisation of waste collection routes, leading to a decrease in fuel consumption and carbon emissions of garbage trucks. Furthermore, the upgraded classification of reusable materials on the outskirts has contributed to the progress of recycling initiatives.

5. Monitoring and reducing pollution in the environment: To monitor the air quality of the city, Barcelona has built a fog computing system. Air quality data was gathered from sensors deliberately placed at different locations and investigated on-site using fog devices. The system reports on pollution levels expedited the response from municipal authorities, which enabled actions such as traffic diversion and implementation of emission regulations, thus mitigating pollution.

10.3 FOG COMPUTING FOR SDG 2 IN SMART AGRICULTURE

Fog computing is essential in revolutionising agriculture and supporting the attainment of SDG 2 [22], which aims to eliminate hunger, and food security, enhance nutrition, and foster sustainable agriculture. The utilisation of fog computing in the field of smart agriculture:

1. Precision Farming and Data Analytics: Sensors, unmanned aerial vehicles, and IoT devices gather data pertaining to soil moisture, temperature, crop vitality, and pest invasions. Fog computing devices locally process this data, carrying out real-time analytics. It allows farmers to make informed decisions based on data, such as implementing accurate irrigation, optimising fertiliser consumption, and implementing tailored pest control measures.

2. Remote Monitoring and Management: Smart agriculture systems employ sensors and cameras to remotely monitor farms. Fog computing devices facilitate the real-time processing of monitoring data,

allowing farmers to remotely oversee and control their fields, even in distant areas. This enables prompt interventions and enhances farm management.

3. Improved Crop Yield and Quality: Constant monitoring of environmental factors that impact crop growth. Fog computing utilising real-time analysis enables prompt reactions to fluctuating conditions, leading to increased agricultural productivity [23], enhanced crop standards, and minimised crop wastage.

4. Decision Support Systems: By integrating meteorological forecasts, historical data, and agronomic models. Computing-based decision support systems deliver farmers with practical advice, allowing them to organise their planting, foresee the optimal time for harvesting, and lessen risks related to weather fluctuations.

The fog model for smart agriculture is economically effective, energy-saving, and capable of accommodating different IoT devices, including basic temperature sensors, water sprinklers, advanced cameras, and aerial drones.

The role of fog computing in smart agriculture is to optimise resource utilisation, enrich agricultural effectiveness, raise crop production, and encourage the use of sustainable farming systems. Fog computing plays a vital role in supporting sustainable and effective farming practices universally by enabling real-time decision-making and offering data-driven insights, hence contributing to the accomplishment of SDG 2.

10.4 FOG COMPUTING FOR SDG 3 IN IMPROVING HEALTHCARE

Fog computing has considerable consequences in improving healthcare, in line with SDG 3 of assuring health and nurturing well-being [24] for individuals of all age groups. Fog computing plays an important role in advancing healthcare.

1. Fog computing allows the real-time collection and processing of patient data via wearable devices, sensors, and medical equipment, allowing for remote monitoring. Healthcare providers have the capability to monitor patient's vital marks, medication, and health issues from a distance.

2. Fog computing assists telemedicine [25] by offering high-definition video sessions, remote diagnostics, and virtual care services. Healthcare experts and specialists are made accessible to patients residing in remote or underdeveloped locations.

3. Fog computing permits the employment of portable and fog-based healthcare solutions in regions with restricted access to healthcare. Fog computing plays a key role in constricting the healthcare access

disparity by bringing medical knowledge and diagnostic skills closer to rural areas. This facilitates the early identification, prompt actions, and enhanced health results.

4. Utilising real-time data analysis to prevent the spread of diseases: Fog computing systems use fog analytics to process healthcare data and illness trends. Real-time analysis facilitates the detection of disease outbreaks, monitoring of patterns, and implementation of preventive actions. Various cloud based ML methods can also be considered the most effective and efficient for disease detection at an early stage [44].

The utilisation of fog computing in healthcare plays a substantial role in the realisation of Sustainable Development Goal 3. By harnessing know-how, fog computing enhances the accessibility of healthcare, enriches diagnostic capabilities, and simplifies remote monitoring. Subsequently, this promotes improved health outcomes and well-being for both individuals and communities, irrespective of geographical limits or resource limits.

10.5 FOG COMPUTING FOR SDG 11 IN SMART CITIES

Fog computing is crucial for the advancement of smart cities, which are integral to the realisation of SDG 11. SDG 11 seeks to create cities and human settlements that are inclusive, safe, resilient, and sustainable. Fog computing provides several benefits to smart cities:

1. Traffic management and optimisation involve the collection of real-time data on traffic flow and congestion through the use of sensors, cameras, and traffic monitoring equipment [26]. The data computational procedure permits real-time analysis. Through the dynamic management of signal timings, rerouting of vehicles, and provision of information to drivers, traffic optimisation becomes more efficient, reducing congestion and improving traffic flow.

2. Public Safety and Emergency Response Systems: IoT devices, sensors, and surveillance cameras are utilised to monitor and identify crises and safety issues in public spaces. Fog computing analytics exploit local data processing to detect anomalies and possible risks. It permits prompt measures, such as the deployment of emergency services [27], enhancements to public safety, and quicker response times in critical situations.

3. Waste management: By employing sensors and intelligent waste bins, the levels of garbage are monitored and collection routes [28] are adjusted. Through data analysis, fog computing devices enhance waste collection schedules and routes, reducing unnecessary trips and promoting efficient waste management. In addition, fog computing enhances resource allocation, identifies maintenance requirements, and monitors the condition of urban infrastructure.

4. Environmental Monitoring: Smart metres and environmental sensors are utilised to monitor energy consumption and environmental conditions. Environmental data computing enables cities to monitor water management, energy consumption, and air quality in real-time. This knowledge enables improved resource allocation, energy conservation, and the adoption of ecologically beneficial practices.

5. Integration of IoT devices and smart metres with the city's energy infrastructure to enable smart grids and efficient energy management. Within smart grids, the computational process facilitates load balancing, identifies patterns in energy consumption, and optimises the distribution of energy.

Fog computing enables the efficient management of resources in smart cities, improves public services, enhances safety, and promotes the establishment of sustainable urban environments. This is achieved through its decentralised architecture, real-time analytics, and localised data processing [29]. Fog computing plays an important role in promoting the development of inclusive, secure, robust, and sustainable cities by using edge technologies, hence contributing to the attainment of Sustainable Development Goal 11.

10.6 FOG COMPUTING FOR SDG 13 IN DROUGHT PREDICTION AND FORECASTING

Fog computing is a distributed computing model that aims to bring computational and storage resources in close proximity to the spot where they are required. This is especially advantageous for applications that necessitate minimal delay, ample data transfer capacity, and instantaneous data processing [30]. Fog computing is particularly suitable for applications that require operation in remote or resource-limited settings.

SDG 13 of the SDGs emphasises the need for immediate measures to address climate change and its consequences. Drought is a significant manifestation of climate change that can result in catastrophic repercussions for agriculture [31–32], water resources, and human well-being. Fog computing can facilitate the creation and implementation of drought prediction and forecasting systems, which can aid in mitigating the adverse effects of drought.

Drought prediction and forecasting involve the deployment of historical data, present observations, and climate models to antedate the intensity of forthcoming droughts. Fog computing enables the combination and examination of vast quantities of data obtained from sensors, satellites, and various other sources. Subsequently, this data can be utilised to train machine learning models capable of accurately forecasting drought conditions.

Examples of fog computing utilised for the drought prediction and forecasting

1. Fog computing is utilised in several instances to predict and forecast drought conditions. A case of such is the Drought Early Warning System (DEWS), currently under development by the World Meteorological Organisation (WMO). DEWS works on fog computing to gather and analyse data from sensors, satellites, and other sources. Subsequently, this data is utilised to forecast drought situations with exceptional precision.
2. Fog Drought is creating a fog computing platform specifically designed for deploying drought prediction and forecasting models in secluded regions.

Fog computing is a very promising know-how that enables the development and implementation of drought prediction and forecasting systems, which in turn aid in mitigating the adverse effects of drought. Fog computing offers numerous advantages for drought prediction and forecasting, such as reduced latency, increased bandwidth, instantaneous data processing, and enhanced security.

10.7 FOG COMPUTING FOR SDG 4 IN HIGHER EDUCATION

Due to the ability to access Internet data and allow users to access, share, and store data in remote servers, cloud computing as a centralised web server cannot be sustained by a massive increase in Internet usage [33]. A company can welcome various users, gadgets like cars, wearables, sensor units, and smart devices using their own Fog facilities. Instead of hindering or stopping educational operations, fog computing technology provides an adaptive platform. It could improve operations in several industries, including education [34].

The architecture depicted in Figure 10.6 is specifically engineered to provide the dependable transfer of all messages received from IoT end-devices to the edge device, hence enabling the establishment of a stream processing pipeline. Once the stream processing pipeline is created, the initial processing of all data received from sensors is performed using generic data functions. During this phase, it is crucial to detect any erroneous data that may impact the study, using historical values and observations as a reference. Subsequently, we need to tackle the temporary interruptions that are causing the loss of data. In order to tackle this form of disconnection, a straightforward approach can be employed to effectively detect the areas where the data is missing, by comparing them to earlier occurrences. After finishing this process, it is crucial to do ongoing data analysis.

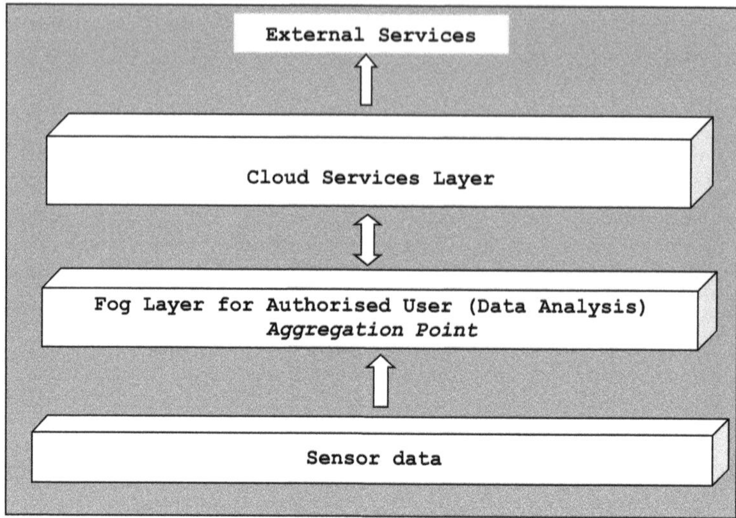

Figure 10.6 Fog computing high-level design for public educational buildings [34].

Fog computing technology enhances educational operations by offering a flexible platform that increases efficiency, rather than impeding or discontinuing them. Fog computing is an emerging technology that is poised for significant growth in the future and is expected to greatly enhance daily operations across various industries, including education [35].

10.8 ETHICAL CONSIDERATIONS AND CHALLENGES

Fog computing presents several ethical considerations when implemented in the context of SDGs. Several examples of these include:

1. Data Security in Fog Computing: It is critical to implement strong security measures to safeguard against unauthorised access, data breaches, and cyber-attacks because of the distribution of data over various edge/ fog devices [36].
2. Resource Allocation in Fog Computing: It is difficult that fog computing resources should be fairly and equally available in various regions with poor infrastructure. Addressing the inconsistency in access to digital resources and ensuring equitable access to these technologies is aligned with SDG 9 (Industry, Innovation, and Infrastructure) [37].
3. Privacy Concerns in Fog Computing: It encompasses the processing of data in close proximity to its origin, which may give rise to privacy concerns. It could involve the storage and processing of data on adjacent equipment [38].
4. Transparency in Fog Computing: It is vigorous to ensure transparency in data processing and various decision-making algorithms [39].

5. Environmental Impact using Fog Computing: It alleviates latency and bandwidth utilisation by locally processing data. It is essential to consider the trade-off between the advantages of decreased data transmission and the higher energy usage and environmental consequences [40]. This aligns with SDG 7 (Affordable and Clean Energy).
6. The ethical employment of data includes that the data assembled and handled via fog computing is engaged in a morally and sensible manner, in accordance with SDG 16 (Peace, Justice, and Strong Institutions). This necessitates the act of evasion biases, discrimination, and assuring impartiality in the procedures of making decisions.
7. Complying with various monitoring frameworks in different locations for data protection, privacy, and security is a challenging task [41, 42]. Fog computing systems must adhere to several legal responsibilities to guarantee ethical conduct.

In order to effectively achieve Sustainable Development Goals, it is crucial to address these ethical considerations and challenges [43] associated with the responsible utilisation of fog computing. This will guarantee that technological advancement positively contributes to societal welfare while maintaining ethical standards.

10.9 CONCLUSION

Fog computing and sustainability are linked in "Fog Computing: Key to SDG Advancement." Fog computing innovations that promote sustainability in various industries are examined in this chapter. The chapter begins with fog computing to improve resource and energy efficiency. It cuts costs, carbon emissions, and environmental impact. It also studies how fog can improve environmental monitoring and conservation. Fog provides real-time information and alarms to aid environmental protection and conservation decisions. Fog-enabled systems assess agricultural data, meteorological trends, and soil conditions to improve farming, minimise water and pesticide use, and boost crop output. The chapter highlights the challenges of fog implementation in sustainability, despite its potential to examine ethical issues like privacy, which is crucial. Research, progress, and cooperation between academic institutions, companies, and governments are needed to fully utilise fog computing to address various global environmental concerns.

REFERENCES

[1] Perera, C., Qin, Y., Estrella, J. C., Reiff-Marganiec, S., & Vasilakos, A. V. (2017). Fog computing for sustainable smart cities: A survey. *ACM Computing Surveys (CSUR), 50*(3), 1–43.

[2] Farooqi, A. M., Hassan, S. I., & Alam, M. A. (2019, February). Sustainability and fog computing: applications, advantages and challenges. In *2019 3rd International Conference on Computing and Communications Technologies (ICCCT)* (pp. 18–23). IEEE.

[3] Yi, S., Li, C., & Li, Q. (2015, June). A survey of fog computing: concepts, applications and issues. In *Proceedings of the 2015 workshop on mobile big data* (pp. 37–42).

[4] Gultepe, I., Tardif, R., Michaelides, S. C., Cermak, J., Bott, A., Bendix, J., ... & Cober, S. G. (2007). Fog research: A review of past achievements and future perspectives. *Pure and applied geophysics, 164*, 1121–1159.

[5] Stojmenovic, I., Wen, S., Huang, X., & Luan, H. (2016). An overview of fog computing and its security issues. *Concurrency and Computation: Practice and Experience, 28*(10), 2991–3005.

[6] Dastjerdi, A. V., Gupta, H., Calheiros, R. N., Ghosh, S. K., & Buyya, R. (2016). Fog computing: Principles, architectures, and applications. In *Internet of things* (pp. 61–75). Morgan Kaufmann.

[7] Mustapha, U. F., Alhassan, A. W., Jiang, D. N., & Li, G. L. (2021). Sustainable aquaculture development: a review on the roles of cloud computing, internet of things and artificial intelligence (CIA). *Reviews in Aquaculture, 13*(4), 2076–2091.

[8] Farooqi, A. M., Hassan, S. I., & Alam, M. A. (2019, February). Sustainability and fog computing: applications, advantages and challenges. In *2019 3rd International Conference on Computing and Communications Technologies (ICCCT)* (pp. 18–23). IEEE.

[9] Kaur, G., Bharathiraja, N., Singh, K. D., Veeramanickam, M. R. M., Rodriguez, C. R., & Pradeepa, K. (2024). Emerging Trends in Cybersecurity Challenges with Reference to Pen Testing Tools in Society 5.0. *Artificial Intelligence and Society 5.0*, 196–212.

[10] Yi, S., Li, C., & Li, Q. (2015, June). A survey of fog computing: concepts, applications and issues. In *Proceedings of the 2015 workshop on mobile big data* (pp. 37–42).

[11] G. Kaur, B. N, M. S, P. K, S. G and V. K. M, te,A Security model with efficient AES and Security Performance Trade-off Analysis of Cryptography Systems with Cloud Computing," *2023 Fifth International Conference on Electrical, Computer and Communication Technologies (ICECCT)*, Erode, India, 2023, pp. 01–08, doi: 10.1109/ICECCT56650.2023.10179752

[12] Kaur, G., & Kaur, S. (2023). Critical analysis of secure strategies against threats on cloud platform. In *Mobile Radio Communications and 5G Networks: Proceedings of Third MRCN 2022* (pp. 443–455). Singapore: Springer Nature Singapore.

[13] Nti, E. K., Cobbina, S. J., Attafuah, E. E., Opoku, E., & Gyan, M. A. (2022). Environmental sustainability technologies in biodiversity, energy, transportation and water management using artificial intelligence: A systematic review. *Sustainable Futures, 4*, 100068.

[14] Feroz, A. K., Zo, H., & Chiravuri, A. (2021). Digital transformation and environmental sustainability: A review and research agenda. *Sustainability, 13*(3), 1530.

[15] Rosário, A. T., & Dias, J. C. (2022). Sustainability and the digital transition: A literature review. *Sustainability, 14*(7), 4072.

[16] Kaur, G., & Kaur, J. (2023, August). Detailed Investigation of the use of Text Mining and Natural Language Processing (NLPs) for Cancer Detection. In *2023 Second International Conference on Augmented Intelligence and Sustainable Systems (ICAISS)* (pp. 1156–1162). IEEE.

[17] Kraemer, F. A., Braten, A. E., Tamkittikhun, N., & Palma, D. (2017). Fog computing in healthcare–a review and discussion. *IEEE Access, 5*, 9206–9222.

[18] Kumari, A., Tanwar, S., Tyagi, S., & Kumar, N. (2018). Fog computing for Healthcare 4.0 environment: Opportunities and challenges. *Computers & Electrical Engineering, 72*, 1–13.

[19] Mutlag, A. A., Abd Ghani, M. K., Arunkumar, N. A., Mohammed, M. A., & Mohd, O. (2019). Enabling technologies for fog computing in healthcare IoT systems. *Future generation computer systems, 90*, 62–78.

[20] Huang, X., Fan, W., Chen, Q., & Zhang, J. (2020). Energy-efficient resource allocation in fog computing networks with the candidate mechanism. *IEEE Internet of Things Journal, 7*(9), 8502–8512.

[21] Chang, Z., Zhou, Z., Ristaniemi, T., & Niu, Z. (2017, December). Energy efficient optimization for computation offloading in fog computing system. In *GLOBECOM 2017–2017 IEEE Global Communications Conference* (pp. 1–6). IEEE.

[22] de Araujo Zanella, A. R., da Silva, E., & Albini, L. C. P. (2020). Security challenges to smart agriculture: Current state, key issues, and future directions. *Array, 8*, 100048.

[23] Qureshi, R., Mehboob, S. H., & Aamir, M. (2021). Sustainable green fog computing for smart agriculture. *Wireless Personal Communications, 121*(2), 1379–1390.

[24] Katal, A. (2023). Leveraging fog computing for healthcare. In *Deep Learning Technologies for the Sustainable Development Goals: Issues and Solutions in the Post-COVID Era* (pp. 51–68). Singapore: Springer Nature Singapore.

[25] Odun-Ayo, I., & Alagbe, O. (2022). Fog computing technology: a review of current trends. *The United Nations and Sustainable Development Goals*, 261–275.

[26] Mishra, P., & Singh, G. (2023). Enabling Technologies for Sustainable Smart City. In *Sustainable Smart Cities: Enabling Technologies, Energy Trends and Potential Applications* (pp. 59–73). Cham: Springer International Publishing.

[27] Sulaiman, A., Nagu, B., Kaur, G., Karuppaiah, P., Alshahrani, H., Reshan, M. S. A., ... & Shaikh, A. (2023). Artificial Intelligence-Based Secured Power Grid Protocol for Smart City. *Sensors, 23*(19), 8016.

[28] Mishra, P., Thakur, P., & Singh, G. (2022). Sustainable smart city to society 5.0: State-of-the-art and research challenges. *SAIEE Africa Research Journal, 113*(4), 152–164.

[29] Siddiqui, E. F., & Nayak, S. K. (2023, March). Agitating sustainability using fog computing for smart cities. In *2023 10th International Conference on Computing for Sustainable Global Development (INDIACom)* (pp. 680–687). IEEE.

[30] Zhang Xiang, Z. X., Chen NengCheng, C. N., Sheng Hao, S. H., Ip, C., Yang Long, Y. L., Chen YiQun, C. Y., ... & Niyogi, D. (2019). Urban drought challenge to 2030 Sustainable Development Goals.

[31] Kaur, A., & Sood, S. K. (2020). Artificial intelligence-based model for drought prediction and forecasting. *The Computer Journal, 63*(11), 1704–1712.

[32] Kaur, A., & Sood, S. K. (2020). Deep learning based drought assessment and prediction framework. *Ecological Informatics*, 57, 101067.

[33] Raman, A. (2019). Potentials of fog computing in higher education. *International Journal of Emerging Technologies in Learning*, 14(18).

[34] Adel, A. (2020). Utilizing technologies of fog computing in educational IoT systems: privacy, security, and agility perspective. *Journal of Big Data*, 7(1), 99.

[35] Abinaya, N., Kumar, A. S., Chaturvedi, A., Musirin, I. B., Rao, M., Kaur, G., ... & Arya, N. (2024). Big Data in Real Time to Detect Anomalies. In *Big Data Analytics Techniques for Market Intelligence* (pp. 372–397). IGI Global.

[36] Guan, Y., Shao, J., Wei, G., & Xie, M. (2018). Data security and privacy in fog computing. *IEEE Network*, 32(5), 106–111.

[37] Chang, Z., Liu, L., Guo, X., & Sheng, Q. (2020). Dynamic resource allocation and computation offloading for IoT fog computing system. *IEEE Transactions on Industrial Informatics*, 17(5), 3348–3357.

[38] Yi, S., Qin, Z., & Li, Q. (2015). Security and privacy issues of fog computing: A survey. In *Wireless Algorithms, Systems, and Applications: 10th International Conference, WASA 2015, Qufu, China, August 10–12, 2015, Proceedings 10* (pp. 685–695). Springer International Publishing.

[39] Kim, Y., Kim, D., Son, J., Wang, W., & Noh, Y. (2018, May). A new fog-cloud storage framework with transparency and auditability. In *2018 IEEE International Conference on Communications (ICC)* (pp. 1–7). IEEE.

[40] Sarkar, S., Chatterjee, S., & Misra, S. (2015). Assessment of the suitability of fog computing in the context of internet of things. *IEEE Transactions on Cloud Computing*, 6(1), 46–59.

[41] Saluja, K., Gupta, S., Solanki, V., Debnath, S. K., & Bansal, A. (2023, May). An EEG-based brain-computer interface for guiding mobile robots. In *2023 3rd International Conference on Advances in Computing, Communication, Embedded and Secure Systems (ACCESS)* (pp. 268–273). IEEE.

[42] Soomro, A. M., Naeem, A. B., Bagchi, S., Sharma, N., Singh, P., & Debnath, S. K. (2023, April). Uncovering Spam in Twitter: A Machine Learning Approach. In *2023 International Conference on Computational Intelligence and Sustainable Engineering Solutions (CISES)* (pp. 993–998). IEEE.

[43] Yakubu, J., Abdulhamid, S. I. M., Christopher, H. A., Chiroma, H., & Abdullahi, M. (2019). Security challenges in fog-computing environment: a systematic appraisal of current developments. *Journal of Reliable Intelligent Environments*, 5(4), 209–233.

[44] Harnal, S., Jain, A., Rathore, A. S., Baggan, V., Kaur, G., & Bala, R. (2023, March). Comparative approach for early diabetes detection with machine learning. In *2023 International conference on emerging smart computing and informatics (ESCI)* (pp. 1–6). IEEE.

Chapter 11

Resource allocation using a hybrid evolutionary model and machine learning

Mandeep Kaur
Chitkara University Institute of Engineering and Technology,
Chitkara University, Chandigarh, India

Rajni Aron
National Forensic Science University, Ponda, India

*Heena Wadhwa, Righa Tandon, Htet Ne Oo,
and Gagandeep Kaur*
Chitkara University Institute of Engineering and Technology,
Chitkara University, Chandigarh, India

11.1 INTRODUCTION

The practice of allocating resources, such as time, money, and labour, to accomplish particular goals is referred to as resource allocation. In several fields, including manufacturing, logistics, and project management, among others, resource allocation is crucial. The optimum resource allocation achievable under a given set of restrictions and objectives has traditionally been sought after using mathematical optimization approaches. Scholars have recently studied resource allocation using hybrid evolutionary models and machine learning methods. Evolutionary algorithms, which draw their inspiration from natural selection and genetic algorithms, are combined with standard optimization approaches in a hybrid evolutionary model. The fittest members of a population live and procreate, while the less fit members are eliminated. These algorithms replicate the process of natural selection [1–3].

Resource allocation is effectively allocating scarce resources among conflicting demands has long been a significant problem in computing, telecommunications, logistics, and other areas. Traditional resource allocation techniques frequently fail to meet the needs for optimization, adaptability, and efficiency in today's dynamic, diversified, and frequently unpredictable technological context. The novel idea of this chapter is the investigation of state-of-the-art technology for resource allocation, the fusion of hybrid

DOI: 10.1201/9781003494430-11

evolutionary models with machine learning methods. With solutions that not only optimize resource utilization but also adapt to and learn from changing situations, this novel paradigm represents a substantial advancement in resolving the difficulties of resource allocation [4].

Techniques from machine learning are also employed in resource allocation, especially when generating predictions and decisions. Machine learning algorithms can analyze historical patterns of resource allocation and forecast future resource requirements. The resource allocation process can be automated with the help of these models, requiring less human involvement. Resource allocation in many domains has shown promise when combined with evolutionary algorithms and machine-learning approaches. The evolutionary algorithm can learn from the data and modify its search strategy by including machine learning models, which increases the system's effectiveness in locating the best solutions. This strategy may result in better resource management, more production, and cost reductions.

11.1.1 Background and motivation

In today's systems and technology, resource allocation is crucial. The Internet of Things (IoT) resource allocation, network bandwidth management, and cloud computing resource optimization are just a few examples where efficient allocation is crucial to achieve the best performance and cost-efficiency. The difficulties relating to resource allocation have significantly increased as technology keeps developing and systems become more complicated. Resource allocation issues can appear in many ways in today's linked society. To ensure service quality and save operating expenses, for instance, allocating computing resources efficiently to various virtual machines in the cloud is essential. For seamless connectivity to be possible in wireless networks, spectrum and bandwidth management are crucial. In addition, delivering low-latency and high-performance applications has become a focus in developing sectors like edge computing through the dynamic allocation of computing resources at the network's edge [5–7].

This chapter's inspiration came from the realization that there were resource allocation problems and sustainable solutions were required. There is growing interest in utilizing the strength of both evolutionary algorithms and machine learning to successfully solve the problems posed by the complexity of modern systems because traditional methods frequently fall short in doing so.

11.1.2 Research objectives

The primary objective of this chapter is to explore the synergy between evolutionary algorithms and machine learning techniques in the context of resource allocation. We aim to explore how these two powerful paradigms can be integrated to create hybrid models offering superior resource allocation capabilities. Our goal is to provide insights into such hybrid models' principles, advantages, and challenges.

We will also investigate real-world applications and case studies that demonstrate the practicality and effectiveness of these hybrid approaches. By doing so, we seek to equip researchers, practitioners, and decision-makers with a deeper understanding of leveraging hybrid evolutionary models and machine learning for resource allocation tasks.

11.1.3 Scope of the chapter

As we allocate resources utilizing hybrid evolutionary models and machine learning, the chapter will concentrate on a few important features of the process. The chapter mainly focuses on these topics:

- Investigating the principles of evolutionary algorithms and machine learning in the context of resource allocation.
- Looking into the benefits and restrictions of each strategy when used alone.
- offering a theoretical foundation for combining machine learning and evolutionary algorithms to create hybrid models.
- demonstrating use cases from many fields, such as edge computing, cloud computing, and the Internet of Things, to show how effective these hybrid models are in the real world.

11.1.4 Methodology

Utilizing cloud computing and industrial applications, the technique entails a methodical investigation of resource allocation problems and solutions. After providing an overview of the challenges involved in allocating resources, it moves on to talk about the fundamental ideas of sustainable distribution. Then it is analyzed that how well evolutionary algorithms distribute cloud resources, network bandwidth, and edge computing resources.

Next, several machine learning approaches are examined to see if they can be used to predict resource use and improve decision-making. These techniques' advantages and disadvantages are thoroughly examined. Its sustainability attributes are evaluated using a hybrid methodology that combines machine learning and evolutionary techniques. The significance of sustainability in resource allocation procedures is emphasized through the evaluation of research opportunities and success metrics. This methodology sheds light on the dynamic field of resource allocation, emphasizing areas for further investigation and the need for using sustainable practices. The flow chart in Figure 11.1 shows the methodology used in this chapter.

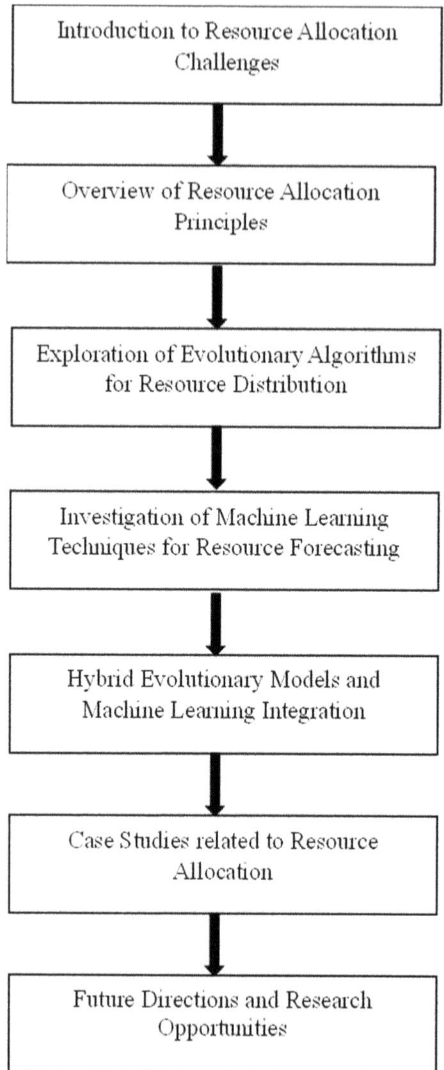

Figure 11.1 Methodology used.

11.2 RESOURCE ALLOCATION IN CONTEMPORARY SYSTEMS

Resource allocation, the art and science of distributing finite resources to meet various demands, forms the backbone of modern systems across industries. In this section, we delve into the fundamental principles, pressing challenges, and paramount importance of resource allocation in today's dynamic and interconnected world. This section introduces the main theme of resource allocation in contemporary systems, which acts as the cornerstone of the chapter. It gives the background for appreciating the difficulties and significance of resource management in the modern technology environment. The varied nature of resources—from network bandwidth and edge computing resources to processing resources in cloud environments.

11.2.1 Resource allocation fundamentals

Resource allocation is not a novel idea; it has been a crucial procedure ever since complex systems first emerged. But in order to successfully navigate the complexities of modern resource allocation, it is crucial to comprehend its underlying principles [8]. This section explores the fundamental ideas and concepts of resource allocation throughout the chapter. The course may encompass fundamental ideas, approaches, and theories that serve as the foundation for comprehending the distribution of resources in systems. Talks about allocation algorithms, optimization strategies, or the underlying ideas governing resource distribution may fall under this category. The objective of this chapter is to present a thorough grasp of the essential elements that impact resource allocation techniques by outlining the principles.

The basic ideas that guide resource allocation in the complex systems of today are examined in this section:

Basic Information and Definition: To efficiently accomplish particular goals, resource allocation is the practice of dividing limited resources among conflicting demands. It includes different resources involved (such as computing power, bandwidth, and human resources), as well as the fundamental allocation rules [9].

Resource Restrictions: Recognizing that resources are finite by nature is essential to resource allocation. Resource constraints are important factors to take into account when designing, running, and optimizing a variety of systems, such as distributed computing environments, computer systems, and networks. These limitations could be brought about by financial limitations, technological barriers, physical limitations, or other issues that affect a system's capacity and capabilities [10].

Resource Allocation Goals: It is critical to understand the underlying objectives of resource allocation. The objectives of resource allocation include reaching scalability, dependability, and user satisfaction, as well as maximizing system performance and making sure resources are used efficiently. These objectives focus on finding economical solutions, being flexible in changing circumstances, allocating resources fairly, adhering to security protocols, and using less energy. It is imperative to strike a balance between these goals since effective resource allocation necessitates trade-offs in order to build a system that is high-performing, dependable, and long-lasting and satisfies organizational and user requirements.

11.2.2 Resources allocation challenges

Allocating resources does not come without challenges, especially in modern systems where needs are dynamic and varied [11–13]. Most likely, this subsection summarizes the different difficulties and complexities involved in allocating resources in modern systems. The dynamic nature of modern systems, fluctuating workloads, or the requirement for effective resource optimization can all present challenges. The development of efficient resource allocation models depends on recognizing and resolving these issues. Comprehending the obstacles and complexities involved facilitates the development of flexible and resilient solutions that improve system performance as a whole. A few challenges are covered in Figure 11.2.

Dynamic Workloads: The unpredictability and fluctuation of the computational demands imposed on a system over time present a challenge for dynamic workloads. It can be difficult to allocate and manage resources efficiently in dynamic workloads due to the large variations in task and process volume and quality. Maintaining optimal performance requires systems to adjust to abrupt spikes or drops in workload demands. Effective resource allocation strategies are necessary to meet this challenge. These strategies must be able to dynamically scale resources in response to changes in workload patterns in real-time. This will ensure that resources are used efficiently and can adapt to changing computational demands without under- or over-provisioning them. For systems to continue being responsive and resilient in dynamic and changing environments, the dynamic workload challenge must be solved.

Resource Fragmentation: When available resources in a system are dispersed or split into smaller, non-contiguous segments, it can be difficult to allocate them effectively. This is known as resource fragmentation. Resources of all kinds, including memory, storage, and network bandwidth, can become fragmented in this way. Resources that are dispersed can be used less effectively, which can result in slower system response times and lower overall performance. This problem is especially pertinent to systems that have dynamic workloads or frequently allocate and deallocate resources, as these operations may cause resource fragmentation. Effective resource management tactics, such as adaptive

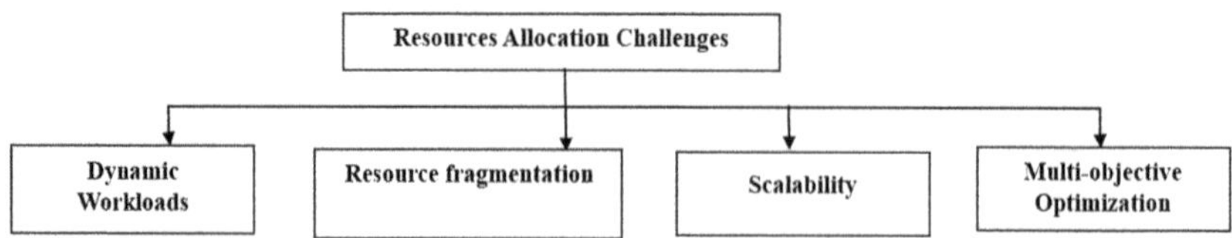

Figure 11.2 Resource allocation challenges.

allocation algorithms or defragmentation methods, are necessary to minimize the effects of resource fragmentation and preserve peak system performance.

Scalability: Scalability is the capacity, performance, and resource adaptation of a system that allows it to efficiently and effectively handle an increasing workload or demand. A scalable system can handle expansion without sacrificing user experience, performance, or responsiveness. Two common categories for scalability are vertical and horizontal. Vertical scalability refers to expanding the capacity of current resources, like upgrading hardware components, while horizontal scalability refers to adding more resources, like servers or nodes, to spread the workload. In dynamic and changing environments with fluctuating workloads, achieving scalability is essential because it guarantees that a system can easily grow or shrink to meet demand without compromising efficiency. In the rapidly evolving world of today, creating scalable architectures and utilizing scalable practices are essential to the design and management of systems.

Multi-objective Optimization: When there are several competing objectives that must be simultaneously optimized, multi-objective optimization is a problem-solving strategy that can be used. Multi-objective optimization takes several, frequently conflicting, objectives into account as opposed to traditional single-objective optimization, which only looks at one. These goals might be different from one another and could include optimizing for different aspects such as performance, cost, energy efficiency, or other standards.

11.2.3 Resource allocation's importance in contemporary systems

In modern systems, it is impossible to emphasize the importance of resource allocation. The importance of resource allocation in the current technological environment is emphasized in this subsection. It might go into detail about how cost-effectiveness, scalability, and system performance are all directly impacted by effective resource allocation. The section also addresses how resource allocation helps to support emerging technologies, guarantee system reliability, and satisfy user demands. The chapter highlights the significance of resource allocation in modern systems and the need for sophisticated models and approaches, like machine learning and hybrid evolutionary algorithms.

Performance Optimization: Efficiency in resource allocation has a direct impact on performance optimization. The systematic process of improving a system, application, or process's overall functionality, speed, and efficiency is known as performance optimization. To attain the best outcomes, the main goal is to enhance performance metrics like response time, throughput, or resource utilization. Numerous domains, such as software development, computer systems, networks, and algorithms, can benefit from this optimization. The procedure entails locating bottlenecks, examining how the system behaves in various scenarios, and putting specific improvements into practice.

Cost-Effectiveness: In order to maximize value or benefit in relation to incurred costs, resources must be allocated and utilized efficiently. This is known as cost-effectiveness. A cost-effective strategy aims to maximize results while lowering costs in a variety of situations, including project management, business operations, and technology deployment. To make sure that the advantages of an investment or action exceed the costs involved, it entails carefully weighing resource allocation, strategic decision-making, and efficiency gains. In order to achieve cost-effectiveness, performance, quality, and financial factors must be balanced. This will ultimately result in the most resource-efficient use of resources without sacrificing intended results. Retaining cost-effectiveness in dynamic and changing environments requires ongoing process evaluation and improvement.

Quality of Service (QoS): The parameters and metrics used to evaluate and guarantee the dependability and efficiency of a system or service are referred to as Quality of Service (QoS) metrics. Quality of Service (QoS) is defined as the ability to satisfy certain needs and standards about things like latency, bandwidth, reliability, and overall user experience in a variety of technological domains, such as networking, telecommunications, and cloud computing. When resources are scarce or network conditions are unstable, QoS mechanisms give priority to specific kinds of traffic or services in order to ensure a certain degree of performance. As a result, a consistent and acceptable level of service quality is maintained, and critical applications are given the resources and bandwidth they require. Effective implementation of Quality of Service (QoS) leads to improved system stability and user experience in environments where different applications or users compete for shared resources.

Adaptability and Resilience: A system's ability to respond and recover efficiently in the face of shifting circumstances, interruptions, or unanticipated obstacles is characterized by its resilience and adaptability. Adaptability pertains to the capacity of a system to modify and progress in reaction to novel demands, ever-changing surroundings, or developing patterns. However, resilience describes a system's capacity to withstand shocks, recover swiftly from adversity, and absorb failures while still preserving critical functionality. When combined, adaptability and resilience help a system be robust and sustainable by allowing it to navigate uncertainty and keep operating at its best even in the face of disruptions. In dynamic and complex environments where unpredictability is ingrained, these attributes are especially crucial for ensuring that systems can recover quickly from setbacks and adapt flexibly to changing needs in order to maintain operational integrity.

The fundamental ideas, difficulties, and crucial functions of resource allocation in contemporary systems are explored in order to give readers a thorough understanding of the topic. These revelations serve as the starting point for talks about cutting-edge methods for solving problems with resource allocation, such as the use of hybrid evolutionary models and machine learning [14].

11.3 EVOLUTIONARY ALGORITHMS FOR RESOURCE ALLOCATION

Modern systems' resource allocation problems have prompted the investigation of cutting-edge methods, such as the use of evolutionary algorithms. In order to solve optimization problems in the context of resource allocation, evolutionary algorithms for resource allocation represent a computational approach motivated by the concepts of natural selection and evolution. These algorithms mimic the process of biological evolution to iteratively evolve a population of potential solutions toward optimal or nearly optimal solutions. They are frequently derived from the field of evolutionary computation. Evolutionary algorithms are utilized in the resource allocation domain to determine optimal resource distributions that meet predetermined goals or constraints. They include a variety of methods, including evolutionary strategies, genetic programming, and genetic algorithms. These algorithms can adjust to changing circumstances and maximize the distribution of resources by iteratively developing a population of viable resource allocation solutions through mechanisms like crossover, mutation, and selection.

11.3.1 Evolutionary algorithms: a quick overview

For the purpose of solving optimization problems, evolutionary algorithms offer a computational strategy based on biological evolution. Evolutionary algorithms are briefly reviewed here, along with their basic ideas and elements. Through iterative processes like crossover, mutation, and selection, these algorithms work with a population of possible solutions to gradually refine and enhance them over successive generations. Whereas crossover brings together components of chosen solutions to form new ones, the selection mechanism imitates the natural selection of fit individuals. Changes brought about by mutation are arbitrary. Until a workable solution is found, these iterative processes keep going. Resource allocation problems in systems like cloud computing and network management can be addressed with evolutionary algorithms' robust optimization approach, which is adaptable and applicable in a variety of domains.

11.3.1.1 Evolutionary algorithms' basic principles

Natural Selection: The fundamental idea behind evolutionary algorithms is the idea of natural selection, in which solutions develop through generations. Natural selection in biology and evolutionary algorithms share a fundamental principle. In this context, the evolutionary process involves successive generations, and solutions to a problem are compared to biological entities. People who possess characteristics that help them adapt better or solve problems more successfully are more likely to stick around and leave a legacy for future generations. In biological systems, where the fittest individuals have a higher chance of surviving and passing on their traits to the following generation, this iterative cycle mimics the evolutionary dynamics seen in those systems.

Population-Based Optimization: Evolutionary algorithms operate on populations of potential solutions. Instead of concentrating on a single solution, evolutionary algorithms work on populations of possible solutions, which sets them apart from conventional optimization techniques. Through the use of a population-based approach, diversity is introduced and a wider range of solutions can be explored. Important mechanisms are crossover, in which features from several solutions are merged to form new individuals, and mutation, in which arbitrary modifications are added to individual solutions. The population evolves over generations, convergent towards optimal or nearly optimal solutions, through these mechanisms and the concept of selection, where individuals are chosen based on their fitness.

Fitness Evaluation: Evolutionary algorithms rely on fitness, which is assessed through the use of a fitness function, to determine the efficacy of solutions. How well a solution tackles the current issue is measured by this function. Better fitness scores indicate greater problem-solving ability, which gives them an advantage during the selection process. The optimization process is heavily dependent on fitness evaluation, which also affects the population's evolutionary trajectory. The search for solutions that meet the specified optimization goals is guided by the fitness landscape as the algorithm iterates, eventually producing better and more suited solutions over time.

11.3.2 Resource allocation using evolutionary algorithms

Different resource allocation scenarios have been discovered to be applicable to evolutionary algorithms. This section emphasizes how useful they are.

11.3.2.1 Allocating cloud resources

Resource Pool Management: Allocating resources is a major problem in the cloud computing space, and evolutionary algorithms provide a useful method. Under "Resource Pool Management," these algorithms maximize how computing resources are distributed in cloud environments. To effectively respond to shifting workloads and fluctuating resource demands, resources are allocated in a dynamic manner, achieving this optimization. Evolutionary algorithms are particularly good at adjusting to the changing conditions of cloud environments, making the most efficient use of computational resources, and maximizing system performance.

Virtual Machine Placement: One particular area in which evolutionary algorithms are used in cloud resource allocation is "Virtual Machine Placement." In this case, the algorithms are essential in determining where in the cloud infrastructure to put virtual machines (VMs). Putting virtual machines (VMs) in the right places will reduce resource contention, improve performance, and guarantee effective resource use. Evolutionary algorithms evolve solutions that address the complexities of VM placement in dynamic cloud environments through an iterative process of evaluation and adaptation, taking into account multiple factors like workload distribution, resource availability, and performance metrics. In cloud computing scenarios, the optimal balance between resource allocation and system performance is achieved through the significant contribution of evolutionary algorithms [15].

11.3.2.2 Allocating network bandwidth

Dynamic Bandwidth Allocation: Through "Dynamic Bandwidth Allocation," evolutionary algorithms are essential to the optimization of network bandwidth allocation. In this case, these algorithms respond to shifting demands and priorities within a network by dynamically adjusting the bandwidth distribution in real-time. Evolutionary algorithms guarantee optimal and efficient use of available bandwidth resources by adjusting to changing traffic patterns and user demands. By adapting to the changing requirements of various applications and services, this dynamic allocation strategy helps to improve network responsiveness and performance.

Quality of Service (QoS) Management: "Quality of Service (QoS) Management" is a key area in which evolutionary algorithms are used in network bandwidth allocation. Through dynamic resource allocation of the network according to traffic patterns and priority levels, these algorithms help to provide QoS. Evolving algorithms contribute to consistent and dependable service levels by taking into account the unique needs of various data and application types. This guarantees that important tasks have the bandwidth they require to meet performance standards. Evolutionary algorithms support a variety of applications with different network demands and improve the overall user experience with QoS management.

11.3.2.3 Allocation of resources for edge computing

Real-time Resource Management: Evolutionary algorithms play a key role in "Real-time Resource Management" in the context of edge computing. To facilitate low-latency services and applications, these adaptive algorithms dynamically allocate resources at the network's edge. Evolutionary algorithms used in edge computing help to minimize latency and maximize the performance of applications that demand quick responses and data processing by quickly modifying resource allocations in response to real-time demands [16].

Energy-Efficient Allocation: The "Energy-Efficient Allocation" of resources in edge computing environments is another area in which evolutionary algorithms really shine. These algorithms help ensure sustainability and economy of scale by maximizing the distribution of computational resources and reducing superfluous energy usage. Evolutionary algorithms satisfy the inherent constraints and demands of edge computing infrastructures by constantly adapting and optimizing themselves to strike a balance between resource efficiency and energy conservation. The business viability and environmental sustainability of edge computing solutions depend heavily on this energy-conscious resource allocation [17–19].

11.4 RESOURCE ALLOCATION USING MACHINE LEARNING

Machine learning has become a transformational force in the dynamic world of resource allocation. By using computational models that recognize patterns in data and forecast future events, resource allocation through machine learning is made possible. By using this method, resource distribution across different domains becomes more flexible and efficient. The purpose of this section is to thoroughly examine how machine learning can be used to address problems with resource allocation. It discusses the fundamental ideas, machine learning methods, real-world applications, benefits, and difficulties of using machine learning in this field.

11.4.1 Machine learning overview

11.4.1.1 Basics of machine learning

Fundamentally, machine learning is the study of how to make computers learn from data and make predictions or judgements without having to

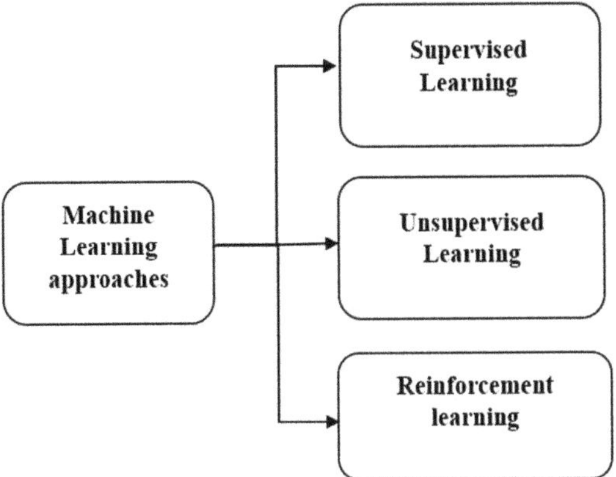

Figure 11.3 Machine learning approaches.

be explicitly programmed. Artificial intelligence includes machine learning, which gives systems the ability to learn from data and gradually get better at what they do without the need for explicit programming. Key concepts, including supervised learning, unsupervised learning, and reinforcement learning, are introduced in the "Basics of Machine Learning" section [20, 21]. To fully grasp how resource allocation can be revolutionized by machine learning, it is imperative to understand the fundamental principles.

There are three major categories of machine learning: supervised, unsupervised, and reinforcement learning, shown in Figure 11.3. To forecast results, supervised learning includes building models on labelled data. Unsupervised learning investigates data structures and patterns without the use of labels. The focus of reinforcement learning is on making decisions as a result of interactions with the environment.

Training and Inference: Machine learning models go through two crucial phases: training and inference. By modifying their internal parameters, models gain knowledge from historical data during training. The stage of inference is when trained models generate forecasts or judgements based on fresh, unforeseen data [22].

Data Preparation: Data is the lifeblood of machine learning, and the preparation of data forms the foundation of any effective model construction. In order to guarantee that the data used for model training and inference is correct and best suited for the task at hand, data preparation entails a number of crucial processes. The quality and efficacy of machine learning models are significantly shaped during this essential step, which includes data cleansing, feature selection, transformation, and engineering.

11.4.1.2 Important Machine Learning Methods

It is critical to comprehend the different machine learning algorithms and how they relate to resource allocation:

Regression and Classification: These techniques are used to forecast categorical and continuous outcomes, respectively. Regression can help predict resource demands in the context of resource allocation, whereas categorization can support decision-making. Both are shown in Figure 11.4.

Clustering: Data points with similar characteristics are grouped via clustering algorithms. Clustering can be used to find resource or demand groupings with similar characteristics, aiding allocation strategies as shown in Figure 11.5.

Deep learning: Deep learning is a branch of machine learning that uses artificial neural networks to recognize complex data patterns. It may be used for resource allocation challenges, such as maximizing neural network setup for available computational resources.

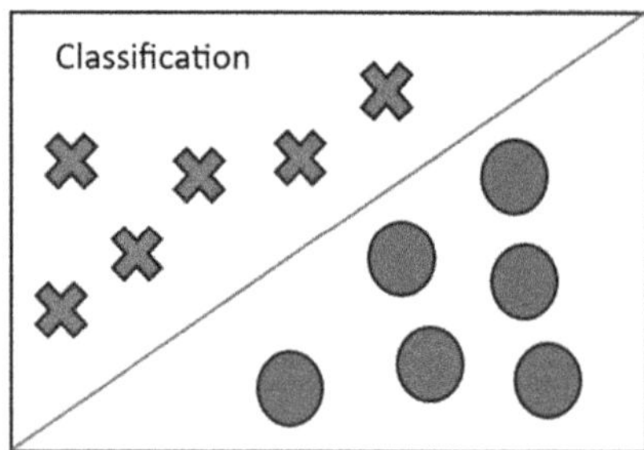

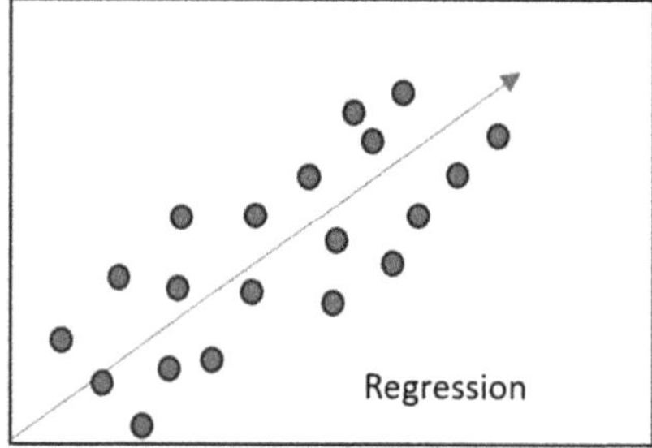

Figure 11.4 Classification and regression.

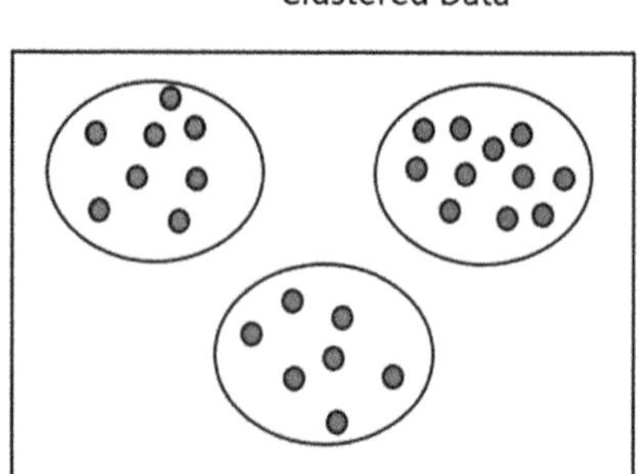

Figure 11.5 Clustering.

11.4.2 Methods of machine learning for resource allocation

11.4.2.1 Predictive resource allocation

Machine learning enables "Predictive Resource Allocation," in which models make predictions about future resource demands based on historical data analysis. This makes it possible to proactively adjust allocation, avoiding under- or over-provisioning and guaranteeing the best possible use of resources.

Predictive resource allocation is an important application of machine learning in resource allocation:

Demand Prediction: Based on past data and patterns, machine learning models can be taught to forecast future resource demands. Systems can allocate resources proactively thanks to this predictive feature, ensuring peak performance [23].

Prediction of Resource Availability: Prediction of resource availability is equally crucial. Resource availability can be predicted by machine learning models, enabling real-time adjustments to allocation strategies to avoid resource contention.

11.4.2.2 Decision-making in optimization and allocation

The chapter examines the use of machine learning to support "Decision Making in Optimization and Allocation." Informed decisions about resource allocation, taking into account workload, priorities, and system constraints, are made possible by machine learning models that learn from large, complex datasets. Allocation decisions can be improved with the use of machine learning:

Resource Allocation Decision Optimization: By taking into account a variety of goals, restrictions, and current circumstances, machine learning models can improve resource allocation choices. Resource-effective allocations result from this adaptive optimization.

Reinforcement Learning for Allocation Rules: Through interactions with the resource allocation environment, reinforcement learning, a subset of machine learning, can be used to build allocation rules. It enables systems to gradually learn the best allocation techniques [24–26].

11.4.3 Machine learning's advantages and drawbacks for resource allocation

11.4.3.1 Advantages

When allocating resources, machine learning has special benefits:

Adaptability and Learning: Machine learning models are adaptable and able to learn from changing situations, which makes them suitable for scenarios involving dynamic resource allocation.

Optimization: Machine learning is excellent at streamlining choices made in response to data-driven insights, ultimately improving performance and efficiency [27].

Scalability: From small-scale systems to large-scale, distributed environments, many machine learning algorithms are scalable and can address resource allocation problems at different scales.

11.4.3.2 Limitations

However, using machine learning for resource allocation also has its drawbacks:

Data Quantity and Quality: For machine learning models to function properly, they require high-quality training data. It can be difficult to find sufficient, high-quality data in some specialized resource allocation sectors.

Complexity: Machine learning models add complexity, and their decision-making procedures occasionally may not be transparent. Model complexity and interpretability must constantly be balanced.

Training and Upkeep: As resource allocation conditions change, machine learning models need constant training and validation. This continual effort may require a lot of resources.

This section gives a thorough grasp of how machine learning can alter resource allocation techniques by examining these aspects of machine learning in resource allocation. It prepares the ground for the chapter's later sections, which will further examine the integration of hybrid evolutionary models and machine learning to solve the difficulties and complexities of contemporary resource allocation.

11.5 HYBRID EVOLUTIONARY MODELS AND MACHINE LEARNING INTEGRATION

Combining evolutionary algorithms' powerful exploration capabilities with machine learning's data-driven intelligence, or hybrid evolutionary models and machine learning integration, is a cutting-edge methodology. This action equips systems to effectively address complex optimization and decision-making challenges, adapt to dynamic environments, accelerate convergence, generalize across various problem domains, and foster innovation in industries ranging from logistics and finance to healthcare and recommendation systems. An important step towards adaptable, effective, and versatile problem-solving in a society that is becoming more and more data-centric is the fusion of these two paradigms [28].

11.5.1 Conceptual framework

Systems that integrate machine learning and evolutionary algorithms can be understood, designed, and put into use using a "Conceptual Framework for Hybrid Models," which offers a theoretical and organized framework. For successfully integrating these many computational methodologies, this framework acts as a road map. Here are the steps that should be followed to create an integrated framework:

Problem Definition and Scope: Start by stating the issue or objective the hybrid model is meant to solve. In order to do this, one must comprehend the goals, limitations, and type of data at hand. To construct a successful hybrid model, a specific problem definition is essential.

Component Models: List each of the distinct evolutionary algorithms and machine learning strategies that will be incorporated into the hybrid system. Genetic algorithms, neural networks, reinforcement learning, and any other pertinent techniques may be among them.

Integration Mechanisms: Describe how these various component models will communicate with one another within the hybrid system. Data exchange, feedback loops, or decision fusion may be necessary for this. Both the advantages of evolutionary algorithms and those of machine learning should be utilized by the integration mechanism.

Data Flow and Preprocessing: Specify the data collection, preprocessing, and sharing procedures across the component models. In order to make sure that the two methodologies work together, data pretreatment activities like cleaning, feature engineering, or normalization may be required.

Decision-Making Strategy: Specify how the hybrid model will decide or forecast the future depending on the results of its component models. Voting systems, weighted averages, or more sophisticated fusion methods may be used to combine the forecasts in this case.

Think about how the hybrid model will adjust to shifting circumstances or data. To make sure the model is useful over time, this can entail online learning tools, model retraining, or parameter changes.

Performance Indicators: Detail the assessment standards and performance indicators that will be applied to judge the viability of the hybrid model. These metrics, which may include measurements of accuracy, precision, recall, or domain-specificity, should be in line with the goals of the issue.

Validation and Testing: Explain the techniques for validating and testing the hybrid model. Included here are the methods for dividing the data into training, validation, and testing sets, as well as any cross-validation or other validation strategies that will be used.

Scalability and Efficiency: Take the hybrid model's scalability and efficiency into account, particularly when working with big datasets or real-time applications. It is necessary to provide methods for parallelization, distributed computing, or optimization [29].

Interpretability and Explainability: Address the question of how the outcomes of the hybrid model may be understood and justified. This is key in applications where user comprehension and trust are essential because transparency is one of them.

Ethical and Regulatory Considerations: Recognize any ethical or regulatory concerns with the hybrid approach, particularly in industries like healthcare or finance. Respect for ethical standards and privacy laws is essential.

Continuous Improvement and Maintenance: Outline techniques for sustaining and enhancing the hybrid model over time. This involves keeping an eye on its performance, updating component models, and handling problems as they come up [30].

11.5.2 Benefits of combining evolutionary algorithms and machine learning

The following benefits can be derived from combining evolutionary algorithms with machine learning:

Enhanced Optimization: Hybrid models combine the local refining skills of machine learning with the global exploration capabilities of evolutionary algorithms to provide more effective optimization processes.

Adaptability: By enabling systems to adapt to changing surroundings and evolving data, the combination makes them appropriate for dynamic scenarios in which conventional methods may struggle.

Data-Driven Decisions: Machine learning extracts knowledge from data to drive evolutionary algorithms, concentrating exploration on promising regions and enhancing convergence speed and solution quality.

Robustness: Hybrid models frequently exhibit higher levels of robustness and resilience when confronted with noisy data or ambiguous circumstances. They are competent at navigating the difficulties of everyday life.

Generalizability: These models may generalize effectively across different problem domains, providing solutions that are effective for a variety of applications.

11.5.3 Limitations of combining evolutionary models with machine learning

Numerous difficulties arise when combining evolutionary models and machine learning [31].

Complexity: Developing and applying hybrid models can be difficult, frequently needing knowledge of both machine learning and evolutionary techniques.

Hyper-Parameter Tuning: Choosing the best hyper-parameters for each component can be difficult because of the potential for complex interactions.

Interpretability: Using multiple sophisticated models together might limit interpretability, making it more difficult to comprehend and defend choices.

Data Compatibility: Making sure that data is compatible with evolutionary and machine learning components can be a difficult effort that necessitates thorough pretreatment.

Computation and Resource Requirements: Hybrid models can be computationally demanding, needing significant computational resources, especially when working with huge datasets or complicated situations.

Selection of Algorithms: Selecting the best evolutionary algorithm and machine learning approach for a given issue is a difficult task that calls for experimentation and domain expertise.

For the benefits of combining evolutionary algorithms with machine learning to be properly utilized and for hybrid models to reach their full potential in a variety of application domains, these hurdles must be successfully overcome.

11.6 CASE STUDIES

This section provides in-depth analyses of real-world applications that highlight the practical applicability and efficacy of hybrid evolutionary models and machine learning integration in addressing challenging resource allocation problems across diverse sectors. These case studies offer useful insights into the adaptability, effectiveness, and versatility of hybrid techniques by providing specific instances of how they might be used to tackle certain issues.

11.6.1 Case study 1: Cloud resource allocation application using a hybrid model

In this case study, we look at the use of machine learning and hybrid evolutionary models in the field of allocating cloud resources. Changing workloads and resource demands are common in cloud computing systems, which are dynamic. Here, a hybrid model brings together machine learning's data-driven insights with the adaptability of evolutionary algorithms to shifting environments. The study demonstrates how this hybrid strategy optimizes the distribution of virtual machines, storage, and network resources, guaranteeing that cloud resources are distributed effectively, lowering costs, and boosting performance. Dynamic resource provisioning, scalability, cost reduction, and flexibility for changing workloads are among the important topics covered.

11.6.2 Case study 2: Using machine learning to improve IoT resource allocation

The Internet of Things (IoT) domain takes the front stage in the second case study. IoT ecosystems include a wide range of interconnected devices, each with its own resource limitations and data processing needs. This case study shows how hybrid models, which use machine learning for predictive analytics and optimization, can improve resource allocation in IoT environments. It shows the distribution of computing resources to Internet of Things (IoT) devices based on data-driven insights, enabling effective data processing, decreased latency, and enhanced system performance. In order to handle the dynamic nature of IoT environments, the study also analyses scalability issues and how adaptable the hybrid method is.

11.6.3 Case study 3: Edge computing resource allocation using hybrid methodologies

The focus of the third case study is edge computing, a paradigm in which computation and data processing take place closer to the data source. In order to address resource allocation issues in edge computing environments, this case study investigates the use of hybrid evolutionary models and machine learning. Given the variety and frequent resource limitations of edge devices, the hybrid model optimizes the allocation of computational resources while taking into account elements like latency, energy efficiency, and data privacy. The research demonstrates the adaptability of hybrid techniques in meeting the unique demands of edge computing scenarios, guaranteeing that computational operations are carried out effectively and quickly at the edge of the network.

Collectively, these case studies demonstrate the flexibility and efficiency of integrating machine learning and hybrid evolutionary models in resource allocation scenarios. They give practical illustrations of how these hybrid techniques can be tailored to satisfy the particular requirements of various domains, assuring optimum resource utilization, cost reductions, and enhanced system performance. Readers learn a lot about the advantages and possible uses of hybrid models in tackling current resource allocation issues across many industries through these real-world examples.

11.7 FUTURE DIRECTIONS AND RESEARCH OPPORTUNITIES

This section focuses on the future of resource allocation and the interesting possibilities for research it offers. Resource allocation techniques must change to keep up with new trends and tackle problems that have not yet been solved as technology continues to advance. This section examines the changing landscape of resource allocation and identifies possible directions for future study.

11.7.1 Emerging trends in resource allocation

The future of this industry is being shaped by new patterns in resource allocation. Important areas of interest are as follows:

Edge and Fog Computing: Resource allocation is being moved closer to the data source as a result of the growth of IoT devices and the demand for real-time data processing. In these edge and fog computing environments, where computational resources are dispersed over the network's edge rather than centralized in data centres, researchers can investigate how resource allocation algorithms need to change [29, 30].

Autonomous Resource Management: The idea of autonomous resource management entails the creation of clever algorithms and systems that are capable of making judgements about the real-time distribution of resources without the need for human participation. To develop resource allocation systems that dynamically adapt to shifting situations, researchers can explore reinforcement learning, self-optimizing algorithms, and AI-driven techniques.

Interoperability in Multi-Cloud Environments: Interoperability in Multi-Cloud Environments: As more businesses use different cloud providers and platforms, it is essential to provide seamless interoperability and effective workload distribution. Standards, norms, and tactics for maximizing resource allocation in intricate multi-cloud environments can all be studied in research.

Quantum Computing: Resource allocation could be revolutionized by quantum computing, especially for extremely complicated optimization tasks. Researchers can look into how resource allocation problems can be solved more effectively using hybrid models and quantum algorithms, possibly using quantum anneals or gate-based quantum computers.

Sustainability and Green Computing: As environmental concerns intensify, there is a pressing need for resource allocation strategies that limit energy use and lower carbon emissions. The development of environmentally friendly resource allocation techniques and algorithms that take into account energy-efficient hardware and data centres might be the subject of research.

11.7.2 Unresolved challenges and open questions

Unresolved Challenges and Open Questions highlight the persistent issues that researchers and practitioners face in resource allocation:

Dynamic Workloads: Addressing dynamic workloads requires resource allocation algorithms that can adapt in real-time to changes in demand. Research can explore techniques for workload prediction, anomaly detection, and adaptive resource provisioning to optimize allocation in rapidly changing environments.

Resource Fragmentation: Resource fragmentation occurs when resources are allocated in a way that leads to suboptimal utilization. Research can investigate strategies to minimize fragmentation, such as resource pooling, load balancing, and efficient task scheduling [32].

Privacy and Security: Ensuring the privacy and security of data during resource allocation processes is crucial, especially in edge and cloud environments. Research can focus on encryption techniques, access control mechanisms, and secure allocation algorithms that protect sensitive information.

Multi-Objective Optimization: Many resource allocation scenarios involve conflicting objectives, such as minimizing costs while maximizing performance. Research can delve into multi-objective optimization techniques, such as Pareto optimization, to find optimal trade-off solutions that balance these objectives effectively.

Ethical Considerations: As resource allocation becomes more automated, ethical considerations arise, particularly in critical domains like healthcare and finance. Researchers can explore ethical frameworks and guidelines for responsible resource allocation, considering fairness, bias, and transparency.

11.8 CONCLUSION

This chapter concludes by analyzing the use of evolutionary algorithms and machine learning to investigate the dynamic terrain of resource allocation in modern systems. The chapter demonstrates the adaptability and efficiency of evolutionary algorithms in scenarios like cloud resource allocation and network bandwidth management through a thorough examination of the principles, difficulties, and significance of resource allocation. Furthermore, decision-making and predictive resource allocation are improved by the incorporation of machine learning techniques. Case studies provide practical examples of how hybrid approaches—which combine machine learning and evolutionary algorithms—can be applied to edge computing resource allocation. They demonstrate resource efficiency, low latency, and adaptability. The chapter emphasizes how important these innovative approaches are to solving the changing problems of resource allocation and offers insights into their advantages, drawbacks, and potential applications in the development of more adaptable and optimized systems.

REFERENCES

[1] Malik, S., Tahir, M., Sardaraz, M., & Alourani, A. (2022). A resource utilization prediction model for cloud data centers using evolutionary algorithms and machine learning techniques. *Applied Sciences*, 12(4), 2160.

[2] Emadi, A., Sobhani, R., Ahmadi, H., Boroomandnia, A., Zamanzad-Ghavidel, S., & Azamathulla, H. M. (2022). Multivariate modeling of river water withdrawal using a hybrid evolutionary data-driven method. *Water Supply*, 22(1), 957–980.

[3] Yadav, A. M., Tripathi, K. N., & Sharma, S. C. (2023). An opposition-based hybrid evolutionary approach for task scheduling in fog computing network. *Arabian Journal for Science and Engineering*, 48(2), 1547–1562.

[4] Obiedat, R., Qaddoura, R., Al-Zoubi, A. M., Al-Qaisi, L., Harfoushi, O., Alrefai, M. A., & Faris, H. (2022). Sentiment analysis of customers' reviews using a hybrid evolutionary svm-based approach in an imbalanced data distribution. *IEEE Access*, 10, 22260–22273.

[5] Mahmoudzadeh, H., Matinfar, H. R., Kerry, R., Eskandari, S., Ebrahimi-Khusfi, Z., & Taghizadeh-Mehrjardi, R. (2022). New hybrid evolutionary models for spatial prediction of soil properties in Kurdistan. *Soil Use and Management*, 38(1), 191–211.

[6] Han, Y., Niyato, D., Leung, C., Miao, C., & Kim, D. I. (2022, May). A dynamic resource allocation framework for synchronizing metaverse with iot service and data. *In ICC 2022-IEEE International conference on Communications* (pp. 1196–1201). IEEE.

[7] Miguel, F. M., Frutos, M., Méndez, M., & Tohmé, F. (2022). Order batching and order picking with 3D positioning of the articles: solution through a hybrid evolutionary algorithm.

[8] Ridha, H. M., Hizam, H., Mirjalili, S., Othman, M. L., Ya'acob, M. E., Ahmadipour, M., & Ismaeel, N. Q. (2022). On the problem formulation for parameter extraction of the photovoltaic model: Novel integration of hybrid evolutionary algorithm and Levenberg Marquardt based on adaptive damping parameter formula. *Energy Conversion and Management*, 256, 115403.

[9] Mahmud, S., Chakrabortty, R. K., Abbasi, A., & Ryan, M. J. (2022). Switching strategy-based hybrid evolutionary algorithms for job shop scheduling problems. *Journal of Intelligent Manufacturing*, 33(7), 1939–1966.

[10] Nwokolo, S. C., Obiwulu, A. U., Ogbulezie, J. C., & Amadi, S. O. (2022). Hybridization of statistical machine learning and numerical models for improving beam, diffuse and global solar radiation prediction. *Cleaner Engineering and Technology*, 9, 100529.

[11] Kumar, Y., Kaul, S., & Hu, Y. C. (2022). Machine learning for energy-resource allocation, workflow scheduling and live migration in cloud computing: State-of-the-art survey. *Sustainable Computing: Informatics and Systems*, 36, 100780.

[12] Bal, P. K., Mohapatra, S. K., Das, T. K., Srinivasan, K., & Hu, Y. C. (2022). A joint resource allocation, security with efficient task scheduling in cloud computing using hybrid machine learning techniques. *Sensors*, 22(3), 1242.

[13] Khan, T., Tian, W., Zhou, G., Ilager, S., Gong, M., & Buyya, R. (2022). Machine learning (ML)-centric resource management in cloud computing: A review and future directions. *Journal of Network and Computer Applications*, 204, 103405.

[14] Chen, X., Yang, L., Chen, Z., Min, G., Zheng, X., & Rong, C. (2022). Resource allocation with workload-time windows for cloud-based software services: a deep reinforcement learning approach. *IEEE Transactions on Cloud Computing*.

[15] Lilhore, U. K., Poongodi, M., Kaur, A., Simaiya, S., Algarni, A. D., Elmannai, H., & Hamdi, M. (2022). Hybrid model for detection of cervical cancer using causal analysis and machine learning techniques. *Computational and Mathematical Methods in Medicine*, 2022(1), 4688327.

[16] Huang, J., Wan, J., Lv, B., Ye, Q., & Chen, Y. (2023). Joint computation offloading and resource allocation for edge-cloud collaboration in internet of vehicles via deep reinforcement learning. *IEEE Systems Journal*, 17(2), 2500–2511.

[17] Soni, D., & Kumar, N. (2022). Machine learning techniques in emerging cloud computing integrated paradigms: A survey and taxonomy. *Journal of Network and Computer Applications*, 205, 103419.

[18] Tuli, S., Gill, S. S., Xu, M., Garraghan, P., Bahsoon, R., Dustdar, S., ... & Jennings, N. R. (2022). HUNTER: AI based holistic resource management for sustainable cloud computing. *Journal of Systems and Software*, 184, 111124.

[19] Shruthi, G., Mundada, M. R., Sowmya, B. J., & Supreeth, S. (2022). Mayfly Taylor optimisation-based scheduling algorithm with deep reinforcement learning for dynamic scheduling in fog-cloud computing. *Applied Computational Intelligence and Soft Computing*, 2022(1), 2131699.

[20] Kaur, G., Goyal, S., & Kaur, H. (2021, December). Brief review of various machine learning algorithms. In *Proceedings of the International Conference on Innovative Computing & Communication (ICICC)*.

[21] Kaur, V., Kaur, G., Dhiman, G., Bindal, R., & Mishra, M. K. (2020). Adaptability of machine learning in cryptography. *Solid State Technology*, 63(4), 2874–2880.

[22] Kruekaew, B., & Kimpan, W. (2022). Multi-objective task scheduling optimization for load balancing in cloud computing environment using hybrid artificial bee colony algorithm with reinforcement learning. *IEEE Access*, 10, 17803–17818.

[23] Pandey, N. K., Diwakar, M., Shankar, A., Singh, P., Khosravi, M. R., & Kumar, V. (2022). Energy efficiency strategy for big data in cloud environment using deep reinforcement learning. *Mobile Information Systems*, 2022(1), 1–11.

[24] Tandon, R., Verma, A., & Gupta, P. K. (2022, April). A secure framework based on nature-inspired optimization for vehicle routing. In *International Conference on Advances in Computing and Data Sciences* (pp. 74–85). Cham: Springer International Publishing.

[25] Tandon, R., Verma, A., & Gupta, P. K. (2024). D-BLAC: A dual blockchain-based decentralized architecture for authentication and communication in VANET. *Expert Systems with Applications*, 237, 121461.

[26] Verma, A., Tandon, R., & Gupta, P. K. (2022). TrafC-AnTabu: AnTabu routing algorithm for congestion control and traffic lights management using fuzzy model. *Internet Technology Letters*, 5(2), e309.

[27] Tandon, R., Verma, A., & Gupta, P. K. (2022, November). Blockchain enabled vehicular networks: A review. In *2022 5th International Conference on Multimedia, Signal Processing and Communication Technologies (IMPACT)* (pp. 1–6). IEEE.

[28] Kaur, M., & Aron, R. (2022). A novel load balancing technique for smart application in a fog computing environment. *International Journal of Grid and High Performance Computing (IJGHPC)*, 14(1), 1–19.

[29] Kaur, M., Aron, R., Wadhwa, H., & Oo, H. N. (2023, April). CNN-based Smart Waste Management System in Fog Computing Environment. In *2023 IEEE 12th International Conference on Communication Systems and Network Technologies (CSNT)* (pp. 774–779). IEEE.

[30] Kaur, M., & Aron, R. (2022). An energy-efficient load balancing approach for scientific workflows in fog computing. *Wireless Personal Communications*, 125(4), 3549–3573.

[31] Kaur, M., & Aron, R. (2022). An energy-efficient load balancing approach for fog environment using scientific workflow applications. In *Distributed computing and optimization techniques: Select proceedings of ICDCOT 2021* (pp. 165–174). Singapore: Springer Nature Singapore.

[32] Islam, M. S. U., Kumar, A., & Hu, Y. C. (2021). Context-aware scheduling in Fog computing: A survey, taxonomy, challenges and future directions. *Journal of Network and Computer Applications*, 180, 103008.

Chapter 12

The cloud frontier

A paradigm shift in sustainable supply chain management through data analytics

Gagandeep Kaur, Harpreet Kaur, and Sonia Goyal
Punjabi University, Patiala, India

12.1 INTRODUCTION

The management of supply chains has undergone a significant transformation in recent years. The current focus is on more than just getting things from one place to another. As of now, the accentuation lies in executing this cycle in a way that decreases natural results, maintains moral guidelines, and improves functional viability [1, 2]. The urgency of incorporating sustainability principles into supply chain operations is emphasized in this chapter, which provides an overview of the dynamic environment under consideration. The essential accentuation in the production network of the executives has been on the decrease of expenses and upgrade of functional productivity. Despite this, there has been a significant shift in focus toward the concept of sustainability in light of the dire circumstances that the global society is confronted with, such as climate change, resource depletion, and social responsibility. Sustainable supply chain management (SSCM) is a comprehensive framework that combines a wide range of goals and ideas to minimize the negative social and environmental effects of supply chain operations and maximize the economic benefits. This section likewise accentuates the vital meaning of cloud innovation in working with this worldview change [3]. Due to its capacity to scale operations, analyze data in real-time, and provide businesses with cost-effective solutions, cloud technology has emerged as a transformative tool for supply chain management. It is an essential part of many sustainability projects because it makes it easier to use data to help make decisions, increases transparency, and encourages supply chain collaboration. The relationship between cloud and supply chain management has been depicted visually in Figure 12.1.

Eventually, the system has been established for comprehending the convergence of cloud innovation and supportability inside supply chains. The ability to evaluate data in real-time, access important information remotely, and facilitate seamless communication among supply chain partners are just a few of the many benefits offered by cloud technology.

DOI: 10.1201/9781003494430-12

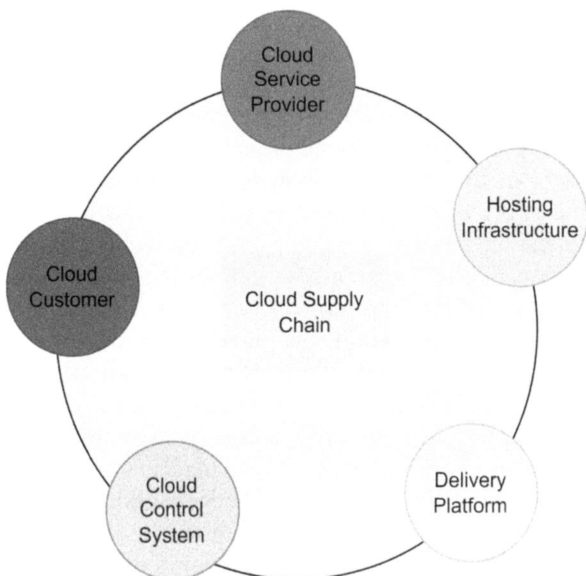

Figure 12.1 Shows the relationship between the supply chain management and Cloud technology.

12.2 RELATED WORK

Vegesna et al. [4] delve into the critical intersection of Information Systems (IS), Supply Chain Management (SCM), and Cloud Computing. The significance of utilizing cloud-based solutions for enhancing multicomponent productivity in supply chain operations is emphasized in the introduction, which appropriately frames the context. Mukherjee et al. [5] highlights the crucial intersection of cloud computing and SCM. The authors aim to address and integrate the challenges posed by the adoption of cloud technologies in the context of industrial production. Dolgui and Sokolov et al. [6] explore the idea of a "Cloud Store network" by coordinating Industry 4.0 and computerized stages in the "Production network as-a-Administration" model. The authors emphasize this strategy's transformative potential for increasing supply chain efficiency. The examination contributes significant bits of knowledge into the convergence of distributed computing, Industry 4.0, and planned operations, preparing for inventive Production network as-a-administration arrangements. The coordination of these advances is ready to rethink ordinary production network standards, offering a promising road for future progressions in strategies and transportation. Chauhan et al. [7] delve into the digitalization of inventory network executives from the perspective of Industry 4.0 empowering innovations, underlining a feasible point of view. The creators give an extensive investigation of the joining of Industry 4.0 innovations in production network processes. They examine the economic ramifications of this digitalization, offering important bits of knowledge into how these mechanical progressions can add to naturally cognizant and socially capable store network rehearses. Liu et al. [8] provide a critical commitment that investigates the cooperative energy between computerized capacities and a roundabout economy with regard to practical store networks. By associating Industry 4.0 with roundabout economy standards, the paper offers a significant point of view on progressing reasonable practices in modern designing and highlights the significance of mechanical reconciliation for an all the more ecologically cognizant future. Ivanov et al. [9] delve into the transformative elements of the digital supply chain, focusing on the pivotal role played by the cloud, platforms, and digital twins. The authors adeptly illustrate the integration of cloud computing, digital platforms, and digital twins, showcasing their collective impact on enhancing supply chain efficiency and responsiveness. Yang et al. [10] present an innovative approach to quality management in perishable supply chain logistics through the integration of edge-cloud, blockchain, and the Internet of Everything (IoE). The authors showcase the synergies between edge computing, cloud infrastructure, blockchain's decentralized ledger, and IoE, providing a comprehensive solution for real-time quality management. This work significantly contributes to the evolving landscape of Internet of Things (IoT) applications in logistics, offering a forward-looking persective on improving perishable supply chain operations through advanced technological integration. Zaman et al. [11] introduce a Grey Decision-Making Trial and Evaluation Laboratory (DEMATEL) model for digital warehouse management within supply chain networks. This research is notable for its application of grey system theory to decision-making processes, offering a methodologically sound approach to evaluating and prioritizing factors in digital warehouse management. The paper contributes to the growing body of literature on decision analytics and digitalization in SCM, offering insights that can inform practical implementations in warehouse operations. Matenga and Mpofu et al. [12] investigate the implementation of a blockchain-based cloud manufacturing SCM system for collaborative enterprise manufacturing, with a specific case study focusing on transport manufacturing. The authors present a case study that exemplifies the practical implementation of this innovative SCM system in the context of transport manufacturing. Key highlights include the utilization of blockchain for secure and transparent transactions, and fostering collaboration among diverse stakeholders in the supply chain.

12.3 THE CLOUD TECHNOLOGY REVOLUTION

Cloud innovative advances, at its center, encompass the arrangement of registering administrations, including capacity, handling, and systems administration, through the Web. With unprecedented levels of scalability, accessibility, and cost-effectiveness, this technology has significantly altered conventional IT structures [13]. Cloud arrangements have become progressively well-known among organizations in different businesses as they look to improve their activities, upgrade efficiency, and cultivate development. The work clarifies the limit of cloud innovation to achieve huge changes. A more equitable environment has emerged for both large and small businesses as a result of the democratization of access to cutting-edge computing resources. The democratization of innovation holds huge ramifications for maintainability attempts since it enables even small endeavors to take on cutting-edge supportability procedures [13].

Moreover, it is possible to argue that cloud technology is intrinsically compatible with sustainability principles. Businesses can adjust their computer resources in response to demand recognition of this technology's scalability, which reduces energy consumption during times of low activity [14]. Moreover, bringing together information stockpiling and handling abilities in the cloud can actually relieve the natural effect frequently connected with on-premises server farms. Moreover, an assessment is led on the effect of cloud innovation on traditional business structures. Software as a Service (SaaS) and Platform as a Service (PaaS) models have made it easier to create subscription-based, cost-effective logistics and SCM solutions. For supply chain reforms driven by sustainability goals, cloud-based solutions are well-suited due to their high degree of flexibility and customization [15].

Furthermore, the study highlights the significance of data within the framework of cloud technology. In the field of SCM, the capacity to gather, evaluate, and respond quickly to large amounts of data is a transformative factor. Cloud-based information examination and AI advancements have arisen as significant apparatuses in propelling supportability endeavors by offering important experiences in different areas, including energy use, squandering the executives, and fossil fuel byproducts decrease [16].

12.4 CLOUD-POWERED DATA AND ANALYTICS

In the domain of supply chains, information assumes a vital part by offering significant bits of knowledge into functional exercises, empowering associations to upgrade processes, limit irregularities, and work on general productivity. However, the task at hand goes beyond merely gathering data; it includes obtaining useful insights from that data. Today's supply chains generate a lot of data from a variety of sources, including customer interactions, IoT devices, sensors, and other sources. The dataset contains a significant measure of data that, when dissected in a capable way, can possibly move manageability exercises.

The subsection places significant emphasis on the importance of real-time data analytics. Cloud technology facilitates the real-time processing and analysis of data, offering prompt insights into the operational aspects of

SCM for enterprises [17–19]. Cloud innovation works with the ongoing handling and investigation of information, offering brief experiences into the functional parts of the production network. A common illustration of the advantages of having quick admittance to stock information is the capacity to forestall both extreme stock levels and stock deficiencies. This ability limits wastage as well as upgrades the usage of assets, bringing about improved productivity. In addition, the application of predictive analytics plays a significant role in sustainability initiatives. Organizations can predict demand in the future, optimize production schedules, and mitigate environmental effects by utilizing previous data and machine learning algorithms [20–23]. It is possible that adopting this proactive strategy will result in significant cost savings and advantages for sustainability.

12.5 SMART INVENTORY MANAGEMENT WITH CLOUD SOLUTIONS

Inventory management plays a pivotal role in the efficient functioning of supply chain activities. In the past, inventory management relied on manual procedures and periodic inspections, both of which were prone to errors and frequently resulted in excessive or inadequate supply. These shortcomings brought about monetary results as well as ecological consequences, as the presence of excess stock can add to wastage.

The use of cloud innovation [3, 24–26], because of its capacity to give continuous information abilities, brings about the change of stock administration into a more adaptable and responsive activity. Cloud-based frameworks, as shown in Figure 12.2, bring the capacity to ongoing admittance to data in regards to the amount of stock at different focuses along the production network, enveloping the two distribution centers and conveyance focuses. The capacity to get to continuous information empowers associations to go with

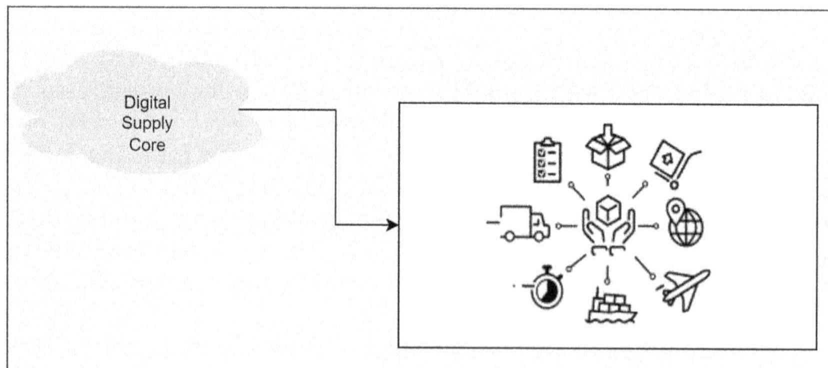

Figure 12.2 Smart inventory management system with cloud solutions.

all-around informed choices on the timing and area of restocking exercises, accordingly limiting wastage and upgrading the productive organization of assets. The chapter also goes into detail about the benefits of using real-time inventory tracking. Cloud innovation empowers firms to convey global positioning frameworks that proposition definite bits of knowledge into the transportation of items [13]. The consideration of the following innovation not only supports safety efforts against robbery, including sack redirection as featured in the part title, yet in addition works with further developed effectiveness in the taking care of and transportation of items.

12.6 SUSTAINABLE LOGISTICS AND TRANSPORTATION

Transportation and logistics play important roles in supply chains and frequently make significant contributions to the overall environmental footprint [27]. This section explains the troubles inherited in ordinary coordinated operations practices and investigates how cloud innovation presents feasible ways of tending to these worries. Cloud innovation works with the examination of broad amounts of information, enveloping traffic designs, atmospheric conditions, and vehicle execution, fully intent on enhancing transportation courses for organizations. Through the essential choice of ideal courses and transportation modes, associations can relieve fuel utilization, reduce emanations, and moderate the biological repercussions related to transportation [28, 29].

Cloud technology enables efficient coordination among many types of transportation, including trucks, trains, ships, and planes [30, 31]. In the field of supply chain logistics, using an intermodal strategy makes it easier to increase flexibility and sustainability. Cloud-based systems can monitor and record emissions data, making it easier for businesses to set and achieve their sustainability goals. The foundation of straightforwardness is urgent regarding natural responsibility as well as satisfying administrative commitments and tending to customer assumptions for ecologically economical labor and products.

12.7 COLLABORATIVE CLOUD-BASED SUPPLY CHAINS

Sustainable supply chain management (SSCM) practices can only be established and maintained through effective collaboration. Supply chains regularly incorporate a huge number of players, enveloping providers, makers, wholesalers, and retailers. The coordination of exercises and the dividing of data between different accomplices are significant to accomplishing supportability targets. The section initiates by giving a prologue to cooperative stages and biological systems that are worked with by cloud innovation. These stages offer a concentrated climate for store network members to participate in correspondence, trade information, and coordinate supportability tries. It has been discovered that cloud-based collaboration tools simplify the process of identifying and resolving sustainability issues by enhancing transparency and easing information flow. Supplier relationships can also benefit from effective collaboration. This section digs into the capability of cloud innovation to increase provider connections through the arrangement of constant perceivability on provider execution, quality, and consistency. Companies are better able to choose and work with suppliers with confidence when they have better visibility, which encourages responsible sourcing practices.

Moreover, the part dives into the thought of inventory network versatility. The establishment of supply chain resilience is crucial to maintaining sustainability in the face of an increasing number of interruptions. The development of robust supply chain networks that are capable of swiftly adapting to a variety of disruptions, such as those brought on by natural disasters, geopolitical events, or other determinants, is made easier by cloud technology. Cloud-based technologies have the potential to improve the efficiency with which sustainability data is collected and presented. This could help businesses meet their regulatory obligations and demonstrate their commitment to ethical business practices.

12.8 CHALLENGES AND FUTURE TRENDS

This study analyzes the different hindrances and imminent improvements in reasonable production network the board, offering important viewpoints on the powerful climate and the utilization of cloud innovation to defeat these deterrents and encourage future headways. Businesses are confronted with a variety of challenges as the incorporation of sustainability into SCM gains prominence. This chapter examines the following significant issues: The utilization of cloud innovation and the assortment of delicate maintainability information lead to critical worries around security and protection. Associations must take on and authorize thorough network safety conventions to protect delicate information. The undertaking of exploring the multifaceted territory of manageability norms and detailing prerequisites relating to administrative consistency can give a considerable test. Cloud-based instruments have the capacity to smooth out and mechanize consistence processes. The execution of supportability rehearses in associations normally requires a groundbreaking social change. The rising globalization and intricacy of supply chains give critical difficulties in actually overseeing supportability all through the entire stock organization. Cloud innovation empowers the consistent trade of information and advances cooperative endeavors among different partners along the inventory network.

The work moreover looks at impending patterns in economical store networks across the board, with a specific accentuation on recognizing key regions where cloud innovation is supposed to have a huge effect. A few patterns can be recognized, including: The idea of round supply chains, which includes the reuse, restoration, or reusing of items and materials, is progressively being perceived and embraced. Cloud innovation can screen and manage the circularity of things. High-level man-made reasoning (computer-based intelligence) and AI (ML) calculations can possibly upgrade maintainability programs through the investigation of broad data and the recognizable proof of regions for development. The developing accentuation of associations lies in the foundation of tough stockpile networks equipped for persevering and alleviating disturbances. Cloud innovation plays a vital part in working with risk evaluation and relief systems. The usage of ecologically reasonable advancements, for example, electric vehicles and environmentally friendly power sources, is encountering a vertical pattern. The integration of various technologies within supply chains could be enhanced by cloud technology.

12.9 CONCLUSION

In a nutshell, this chapter provides a comprehensive summary of the investigation into the role of cloud technology in facilitating SSCM. This work provides an analysis of essential findings and practical insights for professionals in the industry. It summarizes the effects and promise of cloud technology in advancing sustainability within supply chains. In this concluding section, we engage in a retrospective analysis of the transformative process that organizations might undertake when they incorporate cloud technology in alignment with their sustainability objectives. The significance of adopting a comprehensive perspective that encompasses the environmental, social, and economic aspects of supply chain operations is reiterated.

The chapter highlights the perpetual nature of the sustainable journey, underscoring the influence of technical developments, legislative modifications, and evolving consumer expectations on the overall trajectory. It is imperative for companies to maintain a state of agility, adaptability, and unwavering dedication toward the ongoing enhancement of their sustainability endeavors. The study reiterates that cloud technology is a significant catalyst for SSCM, empowering organizations to promote sustainability, improve transparency, and adapt to the dynamic requirements of a swiftly shifting global landscape.

REFERENCES

1. Kaur, G., Goyal, S., & Kaur, H. (2021, December). Brief review of various machine learning algorithms. In *Proceedings of the International Conference on Innovative Computing & Communication (ICICC).*

2. Harnal, S., Sharma, G., Malik, S., Kaur, G., Khurana, S., Kaur, P., ... & Bagga, D. (2022). Bibliometric mapping of trends, applications and challenges of artificial intelligence in smart cities. *EAI Endorsed Transactions on Scalable Information Systems*, 9(4), e8–e8.

3. Harnal, S. et al. (2023). Current and Future Trends of Cloud-based Solutions for Healthcare. In R. Tiwari, D. Koundal, & S. Upadhyay (Eds.), *Image Based Computing for Food and Health Analytics: Requirements, Challenges, Solutions and Practices* (pp. 115–136). Cham: Springer. https://doi.org/10.1007/978-3-031-22959-6_7

4. Vegesna, V. V. (2023). The utilization of information systems for supply chain management for multicomponent productivity based on cloud computing. *International Journal of Management, Technology and Engineering*, 11, 98–113.

5. Mukherjee, S., Chittipaka, V., Baral, M. M., & Srivastava, S. C. (2022). Integrating the challenges of cloud computing in supply chain management. In *Recent Advances in Industrial Production: Select Proceedings of ICEM 2020* (pp. 355–363). Springer Singapore.

6. Ivanov, D., Dolgui, A., & Sokolov, B. (2022). Cloud supply chain: Integrating industry 4.0 and digital platforms in the "Supply Chain-as-a-Service". *Transportation Research Part E: Logistics and Transportation Review*, 160, 102676.

7. Chauhan, S., Singh, R., Gehlot, A., Akram, S. V., Twala, B., & Priyadarshi, N. (2023). Digitalization of supply chain management with industry 4.0 enabling technologies: a sustainable perspective. *Processes*, 11(1), 96.

8. Liu, L., Song, W., & Liu, Y. (2023). Leveraging digital capabilities toward a circular economy: Reinforcing sustainable supply chain management with Industry 4.0 technologies. *Computers & Industrial Engineering*, 178, 109113.

9. Zhang, G., MacCarthy, B. L., & Ivanov, D. (2022). The cloud, platforms, and digital twins—Enablers of the digital supply chain. In *The Digital Supply Chain* (pp. 77–91). Elsevier.

10. Yang, C., Lan, S., Zhao, Z., Zhang, M., Wu, W., & Huang, G. Q. (2022). Edge-cloud blockchain and IoE-enabled quality management platform for perishable supply chain logistics. *IEEE Internet of Things Journal*, 10(4), 3264–3275.

11. Zaman, S. I., Khan, S., Zaman, S. A. A., & Khan, S. A. (2023). A grey decision-making trial and evaluation laboratory model for digital warehouse management in supply chain networks. *Decision Analytics Journal*, 100293.

12. Matenga, A. E., & Mpofu, K. (2022). Blockchain-based cloud manufacturing SCM system for collaborative enterprise manufacturing: a case study of transport manufacturing. *Applied Sciences*, 12(17), 8664.

13. Cigolini, R., Gosling, J., Iyer, A., & Senicheva, O. (2022). Supply chain management in construction and engineer-to-order industries. *Production Planning & Control*, 33(9–10), 803–810.

14. Singh, S., Ramkumar, K. R., & Kukkar, A. (2021, October). Machine learning techniques and implementation of different ML algorithms. In *2021 2nd Global Conference for Advancement in Technology (GCAT)* (pp. 1–6). IEEE.

15. Liu, Y., Kukkar, A., & Shah, M. A. (2022). Study of industrial interactive design system based on virtual reality teaching technology in industrial robot. *Paladyn, Journal of Behavioral Robotics*, 13(1), 45–55.

16. Kaur, M., & Aron, R. (2021). Focalb: Fog computing architecture of load balancing for scientific workflow applications. *Journal of Grid Computing*, 19(4), 40.

17. Kaur, M., & Aron, R. (2022, February). Fog clustering-based architecture for load balancing in scientific workflows. In *Proceedings of International Conference on Computational Intelligence and Data Engineering: ICCIDE 2021* (pp. 213–221). Singapore: Springer Nature Singapore.

18. Tandon, R., & Gupta, P. K. (2019). Optimizing smart parking system by using fog computing. In *Advances in Computing and Data Sciences: Third International Conference, ICACDS 2019, Ghaziabad, India*, April 12–13, 2019, Revised Selected Papers, Part II 3 (pp. 724–737). Springer Singapore.

19. Kaur, M., Aron, R., Wadhwa, H., Tandon, R., Oo, H. N., & Sandhu, R. (2024). A comprehensive exploration of machine learning and IoT applications for transforming water management. In *Innovations in Machine Learning and IoT for Water Management* (pp. 25–50). IGI Global.

20. Bindal, R., Sarangi, P. K., Kaur, G., & Dhiman, G. (2019). An approach for automatic recognition system for Indian vehicles numbers using k-nearest neighbours and decision tree classifier.

21. Reddy, K., Reddy, M. A., Kaur, V., & Kaur, G. (2022). Career guidance system using ensemble learning In Akhila, Kaur, Veerpal, Kaur, and Gagandeep, eds., *Career Guidance System Using Ensemble Learning, Proceedings of the Advancement in Electronics & Communication Engineering* (pp. 33–39) (July 14, 2022).

22. Sarangi, P. K., Sahoo, A. K., Kaur, G., Nayak, S. R., & Bhoi, A. K. (2022). Gurmukhi numerals recognition using ann. In *Cognitive Informatics and Soft Computing: Proceeding of CISC 2021* (pp. 377–386). Singapore: Springer Nature Singapore.

23. Kaur, V., Kaur, G., Dhiman, G., Bindal, R., & Mishra, M. K. (2020). Adaptability of machine learning in cryptography. *Solid State Technology*, 63(4), 2874–2880.

24. Kaur, V., Gupta, K., Baggan, V., & Kaur, G. (2019). Role of cryptographic algorithms in mobile ad hoc network security: an elucidation.

25. Harnal, S., Jain, A., Rathore, A. S., Baggan, V., Kaur, G., & Bala, R. (2023, March). Comparative Approach for Early Diabetes Detection with Machine Learning. In *2023 International Conference on Emerging Smart Computing and Informatics (ESCI)* (pp. 1–6). IEEE.

26. Kaur, V., Kumar, P., Kaur, G., & Kaur, A. (2023, April). Improved Facial Biometric Authentication Using MobileNetV2. In *2023 International Conference on Computational Intelligence and Sustainable Engineering Solutions (CISES)* (pp. 376–380). IEEE.

27. Kaur, G., Kaur, H., & Goyal, S. (2023). Correlation analysis between different parameters to predict cement logistics. *Innovations in Systems and Software Engineering*, 19(1), 117–127.

28. Tandon, R., & Gupta, P. K. (2023). A hybrid security scheme for inter-vehicle communication in content centric vehicular networks. *Wireless Personal Communications*, 129(2), 1083–1096.

29. Wadhwa, H., & Aron, R. (2018, December). Fog computing with the integration of internet of things: Architecture, applications and future directions. In *2018 IEEE Intl Conf on Parallel & Distributed Processing with Applications, Ubiquitous Computing & Communications, Big Data & Cloud Computing, Social Computing & Networking, Sustainable Computing & Communications (ISPA/IUCC/BDCloud/SocialCom/SustainCom)* (pp. 987–994). IEEE.

30. Wadhwa, H., & Aron, R. (2021, December). Fog computing framework for multi constraints based resource utilization. In *2021 2nd International Conference on Computational Methods in Science & Technology (ICCMST)* (pp. 20–23). IEEE.

31. Wadhwa, H., & Aron, R. (2022). A clustering-based optimization of resource utilization in fog computing. In *Proceedings of International Conference on Advanced Computing Applications: ICACA 2021* (pp. 343–353). Springer Singapore.

Chapter 13

Waste management sustainability using cloud and fog computing

Shantanu Bindewari, Anand Singh, and Rajeev Tiwari
IILM University, Greater Noida, India

Priyanka Chawla
NIT Warangal, Warangal, India

13.1 INTRODUCTION

13.1.1 Brief overview of waste management challenges

Waste management poses a multifaceted set of challenges that require innovative solutions for a sustainable future. As urbanization and population growth accelerate, the volume and complexity of waste generated have intensified, placing significant strain on traditional waste management systems. The influx of people into urban areas leads to increased waste generation, overwhelming existing waste infrastructure. War Urbanization exacerbates the need for efficient waste collection, disposal, and recycling mechanisms. Modern societies produce diverse waste streams, including electronic waste, hazardous materials, and non-biodegradable plastics. Effective waste management requires specialized treatment for different waste categories, demanding advanced technologies and processes. Many regions face resource constraints and inadequate funding for waste management infrastructure [1]. Inadequate resources hinder the establishment of efficient waste collection and disposal systems, contributing to environmental degradation. Improper waste disposal leads to environmental pollution, soil contamination, and harm to ecosystems. The degradation of natural habitats and the release of harmful substances into the environment pose long-term threats to biodiversity and human health. Conventional waste collection systems often lack optimization, leading to inefficient routes and increased operational costs [2]. Inefficient collection systems result in delayed waste pickup, increased fuel consumption, and heightened greenhouse gas emissions. Many regions lack adequate infrastructure for recycling, leading to a significant portion of recyclable materials ending up in landfills [3]. Failure to implement effective recycling measures contributes to resource depletion and missed opportunities for sustainable material use. Technological disparities hinder the adoption of advanced waste management solutions in certain regions. The inability to leverage technology impedes the implementation of smart waste management practices and data-driven decision-making. The public's awareness and involvement in waste reduction and recycling initiatives are inadequate. Inadequate public participation hampers the success of waste management initiatives, as behavioral changes are crucial for sustainable waste practices [4].

13.1.2 Importance of sustainable development in waste management

Sustainable development [5] is crucial for the effective and responsible management of waste. It encompasses practices that balance environmental, economic, and social considerations to meet the needs of the present without compromising the ability of future generations to meet their own needs. The importance of sustainable development in waste management can be understood through several key aspects (Figure 13.1).

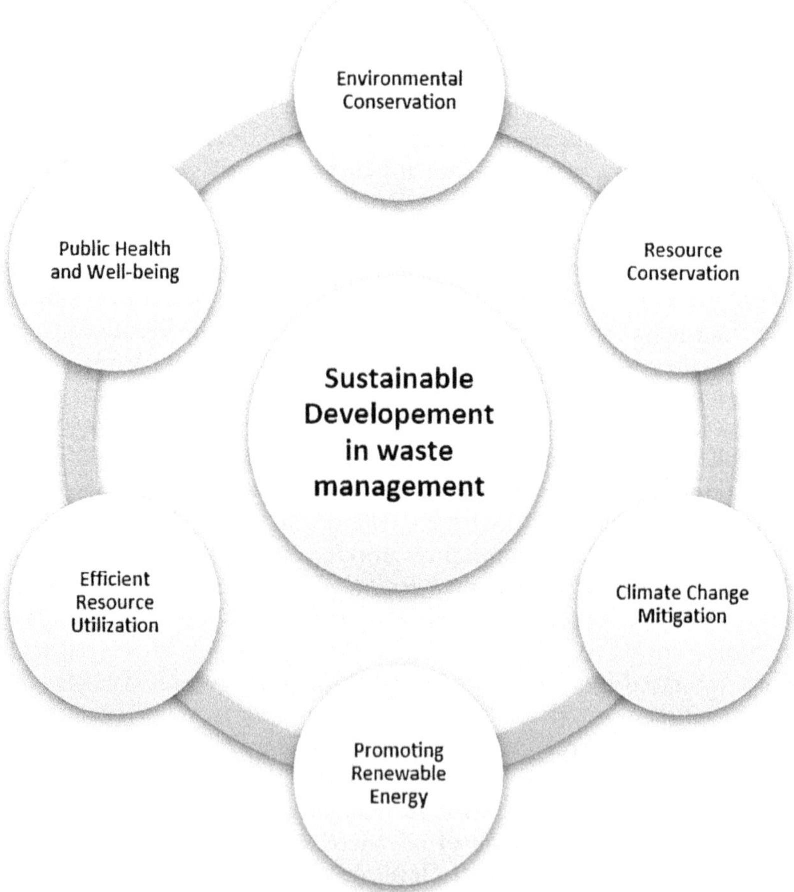

Figure 13.1 Sustainable development in waste management.

13.1.3 Environmental conservation

Preserving Ecosystems: Sustainable waste management minimizes the impact of waste on ecosystems, preventing pollution, soil degradation, and harm to wildlife [6].

Resource Conservation: By prioritizing recycling and waste reduction, sustainable practices help conserve natural resources and reduce the need for raw material extraction.

Climate Change Mitigation: Reducing Greenhouse Gas Emissions: Proper waste management, including the reduction of organic waste in landfills, helps mitigate greenhouse gas emissions such as methane, contributing to climate change mitigation efforts [7].

Promoting Renewable Energy: Sustainable waste practices can include the extraction of energy from waste materials, and supporting the development of renewable energy sources.

Efficient Resource Utilization: Sustainable waste management [5] encourages the adoption of a circular economy, where materials are reused, recycled, and repurposed to minimize waste and maximize resource efficiency. Sustainable practices aim to minimize reliance on landfills, promoting alternative waste treatment methods that are more environmentally friendly [6].

Public Health and Well-being: Sustainable waste management practices minimize the risk of water and air pollution, reducing the potential health hazards associated with improperly managed waste. Clean and well-managed environments contribute to improved quality of life, promoting the well-being of communities.

13.2 INTRODUCTION TO CLOUD AND FOG COMPUTING TECHNOLOGIES

In the ever-evolving landscape of information technology, cloud and fog computing have emerged as transformative paradigms that revolutionize the way data is processed, stored, and managed. These technologies play pivotal roles in addressing the increasing demand for computational power, storage, and real-time processing in diverse applications. This introduction provides a brief overview of cloud and fog computing, highlighting their key characteristics and applications (Figure 13.2) [8].

13.2.1 Cloud computing

Cloud computing [9] refers to the delivery of computing services, including storage, processing power, and applications, over the Internet. These services are hosted and managed by remote servers, allowing users to access and

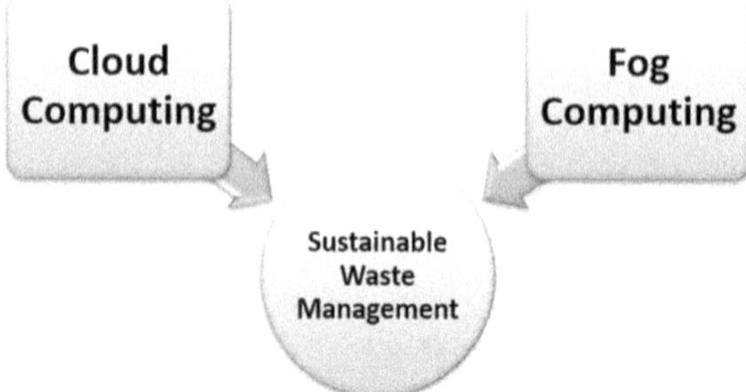

Figure 13.2 Cloud and Fog Computing for sustainable management.

utilize computing resources on a scalable and pay-as-you-go basis. Users can access computing resources as needed, without the need for significant upfront investments. Scalability: Cloud services can easily scale up or down based on demand, providing flexibility and cost efficiency. Resources are shared among multiple users, optimizing utilization and maximizing efficiency. Services are accessible over the internet from anywhere, enabling remote collaboration and global access. Cloud storage solutions provide a secure and scalable way to store and retrieve data. Cloud platforms offer powerful computing capabilities for data processing, analysis, and complex computations. Users can access software applications over the Internet without the need for local installations (Figure 13.3) [8, 9].

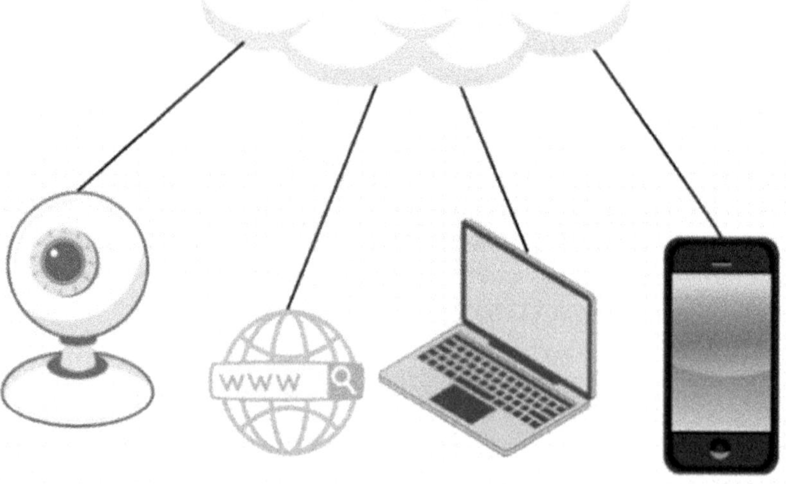

Figure 13.3 Cloud computing.

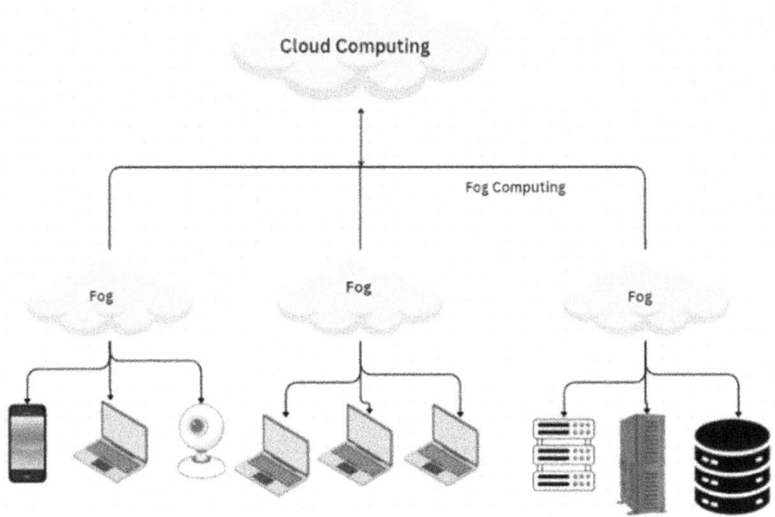

Figure 13.4 Fog computing.

13.2.2 Fog computing

Fog computing extends the capabilities of cloud computing by bringing computation and data storage closer to the edge of the network, closer to the data source or user device. This decentralized approach reduces latency, enhances real-time processing, and improves overall efficiency. Proximity to Edge Devices: Fog computing [10] places computing resources closer to where data is generated, reducing latency and enhancing responsiveness. Resources are distributed across the network, allowing for localized processing and decision-making. Fog computing can scale horizontally by adding computing resources at the edge, supporting the growth of edge devices. Applications in fog computing can provide real-time responses critical for time-sensitive processes. Fog computing is integral in Internet of Things (IoT) environments, processing data from connected devices at the edge [11]. Real-time analytics and decision-making for time-sensitive applications, such as autonomous vehicles and industrial automation [12]. Fog computing supports the development of smart city infrastructure by enabling localized data processing for various applications (Figure 13.4).

13.3 CLOUD COMPUTING IN WASTE MANAGEMENT

13.3.1 Cloud computing in waste management

Cloud computing has been a disruptive force in several industries, including waste management, with numerous advantages that range from increased sustainability to increased efficiency. Waste management organizations can reduce their environmental impact by optimizing resource allocation, streamlining processes, and making data-driven decisions by utilizing cloud-based technology.

The capacity of cloud computing to consolidate and analyze massive volumes of data created during the trash disposal process is one of its main benefits for waste management. Data about waste collection, transportation, and disposal can be gathered in real-time and sent to cloud servers using IoT sensors, GPS tracking systems, and other monitoring equipment. In order to find patterns, improve routes, and forecast maintenance requirements, this data can then be examined using sophisticated analytics tools, which will increase the overall effectiveness of waste management [12] operations.

Additionally, cloud-based platforms make it easier for government organizations, waste collection businesses, recycling facilities, and end users to collaborate and communicate with one another and with each other regarding waste management. Cloud computing encourages more accountability and transparency in the waste management ecosystem by offering a centralized platform for data sharing and coordination, allowing stakeholders to collaborate on shared sustainability objectives.

Furthermore, the use of cutting-edge solutions like garbage sorting technology and smart bins is made possible by cloud computing. These technologies automatically sort and classify various trash kinds using cloud-based analytics, which raises recycling rates and lowers the quantity of waste dumped in landfills. Furthermore, cloud-based systems can make it easier to include sustainable practices [13] and renewable energy sources into waste management procedures, therefore lessening their negative effects on the environment.

All things considered, cloud computing has the ability to completely transform waste management by offering the infrastructure and technologies required to maximize efficiency, cut expenses, and lessen environmental impact. Waste management [8] organizations are increasingly relying on cloud-based solutions as a means of addressing the intricate issues confronting the sector and constructing a more sustainable future, as technology advances (Figure 13.5).

13.3.2 Data collection and monitoring using cloud platforms

Waste management is undergoing a revolution thanks to cloud-based data collection and monitoring systems that improve sustainability initiatives, streamline operations, and provide real-time information. Trash management firms can obtain extensive data on many components of the trash disposal process, such as collection, transportation, and disposal, by integrating IoT sensors [13], GPS tracking systems, and other monitoring equipment.

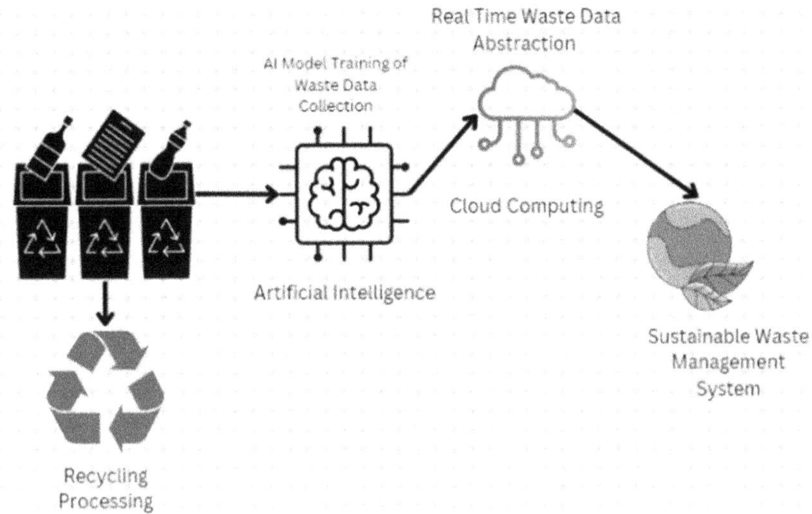

Figure 13.5 Real-time waste management system using cloud computing with artificial intelligence.

These sensors are installed in recycling centers [14], collection trucks, and trash cans. They are always gathering information on temperature, location, fill levels, and other pertinent factors. After that, this data is sent to cloud servers for processing, analysis, and storage. Cloud systems provide a number of benefits for waste management data collection and monitoring, including:

Real-Time Visibility: Stakeholders can remotely monitor the state and performance of assets and processes thanks to cloud-based technologies' real-time visibility into waste management activities. This makes it possible to make preventive decisions and act quickly in the event of problems like equipment failure, bin overflows, or route optimization.

Scalability: Waste management businesses can effortlessly scale up or down their infrastructure in response to shifting demands and needs thanks to the scalability provided by cloud platforms. Expanding data storage capacity or adding more sensors—cloud solutions can handle growth without requiring large upfront investments in hardware or infrastructure.

Data Analytics: To handle and examine the enormous volumes of data produced by IoT [13] sensors and monitoring devices [15], cloud platforms make use of sophisticated analytics tools. Waste management organizations can enhance efficiency and save costs by identifying trends, gaining important insights, and optimizing operations through the application of machine learning algorithms and predictive analytics.

Remote Management and Control: By enabling remote management and control of waste management assets and procedures, cloud-based systems do away with the necessity for manual intervention and on-site oversight. This lowers operating expenses while minimizing human error and improving overall effectiveness and reliability.

Collaboration and Interaction: Cloud platforms make it easier for parties involved in waste management to work together seamlessly by facilitating interaction with other systems and data sources. Cloud-based solutions facilitate interoperability and data interchange to meet shared sustainability goals, whether those goals be data sharing with recycling facilities [14], government agencies, or public stakeholders.

In general, cloud-based platforms for data collection and monitoring provide waste management businesses with the instruments and capacities required to enhance resource efficiency, streamline operations, and propel sustainability programs. Waste management may move toward more intelligent, data-driven processes that improve environmental stewardship and help create a cleaner, more sustainable future by utilizing cloud computing.

13.3.3 Predictive analytics for optimizing waste collection routes

The utilization of complex algorithms, real-time information, and historical data to forecast future trends and patterns is how predictive analytics plays a critical part in waste collection route optimization. Here's how predictive analytics can be utilized for this purpose:

Analysis of Historical Data: Analyzing historical data on garbage generation patterns, collection schedules, route efficiency, and other pertinent variables is the first step in predictive analytics. Waste management firms can pinpoint areas for improvement and create more successful route optimization strategies by analyzing historical patterns and performance indicators.

Integration of Data in Real-Time: Predictive analytics [16] uses real-time data from IoT sensors, GPS tracking systems, and other monitoring devices placed in trash cans and pickup trucks in addition to historical data. Insights regarding fill levels, traffic patterns, weather predictions, and other factors influencing waste collection operations are available from this real-time data.

Route Optimization: Waste management organizations can use predictive models to optimize waste collection routes in a number of ways after they are created. Predictive analytics makes it possible for dynamic route optimization, which enables collection vehicles to modify their routes in real-time in response to unforeseen events or changing traffic conditions.

Demand Forecasting: Businesses can more effectively manage resources and modify collection schedules by using predictive models [17] to predict future demand for garbage collection services in various locations.

Effective Resource Allocation: Waste management firms can minimize fuel consumption, lower vehicle wear and tear, and enhance overall operational efficiency by optimizing routes based on anticipated demand and fill levels.

Service Level Improvement: By achieving service level agreements and customer expectations, predictive analytics may assist in guaranteeing that waste collection [18] services are provided consistently and reliably.

Continuous Improvement: Applying predictive analytics is a continuous process rather than a one-time fix. Waste management firms constantly gather and examine data, improve predictive models, and modify their route optimization tactics in response to input and performance indicators. Routes are kept flexible to changing conditions and optimized over time thanks to this iterative methodology.

Overall, by utilizing data-driven insights and forecasting approaches, predictive analytics [19] helps waste management organizations to optimize collection routes, lower costs, and improve service quality. Waste management operations can improve their efficiency, sustainability, and responsiveness to consumer and environmental demands by utilizing predictive analytics.

13.4 RESOURCE OPTIMIZATION AND CENTRALIZED DECISION-MAKING

The combination of cloud and fog computing technologies can greatly improve resource optimization and centralized decision-making for waste management. Fog computing brings this capability closer to the edge of the network, where data is generated, while cloud computing serves as the foundation for centralized data processing, storage, and management. When combined, they allow for better operational efficiency in waste management systems, timely decision-making, and the effective allocation of resources.

Cloud Computing for Centralized Data Management: The utilization of cloud computing platforms facilitates centralized data management by acting as a repository for massive volumes of data gathered from different sources within the waste management ecosystem, such as GPS devices, IoT sensors, and operational databases. Waste management organizations can obtain a full perspective of their operations through cloud-based data aggregation [20] and storage, which facilitates improved analysis, planning, and decision-making.

Prediction Modeling and Data Analytics: Waste management firms can create prediction models to optimize resource allocation by analyzing past data using cloud-based analytics technologies. Businesses can forecast trash creation patterns, anticipate the need for equipment maintenance, and optimize collection routes to reduce expenses and environmental effects by utilizing machine learning algorithms and statistical methodologies [21].

Real-Time Monitoring and Control via Fog Computing: By moving data processing and analysis closer to the network's edge, or at or close to the location of data generation, fog computing expands the possibilities of cloud computing. Fog computing makes it possible to monitor and operate collection trucks, sorting facilities, and other vital infrastructure in real-time for trash management. Businesses can automate decision-making processes [22], including route modifications based on traffic conditions or bin fill levels, by deploying edge computing devices with sensors and actuators. This allows them to avoid depending exclusively on cloud-based resources.

Dynamic Resource Allocation: Waste management organizations can put into practice dynamic resource allocation plans that instantly adjust to changing circumstances by combining cloud and fog computing technologies. Fog computing devices mounted on collection vehicles, for instance, can study local traffic patterns and modify routes accordingly; in the meantime, fleet management is optimized overall by cloud-based algorithms that take into account demand projections and larger trends.

Centralized Decision-Making and Coordination: By offering a single platform for data exchange and collaboration, cloud computing makes centralized decision-making and coordination easier throughout the waste management ecosystem. Stakeholders can monitor key performance metrics, get real-time insights, and plan actions to optimize resource use and enhance service delivery through cloud-based dashboards and management tools.

Waste management businesses may increase operational resilience, sustainability [14, 21, 22], and efficiency by utilizing the combined powers of cloud and fog computing. Through the use of centralized data management, analytics, real-time monitoring, and dynamic resource allocation, these technologies enable businesses to streamline workflows, make well-informed decisions, and continuously enhance waste management procedures.

13.5 REMOTE MONITORING AND CONTROL THROUGH CLOUD PLATFORMS

Cloud-based remote monitoring and control have several benefits for waste management operations, including proactive decision-making, real-time oversight, and increased efficiency. The following are some ways that cloud platforms help with remote waste management monitoring and control:

13.5.1 Remote data gathering and integration

Within the waste management architecture, cloud platforms act as centralized hubs for gathering and integrating data from diverse sources. This comprises data from IoT sensors placed in garbage cans, trucks used for collection, facilities for sorting, and other crucial locations in the waste management procedure. Waste management organizations can improve monitoring and analysis by gaining a comprehensive perspective of their operations through cloud-based data aggregation.

Real-Time Monitoring: From any location with internet access, cloud solutions allow for real-time monitoring of waste management assets and operations. Through web-based interfaces, managers and operators may monitor key performance indicators including garbage bin fill levels, vehicle whereabouts, and equipment status by using dashboards and analytics tools.

Automated Control and Optimization: By utilizing cutting-edge algorithms and machine learning techniques, cloud platforms provide the automated control [23] and optimization of waste management procedures. Predictive analytics, for instance, can identify trends in the production of waste and instantly adjust collection routes to reduce fuel usage and idle time for vehicles. Additionally, automated sorting systems in recycling facilities can adjust settings based on incoming waste composition data to improve recycling efficiency.

Scalability and Flexibility: Cloud platforms provide scalability and flexibility, enabling waste management firms to adjust their monitoring and control systems in response to shifting operational demands. Without requiring large upfront investments in hardware or infrastructure, cloud-based solutions may readily support growth, whether it's by expanding operations to new locations or adopting new technologies.

13.5.2 Data security and compliance

Cloud computing solutions offer strong security features to safeguard confidential information and guarantee adherence to laws like GDPR and HIPAA. Stakeholders are reassured about the dependability and privacy of remote monitoring and control systems by the use of encryption, access limitations, and frequent security audits.

In conclusion, cloud solutions that enable remote monitoring and control enable waste management firms to maximize efficiency, cut expenses, and improve sustainability by offering real-time information, proactive management tools, and scalability to accommodate changing requirements. Waste management operations can improve their resilience, efficiency, and ability to adapt to changing environmental and regulatory requirements by utilizing cloud-based solutions.

13.6 ENERGY EFFICIENCY IMPROVEMENTS THROUGH CENTRALIZED ANALYSIS

By utilizing cloud and fog computing technologies for centralized analysis, waste management can realize increases in energy efficiency. These technologies can be used in the following ways:

13.6.1 Data collection and aggregation

Within waste management systems, cloud platforms act as centralized repositories for gathering and combining data from many sources. Information from IoT sensors placed in garbage cans, collection trucks, and processing plants is included in this. By processing data closer to the source, fog computing expands this potential and allows for quicker analysis and reaction times.

Real-time energy consumption monitoring is possible thanks to IoT [12] sensors positioned throughout the waste management infrastructure [15]. This involves monitoring the amount of electricity used by transportation vehicles, sorting facilities, and other operational elements. This data can be analyzed using cloud-based analytics to find trends, patterns, and areas that need improvement.

> **Predictive Analytics:** Using previous data and outside variables like trash creation rates and weather, cloud-based predictive analytics [22, 24] models are able to estimate energy use. Waste management organizations can plan maintenance activities during off-peak hours, improve resource allocation, and proactively implement energy-saving measures by forecasting future energy demand.

13.6.2 Route optimization

In order to reduce energy usage, waste collection [24] routes can be optimized using cloud and fog computing. Predictive analytics algorithms are capable of identifying the most energy-efficient routes for collection vehicles by evaluating previous data on waste generation patterns, traffic patterns, and fuel efficiency. On collection trucks, fog computing devices are mounted to enable real-time route adjustments based on the amount of rubbish in the bins and traffic conditions.

> **Monitoring Equipment Efficiency:** Cloud platforms are able to keep an eye on waste management equipment's performance and spot areas where energy efficiency can be increased. For instance, sensors can monitor the performance of shredders, compactors, and other equipment to make sure they are operating as efficiently as possible. In order to increase energy efficiency, cloud-based analytics can identify anomalies and suggest modifications or maintenance.

Integration of Renewable Energy: Cloud platforms can make it easier to incorporate renewable energy sources—like wind turbines or solar panels—into waste management processes. Predictive analytics may optimize the use of renewable energy sources to power waste processing plants and transportation fleets, reducing reliance on fossil fuels and minimizing carbon emissions. This is accomplished by analyzing weather data and energy output estimates.

Continuous Improvement: Energy-efficiency programs in waste management may be continuously monitored and optimized thanks to cloud-based analysis. Waste management firms may monitor the success of energy-saving initiatives, pinpoint areas for more development, and hone their tactics to meet long-term sustainability objectives by gathering and evaluating data over time.

Waste management organizations can find ways to limit environmental effects, cut operating costs, and enhance energy efficiency by utilizing centralized analysis via cloud and fog computing. By enabling data-driven decision-making, predictive analytics, and real-time monitoring, these technologies help waste management operations become more robust and sustainable in the face of changing obstacles.

13.7 COST REDUCTION AND SCALABILITY IN WASTE MANAGEMENT OPERATIONS

Noisy and hazy significant prospects for cost savings and scalability in waste management operations are provided by computing technology. Waste management organizations can decrease costs associated with on-premises equipment by streamlining their operations and utilizing cloud platforms as central hubs for data processing, analysis, and storage. Through real-time data analytics and effective resource allocation and optimization made possible by cloud computing, businesses may detect inefficiencies and take proactive steps to save costs.

Furthermore, fog computing brings cloud computing's capabilities to the network's edge, allowing for real-time asset and process monitoring and control in waste management. Through the reduction of latency and bandwidth usage, this distributed strategy allows for faster reaction times and more effective resource utilization. Furthermore, scalability is provided by cloud and fog computing, which enables waste management operations to adjust their operations in response to fluctuating demands without having to make large upfront expenditures in hardware or equipment. Waste management firms may adjust to changes in waste generation rates, seasonal fluctuations, and other factors with this flexibility, which guarantees cost-effectiveness and optimal resource utilization. In general, waste management operations can realize cost savings and scalability while enhancing efficiency and sustainability by utilizing the power of cloud and fog computing.

13.7.1 Cost savings by minimizing reliance on centralized cloud resources

By strategically deploying fog computing, waste management operations can reduce their dependency on centralized cloud resources and save costs. By bringing computational power closer to the data source, fog computing lessens the requirement for continuous data transfer to and from centralized cloud servers. Here's how this strategy can result in financial savings:

Lower Data Transmission Costs: Large amounts of data produced by sensors in garbage receptacles, collection trucks, and processing centers are transferred to centralized cloud servers in traditional cloud-based systems so they can be analyzed. Significant data transmission costs may result from this, particularly in situations when bandwidth is expensive or scarce.

Reduced Infrastructure Costs: By spreading processing resources closer to the point of data generation, fog computing eliminates the requirement for a large centralized cloud infrastructure. Deployed across the waste management infrastructure, edge devices and gateways may handle data processing, analytics, and decision-making locally, eliminating the need for centralized cloud servers. The decentralization of computer resources has the potential to reduce infrastructure expenses related to processing, data storage, and cloud service subscriptions.

Enhanced Operational Efficiency: By permitting real-time monitoring, control, and decision-making at the network's edge, fog computing improves operational efficiency in waste management. Local data processing minimizes operational overhead and downtime by enabling quicker response times and more effective resource allocation. By optimizing route planning, equipment utilization, and maintenance scheduling, waste management companies can achieve cost savings through improved efficiency and productivity.

Scalability and Flexibility: The scalability and flexibility provided by fog computing enable waste management organizations to adjust their computing infrastructure in response to changing requirements. It is simple to deploy or move edge devices and gateways to adapt to changes in waste generation patterns, population density, or operating requirements. This scalability guarantees that computer resources are utilized efficiently, eliminating excessive expenditures associated with over-provisioning centralized cloud resources.

Improved Dependability and Resilience: Fog computing improves waste management operations' dependability and resilience by lowering dependency on centralized cloud resources. Critical operations, such as route optimization and equipment maintenance, can continue even in the case of network failures or connectivity problems thanks to local data processing. This improved resilience minimizes the risk of service disruptions and associated costs, ensuring uninterrupted waste management operations.

Overall, by minimizing reliance on centralized cloud resources and leveraging the capabilities of fog computing, waste management operations can achieve significant cost savings while enhancing efficiency, reliability, and scalability. This distributed computing approach optimizes resource utilization, reduces operational overhead, and ensures continuous operation, ultimately driving down the overall cost of waste management.

13.8 FOG COMPUTING IN WASTE MANAGEMENT

By bringing computing power closer to the point of data collection, fog computing is gradually transforming waste management by facilitating real-time decision-making and maximizing operational effectiveness. Fog computing in waste management refers to the placement of computer resources, such as gateways and edge devices, at different locations along the waste management infrastructure, such as processing facilities, collection trucks, and waste bins. Here's how waste management is changing as a result of fog computing:

13.8.1 Real-time monitoring and control

By processing data locally at the network's edge, fog computing makes it possible to monitor and control waste management assets and procedures in real-time. Waste containers and collection trucks may gather information on fill levels, temperature, humidity, and other factors thanks to sensors built inside them.

Decreased Latency: Fog computing eliminates dependency on centralized cloud servers and reduces latency by processing data locally. This is especially important for trash management, where prompt judgments are needed to deal with problems like overflowing bins, broken equipment, or optimized routes. Faster reaction times are ensured by lower latency, which helps waste management organizations increase customer satisfaction and operational effectiveness.

Bandwidth Optimization: Fog computing helps to optimize bandwidth utilization and lessen network congestion by reducing the need to transfer massive amounts of data to centralized cloud servers for analysis. Instead, bandwidth is conserved and data transmission expenses are decreased by sending only pertinent data or insights to the cloud for additional analysis or storage.

Fog computing enables offline operation and local data processing in settings with sporadic or restricted connectivity, such as underground facilities or rural places. Reliability and resilience are increased as a result of the continuous monitoring and control of waste management activities even in the absence of connectivity.

13.8.2 Edge analytics

Edge analytics is the process of locally analyzing data at the network's edge before sending it to centralized cloud servers. Fog computing makes this possible. This enables proactive decision-making and operational process improvement by waste management organizations without depending exclusively on cloud-based analytics. It does this by enabling them to extract valuable insights from data in almost real-time.

> **Scalability and Flexibility:** The scalability and flexibility provided by fog computing enable waste management organizations to adjust their computing infrastructure in response to changing requirements. In order to ensure optimal performance and resource usage, edge devices and gateways can be readily deployed or relocated to match fluctuations in waste generation patterns, population density, or operational demands.

Fog computing allows for real-time monitoring, analysis, and control at the network's edge, which improves waste management operations' overall efficiency, dependability, and responsiveness. Fog computing makes trash management smarter, more agile, and more sustainable by combining dispersed edge computing capabilities with centralized cloud computing.

13.9 INTEGRATION OF CLOUD AND FOG COMPUTING

By fusing the benefits of dispersed edge computing with centralized cloud resources, cloud and fog computing integration provides a potent way to optimize waste management operations. The following are some practical ways to use this integration in trash management:

13.9.1 Centralized data management in the cloud

In waste management, cloud computing acts as a centralized location for data administration, storage, and analysis. For consolidation and processing, data from IoT sensors placed in trash cans, collection trucks, and processing plants is sent to the cloud. Stakeholders have access to a comprehensive picture of waste management activities via cloud-based technologies. Real-time insights into variables like fill levels, collection routes, and equipment performance are made possible by this complete picture. Stakeholders can optimize resource allocation, improve operational efficiency, and advance sustainability initiatives in waste management by utilizing cloud-based analytics tools to make data-driven decisions. Thus, cloud-based centralized data management enables ongoing waste management practice improvement and supports well-informed decision-making.

13.9.2 Real-time monitoring and control with fog computing

By processing and analyzing data closer to the data source—such as trash cans or collection trucks—within the waste management infrastructure, fog computing improves cloud capabilities. The system's edge devices and gateways carry out local data processing and analytics, allowing for real-time waste management operation monitoring and control. By processing data locally, this decentralized strategy reduces latency and conserves bandwidth, enabling faster response times and maximizing resource utilization. Waste management systems can benefit from increased responsiveness, faster decision-making, and more efficient operations by utilizing fog computing, which will ultimately result in increased efficacy and cost savings.

13.9.3 Dynamic resource allocation and optimization

Waste management resource allocation and optimization are revolutionized by the combination of cloud and fog computing. Optimized resource allocation is made possible by cloud-based analytics tools that process historical data to create forecast models. At the network's edge, fog computing devices carry out real-time monitoring and control functions to guarantee prompt reaction to shifting circumstances. In waste management operations, this combination of distributed decision-making and centralized analysis guarantees effective resource usage. Waste management organizations may maximize operating efficiency while minimizing costs and environmental effects by dynamically allocating resources based on real-time data and predictive insights. Whether optimizing collection routes, regulating equipment utilization, or scheduling maintenance activities, the combined power of cloud and fog computing enables agile and informed decision-making, eventually promoting sustainable practices and boosting overall performance in waste management.

> Scalability and Flexibility: Waste management operations can now successfully adapt to shifting needs thanks to the scalability and flexibility offered by cloud and fog computing. Cloud solutions provide seamless support for larger volumes of data and computing operations. Because of their scalability, waste management systems are guaranteed to be able to adapt to changes in population density, trash generation rates, and operational needs without compromising their dependability or performance. Furthermore, fog computing equipment is easily deployable or movable to locations that need more processing power or monitoring. Because of this adaptability, waste management organizations can dynamically modify their computing infrastructure to meet changing needs while maintaining optimal operations and resource allocation. By embracing the scalability and flexibility of cloud and

fog computing, waste management operations can successfully satisfy current demands while also being prepared for future challenges and growth opportunities.

Enhanced Reliability and Resilience: The resilience and dependability of waste management [25] operations are strengthened by the incorporation of cloud and fog computing. Redundancy and failover measures are built into cloud-based platforms to provide uninterrupted operation even in the event of hardware malfunctions or network outages. Fog computing devices, on the other hand, support local data processing and decision-making by operating independently in the event of network failures or connectivity problems. By reducing the possibility of service interruptions and ensuring that vital operations continue without interruption, this dual-layered strategy improves the resilience of the system. Consequently, waste management companies can continue to provide uninterrupted services while satisfying legal and customer demands. The amalgamation of cloud and fog computing results in greater operational stability, customer satisfaction, and overall efficacy in waste management.

All things considered, the combination of distributed edge computing's real-time monitoring and control capabilities with the scalability and flexibility of centralized cloud resources provides a complete solution for optimizing waste management [25] operations through cloud and fog computing integration. Waste management businesses may increase operational efficiency, cut costs, and promote environmental sustainability by utilizing the advantages of both paradigms.

13.10 CASE STUDIES AND EXAMPLES

Real-world applications of cloud computing in waste management A wealth of practical applications that change waste management [25] techniques and drive efficiency, sustainability, and cost savings are available when cloud computing and fog computing are combined. Waste management firms can collect and handle data from a variety of sources, such as IoT sensors integrated into waste bins, pickup vehicles, and processing facilities, using cloud-based systems. By enabling local data processing at the network's edge, fog computing expands this capability and makes it easier to monitor and make decisions in real-time. Real-time monitoring and optimization is one useful application. Fog computing devices provide quick edge responses to ensure effective waste collection and transportation, while cloud-based analytics tools analyze data to optimize waste collection routes based on variables like fill levels and traffic conditions.

Furthermore, the use of intelligent trash cans with sensors to track temperature, pollution, and fill levels is made possible by cloud computing.

By processing sensor data locally, fog computing enhances this and enables immediate actions, such as scheduling collection pickups or rerouting vehicles to bins with higher fill levels. This clever strategy reduces gasoline usage, maximizes waste collection routes, and improves operational effectiveness. Furthermore, cloud-based predictive analytics, which predict equipment breakdowns by analyzing past data, make predictive maintenance easier. Fog computing devices perform local monitoring and analysis on waste management equipment. They identify irregularities and provide real-time predictions of probable failures. This proactive strategy lowers maintenance costs, increases equipment lifespan, and minimizes downtime.

Moreover, cloud computing makes it possible to remotely manage and control waste management assets and procedures, giving stakeholders the ability to keep an eye on things and make informed decisions from any location. This capacity is increased by fog computing, which offers local control and decision-making at the edge of the network, allowing for real-time modifications to equipment settings, sorting procedures, and collection routes. This improves waste management [25] overall efficacy, reactivity, and operational efficiency. In general, the amalgamation of cloud and fog computing enables waste management enterprises to enhance their operations, curtail expenses, and alleviate ecological consequences, propelling sustainable and effective waste management methodologies in the contemporary epoch.

13.10.1 Success stories of fog computing implementation in waste disposal

The garbage management industry in Barcelona, Spain, is one noteworthy example of a fog computing application that has been successful in utilizing cloud and fog for garbage disposal. The waste management firm optimized garbage collection and disposal procedures throughout the city by implementing a comprehensive system that utilized cloud and fog computing in partnership with technological partners. IoT sensors were installed in garbage cans and pickup trucks to gather data in real-time on temperature, fill levels, and other pertinent variables. Cloud-based solutions receive this data in order to store, analyze, and make decisions centrally. To facilitate local data processing and real-time monitoring, fog computing devices were also mounted on pickup trucks and at trash processing plants.

The waste management company saw notable increases in cost savings, environmental sustainability, and operational efficiency as a result of this integrated approach. The company reduced fuel usage and idle time in vehicles by optimizing waste collection routes based on real-time data, facilitated by the use of cloud-based analytics tools. Fog computing devices made it possible to react quickly to changing circumstances. For example, they may have started repair work based on equipment performance indicators or rerouted collection vehicles to locations with higher fill levels. These changes minimized the negative effects on the environment and reduced operating expenses by streamlining the waste collection procedure.

Additionally, the waste management company was able to identify and fix equipment issues before they happened because of the use of fog computing in predictive maintenance procedures. The organization extended the lifespan of waste management equipment and minimized downtime by proactively scheduling repairs and predicting maintenance needs through the utilization of machine learning algorithms and historical data analysis. This proactive strategy increased operational reliability and customer satisfaction while lowering maintenance costs.

All things considered, Barcelona's effective use of cloud and fog computing for waste disposal shows how revolutionary these technologies may be for streamlining waste management processes. The waste management company established the standard for future advancements in the sector by achieving increased efficiency, cost savings, and environmental sustainability by merging dispersed edge computing capabilities with centralized data management.

13.10.2 Examples of organizations achieving sustainability through integration

Through the use of cloud and fog computing in waste management, several firms have effectively achieved sustainability by putting creative solutions into place to maximize operational efficiency and reduce environmental impact. Waste Management, Inc., one of the top waste management firms in the US, is one such instance. To improve its garbage collection, sorting, and recycling procedures [14], the organization has integrated fog computing and cloud-based platforms.

Waste Management, Inc. collects real-time data on equipment performance, route efficiency, and fill levels using IoT sensors installed in waste bins and collection vehicles. Cloud-based platforms receive this data in order to conduct centralized analysis and make decisions. Furthermore, local data processing and monitoring are made possible by fog computing devices installed at waste processing facilities, enabling quick reactions to changing conditions.

Waste Management, Inc. has significantly improved sustainability measures, such as decreased carbon emissions, greater recycling rates [14], and decreased landfill waste, by using this integrated approach. The organization has minimized fuel consumption and operating expenses while optimizing resource recovery and recycling efficiency through the improvement of sorting procedures and waste collection routes.

Veolia, a multinational provider of environmental services with a focus on resource recovery and waste management, is another illustration. Veolia has improved its waste management operations, especially in metropolitan areas, by utilizing cloud and fog computing technology.

Veolia gathers and analyzes data from IoT sensors placed in trash cans, collection trucks, and recycling facilities using cloud-based systems. Utilizing this data will lessen the environmental impact, increase operational effectiveness, and optimize garbage collection routes. On collection trucks, fog computing devices allow for real-time monitoring and control. This enables dynamic route changes and prompt reaction to problems like bin overflow or equipment failures.

Veolia has achieved notable sustainability benefits, such as decreased greenhouse gas emissions, increased resource recovery, and enhanced environmental stewardship, through its integrated approach to waste management. Veolia has proven its dedication to sustainability by utilizing cloud and fog computing to provide communities worldwide with economical and effective waste management solutions.

13.11 CHALLENGES AND CONSIDERATIONS

Although combining cloud and fog computing technologies can greatly improve waste management, there are a number of issues and concerns that must be taken into account:

13.11.1 Data security and privacy

Tight data security and privacy protections are required in waste management systems because they collect and analyze sensitive data. To ensure data integrity and confidentiality, cloud and fog computing systems need to have strong security measures like encryption, access controls, and secure communication protocols. While access restrictions limit data access to authorized workers exclusively, encryption guarantees that data remains unreadable to unauthorized users. TLS and HTTPS are examples of secure communication protocols that further protect data while it is being transmitted. Cloud and fog computing environments can significantly reduce the danger of unwanted access or breaches by putting these security measures in place, guaranteeing the security and privacy of sensitive data in waste management systems.

13.11.2 Network connectivity and reliability

The efficiency of fog computing depends on network connectivity to enable smooth data transfer between cloud servers and edge devices. However, maintaining dependable communication lines presents a big challenge in isolated or underdeveloped locations with no connectivity. Sufficient infrastructure is essential for maintaining system functionality and ongoing data transfer. This includes strong network coverage and a fast internet connection. Furthermore, putting redundancy measures in place—like backup communication routes or several network providers—can improve dependability and resilience when connectivity problems arise. Organizations can employ fog computing to optimize operations in various environments, such as remote or underdeveloped areas, and overcome network connectivity limits by solving these difficulties and investing in infrastructure enhancements.

13.11.3 Data volume and processing

Large amounts of data are generated by waste management systems from a variety of sources, including GPS units, IoT sensors, and operational databases. In order to effectively manage and process this data and provide real-time insights for well-informed decision-making, cloud platforms are essential. For cloud systems to manage these massive datasets efficiently and provide fast analysis and reaction, they must be scalable and have sufficient processing power. Furthermore, for effective local data processing and analysis, fog computing devices placed at the edge of the network need to have enough processing power and storage. Waste management companies can use cloud and fog computing technologies to optimize operations, boost efficiency, and make data-driven decisions to efficiently handle difficulties by addressing these data volume and processing requirements.

13.11.4 Interoperability and integration

Interoperability issues arise when cloud and fog computing technologies are integrated with the infrastructure and waste management systems that are currently in place. Enabling easy data interchange between cloud and fog computing environments and guaranteeing compatibility between various hardware and software components are necessary for achieving seamless integration. To enable communication and interoperability among various systems and devices, this calls for standardizing protocols and interfaces. Furthermore, in order to find and fix possible compatibility problems early in the integration process, compatibility testing and validation procedures are crucial. Waste management companies may fully utilize cloud and fog computing technologies to optimize operations, boost efficiency, and spur innovation [26] in waste management methods by successfully resolving interoperability issues.

13.11.5 Cost and resource allocation

Significant costs are associated with implementing cloud and fog computing infrastructure in waste management. These costs include upfront expenditures for software licenses, hardware procurement, and installation, as well as continuous operating costs for support and maintenance. To determine if implementing these technologies will be financially feasible in the long run, waste management organizations need to perform extensive cost-benefit assessments. When allocating resources for implementation, factors including scalability, performance requirements, and expected returns on investment should be taken into account. Companies should also look into ways to reduce costs, such as adopting energy-efficient gear or taking advantage of cloud service providers' pricing structures. Waste management firms may make sure that their investments in cloud and fog computing technologies deliver maximum value, promoting operational efficiency and sustainability in waste management practices, by carefully reviewing costs and resource allocation.

13.11.6 Regulatory compliance

Tight rules and compliance requirements pertaining to data management, environmental protection, and health and safety must be followed by waste management operations. Organizations must make sure that cloud and fog computing solutions are implemented in accordance with all applicable industry standards and regulatory frameworks. This includes adhering to environmental laws regulating trash disposal and recycling procedures as well as data privacy laws like GDPR and HIPAA. Strong security measures must be included in cloud and fog computing solutions to protect sensitive data and reduce the possibility of breaches or illegal access. Organizations should also set up procedures for tracking and reporting regulatory compliance, as well as preserve openness in their operations. Waste management companies can reduce legal risks and show their dedication to moral and responsible business practices by giving regulatory compliance top priority when designing and implementing cloud and fog computing solutions.

Using cloud and fog computing to manage trash presents a number of issues and considerations that must be carefully planned for; stakeholder participation is essential; and solutions must be continuously monitored and evaluated. Companies that handle waste management need to carry out thorough risk assessments and create strong plans to deal with concerns related to data security, network connectivity, interoperability, cost distribution, and regulatory compliance. To guarantee compliance with legal requirements and industry best practices, this entails interacting with technology partners, regulatory agencies, and business specialists. Waste management organizations may optimize the potential of cloud and fog computing technologies [26] to improve operational efficiency, encourage sustainability, and strengthen process resilience by taking proactive measures to solve these issues. Organizations can achieve cost-effectiveness, environmental stewardship, and operational excellence by continuously evaluating and refining their waste management systems, identifying areas for improvement, and stimulating continuous innovation.

13.12 FUTURE TRENDS AND INNOVATIONS

The combination of cloud and fog computing technologies is expected to bring about substantial modifications to the waste management sustainability environment in the future. The development of smart garbage bins, which have sensors and IoT devices attached, is one notable achievement. With the help of these smart bins, waste levels can be monitored in real-time, which optimizes waste collection routes for increased operational effectiveness. In this situation, cloud computing is essential since it handles the processing and analysis of the massive volumes of data produced by these sensors, giving waste management authorities useful insights.

The use of predictive analytics models, which make use of cloud resources to estimate waste generation trends, is another trend worth noting. These models improve waste management planning precision by examining past data and external influences. Additionally, fog computing helps by enabling data processing at the edge, which lessens the need to send all data to centralized cloud servers. Fuel consumption and environmental effects are reduced when waste collection routes are optimized on the spot thanks to this localized decision-making capacity.

In order to guarantee openness and traceability throughout the waste management supply chain [16, 23], blockchain technology [27] becomes essential. Blockchain makes garbage disposal and recycling operations easier to track by producing safe, unchangeable records. The foundation for organizing and keeping these blockchain data is cloud computing, which guarantees dependability, security, and accessibility.

The use of Artificial Intelligence (AI) and machine learning algorithms for waste sorting procedures in recycling plants puts automation front and center. The computational capacity needed to train and run these complex machine learning models is made available via cloud platforms, which enables garbage sorting procedures to become more precise and efficient.

Another aspect of future waste management techniques is to include individuals in waste reduction campaigns; mobile applications are anticipated to be effective instruments for this purpose. The data produced by these apps is hosted and managed by cloud services, which provide insights into user behavior and the efficacy of campaigns. The scalability and accessibility provided by cloud technology thereby increase public knowledge and involvement.

Cloud-based platforms will also help projects that promote material reuse and recycling, or the circular economy. These platforms promote a sustainable approach to waste management by facilitating the sharing of resources and information between businesses and consumers. Essentially, waste management procedures are predicted to undergo a radical transformation as a result of the cooperative synergy of cloud and fog computing technologies, bringing in a new era of creativity, effectiveness, and increased sustainability.

13.13 CONCLUSION

To sum up, the amalgamation of cloud and fog computing technologies holds the potential to transform waste management methodologies, providing a comprehensive strategy toward sustainability. These innovations aim to considerably increase operational efficiency, from the use of predictive analytics models backed by cloud resources to enhance trash generation predictions to the deployment of smart garbage bins with IoT sensors for real-time monitoring and optimal collection routes. The waste management supply chain [23, 28] is made transparent and traceable by blockchain technology [27], which is bolstered by cloud computing's storage capabilities. Recycling operations are further streamlined by the automation of waste sorting procedures with AI and machine learning, made possible by cloud platforms. Cloud-hosted mobile applications encourage public awareness and engagement in waste reduction projects by involving citizens. Platforms for the circular economy that are backed by cloud infrastructure encourage resource sharing and environmentally friendly behavior. As these trends develop, the synergistic relationship between fog and cloud computing will likely spur future technical advancements as well as a more ecologically friendly and sustainable approach to waste management.

The convergence of fog computing and cloud computing technologies offers a comprehensive and revolutionary paradigm shift for sustainable waste management in the future. The emergence of intelligent trash cans equipped with IoT sensors is evidence of the possibility for real-time tracking and route optimization, all handled and processed by cloud computing's computational power. With the use of cloud resources and predictive analytics models, waste management is being taken to a new level of strategic planning and accuracy. Blockchain technology plays a pivotal role in guaranteeing transparency [27] and traceability in the complex network of waste management supply chains [28]. It also utilizes cloud computing to enable the safe and convenient storage of this vital information. The processing power offered by cloud platforms enables the automation of garbage sorting procedures, which are supported by advanced AI and machine learning algorithms. Scalability and data management insights provided by cloud-hosted mobile applications make involving individuals in the fight against waste reduction a dynamic undertaking. Cloud-based platforms provide a fertile environment for circular economy efforts that encourage resource exchange and sustainable practices. The symbiotic relationship between cloud and fog computing signals a new era of operational efficiency and demonstrates a strong commitment to more environmentally responsible and sustainable waste management in the wake of these technological breakthroughs.

REFERENCES

[1] Abdel-Shafy, H. I., & Mansour, M. S. M. (2018). Solid waste issue: Sources, composition, disposal, recycling, and valorization. *Egyptian Journal of Petroleum*, 27(4), 1275–1290.

[2] Abubakar, I. R., Maniruzzaman, K. M., Dano, U. L., AlShihri, F. S., AlShammari, M. S., Ahmed, S. M. S., Al-Gehlani, W. A. G., & Alrawaf, T. I. (2022). Environmental Sustainability Impacts of Solid Waste Management Practices in the Global South. *International Journal of Environmental Research and Public Health*, 19(19), 12717.

[3] Siddiqua, A., Hahladakis, J.N. & Al-Attiya, W.A.K.A. (2022). An overview of the environmental pollution and health effects associated with waste landfilling and open dumping. *Environ Sci Pollut Res*, 29, 58514–58536.

[4] Rodríguez-Espíndola, O., Cuevas-Romo, A., Chowdhury, S., Díaz-Acevedo, N., Albores, P., Despoudi, S., Malesios, C., & Dey, P. (2022). The role of circular economy principles and sustainable-oriented innovation to enhance social, economic and environmental performance: Evidence from Mexican SMEs. *International Journal of Production Economics*, 248, 108495.

[5] Sustainable Development. (2012). *In Lees' Loss Prevention in the Process Industries* (pp. 2507–2521). Elsevier.

[6] Abubakar, I. R., Maniruzzaman, K. M., Dano, U. L., AlShihri, F. S., AlShammari, M. S., Ahmed, S. M. S., Al-Gehlani, W. A. G., & Alrawaf, T. I. (2022). Environmental Sustainability Impacts of Solid Waste Management Practices in the Global South. *International Journal of Environmental Research and Public Health*, 19(19), 12717.

[7] Budihardjo, M. A., Humaira, N. G., Ramadan, B. S., Wahyuningrum, I. F. S., & Huboyo, H. S. (2023). Strategies to reduce greenhouse gas emissions from municipal solid waste management in Indonesia: The case of Semarang City. *Alexandria Engineering Journal*, 69, 771–783.

[8] Angel, N. A., Ravindran, D., Vincent, P. M. D. R., Srinivasan, K., & Hu, Y.-C. (2021). Recent advances in evolving computing paradigms: Cloud, edge, and fog technologies. *Sensors*, 22(1), 196. https://doi.org/10.3390/s22010196

[9] Naghavi, M. (2012). Cloud Computing As An Innovation In GIS & SDI: methodologies, services, issues and deployment techniques. *Journal of Geographic Information System*, 04(06), 597–607.

[10] Das, R., & Inuwa, M. M. (2023). A review on fog computing: Issues, characteristics, challenges, and potential applications. *Telematics and Informatics Reports*, 10, 100049.

[11] Alwakeel, A. M. (2021). An overview of fog computing and edge computing security and privacy issues. *Sensors*, 21(24), 8226.

[12] Bathla, G., Bhadane, K., Singh, R. K., Kumar, R., Aluvalu, R., Krishnamurthi, R., Kumar, A., Thakur, R. N., & Basheer, S. (2022). Autonomous vehicles and intelligent automation: Applications, challenges, and opportunities. In M. P. Kumar Reddy (Ed.), *Mobile Information Systems* (Vol. 2022, pp. 1–36). Hindawi Limited.

[13] Fang, B., Yu, J., Chen, Z., Osman, A. I., Farghali, M., Ihara, I., Hamza, E. H., Rooney, D. W., & Yap, P.-S. (2023). Artificial intelligence for waste management in smart cities: a review. *Environmental Chemistry Letters*, 21(4), 1959–1989.

[14] Abis M, Bruno M, Kuchta K, Simon FG, Grönholm R, Hoppe M, Fiore S (2020) Assessment of the synergy between recycling and thermal treatments in municipal solid waste management in europe. *Energies*, 13(23), 6412.

[15] Farjana, M., Fahad, A. B., Alam, S. E., & Islam, Md. M. (2023). An IoT- and Cloud-Based E-Waste Management System for Resource Reclamation with a Data-Driven Decision-Making Process. *IoT*, 4(3), 202–220.

[16] Seyedan, M., Mafakheri, F. (2020). Predictive big data analytics for supply chain demand forecasting: methods, applications, and research opportunities. *J Big Data*, 7, 53.

[17] Srivastava, S. (2023, June 12). *How Predictive Analytics Can Streamline Operations and Close Business Gaps*. Appinventiv. https://appinventiv.com/blog/predictive-analytics-for-growing-business/

[18] *Route Optimization of Vehicle Collecting Municipal Solid Waste: A Review.* (2023, May). https://www.ijres.org/

[19] Fang, B., Yu, J., Chen, Z. et al. (2023). Artificial intelligence for waste management in smart cities: a review. *Environ Chem Lett*, 21, 1959–1989.

[20] Yousefi, S., Karimipour, H., & Derakhshan, F. (2021). Data aggregation mechanisms on the Internet of Things: a systematic literature review. *Internet of Things*, 15, 100427.

[21] Xia W, Jiang Y, Chen X, Zhao R. (2022). Application of machine learning algorithms in municipal solid waste management: A mini review. *Waste Management & Research*, 40(6), 609–624.

[22] Dhingra, S., Madda, R. B., Patan, R., Jiao, P., Barri, K., & Alavi, A. H. (2021). Internet of things-based fog and cloud computing technology for smart traffic monitoring. *Internet of Things*, 14, 100175.

[23] Abdallah T, Diabat A, Simchi-Levi D (2012) Sustainable supply chain design: a closed-loop formulation and sensitivity analysis. *Prod Plan Control*, 23(2–3), 120–133.

[24] Ahmad, I.S., & Kim, D.H. (2020). Quantum GIS based descriptive and predictive data analysis for effective planning of waste management. *IEEE Access*, 8, 46193–46205.

[25] World Bank Group. (2009). *The Solid Waste Management Sector*; Department of Economic Affairs: New Delhi, India.

[26] Gao H, Ding XH, Wu S (2020). Exploring the domain of open innovation: bibliometric and content analyses. *J Clean Prod*, 275, 122580.

[27] Jiang, P., Zhang, L., You, S., Fan, Y. V., Tan, R. R., Klemeš, J. J., & You, F. (2023). Blockchain technology applications in waste management: Overview, challenges and opportunities. *Journal of Cleaner Production*, 421, 138466.

[28] Dyczkowska J, Bulhakova Y, Łukaszczyk Z, Maryniak A (2020) Waste management as an element of the creation of a closed loop of supply chains on the example of mining and extractive industry. *Manag Syst Prod Eng*, 28(1), 60–69.

Chapter 14

Fog computing

Advancing Sustainable Development Goals (SDGs)

Premkumar Chithaluru
Mahatma Gandhi Institute of Technology, Hyderabad, India

Pallati Narsimhulu and N Sudhakar Yadav
Chaitanya Bharathi Institute of Technology, Hyderabad, India

Priyanka Chawla
NIT Warangal, Warangal, India

Rajeev Tiwari
IILM University, Greater Noida, India

14.1 INTRODUCTION

Fog computing, a decentralized computing paradigm that extends cloud computing capabilities to the edge of the network, has the potential to contribute to several SDGs as outlined by the United Nations. Here's how fog computing can be associated with specific SDGs [1]:

- **SDG 3: Good Health and Well-Being:**
  - **Healthcare Data Processing:** Fog computing can process healthcare data from wearable devices and medical sensors at the edge, allowing for real-time monitoring of patients' health. This contributes to improved healthcare services and early disease detection.
- **SDG 4: Quality Education:**
  - **Edge-Based Educational Content:** Fog computing can deliver educational content and applications to remote and underserved areas, enabling access to quality education, especially in regions with limited connectivity.
- **SDG 7: Affordable and Clean Energy:**
  - **Energy Efficiency:** Fog computing can optimize energy consumption by offloading processing tasks to edge devices, reducing the need for centralized data centers. This contributes to energy efficiency and sustainability.

DOI: 10.1201/9781003494430-14

- **SDG 9: Industry, Innovation, and Infrastructure:**
  - **Edge Computing Infrastructure:** Fog computing supports the deployment of edge computing infrastructure, which is essential for enabling Internet of Things (IoT) applications and improving industrial processes, infrastructure management, and logistics.
- **SDG 11: Sustainable Cities and Communities:**
  - **Smart City Applications:** Fog computing is a key enabler of smart city solutions, such as traffic management, waste management, and energy conservation, which contribute to creating more sustainable and efficient urban environments.
- **SDG 13: Climate Action:**
  - **Environmental Monitoring:** Fog computing can support real-time environmental monitoring and data analysis, helping to address climate-related challenges by providing valuable insights and early warnings.
- **SDG 16: Peace, Justice, and Strong Institutions:**
  - **Security and Surveillance:** Fog computing can enhance security and surveillance systems with real-time analytics, contributing to public safety and law enforcement.
- **SDG 17: Partnerships for the Goals:**
  - **Collaborative IoT Ecosystem:** Fog computing encourages collaboration between various stakeholders, including governments, businesses, and communities, to develop and implement IoT solutions for sustainable development.
- **Cross-Cutting Impact:**
  - **Reduced Latency:** Fog computing reduces data transfer latency, making it suitable for applications like disaster response, which requires rapid decision-making and communication.
  - **Data Privacy:** By processing data locally, fog computing can enhance data privacy, aligning with the principles of responsible data handling and GDPR compliance.

It's important to note that fog computing's association with SDGs depends on its effective implementation and integration into various domains and industries. Additionally, the use of fog computing technologies should prioritize environmental sustainability and consider the potential environmental impact of edge devices and infrastructure. By harnessing the power of fog computing, stakeholders can work toward achieving several SDGs while also improving efficiency, responsiveness, and sustainability in various sectors.

While fog computing offers numerous benefits, it also faces several challenges and considerations that require attention for its widespread adoption and effective implementation. Here are some of the key challenges of fog computing [2]:

- Security Concerns:
  - **Data Privacy:** Since fog computing processes data at the edge, there may be concerns about data privacy and security, especially when handling sensitive information.
  - **Device Vulnerabilities:** Edge devices in fog computing environments may have limited security measures, making them vulnerable to attacks.
- Scalability:
  - **Resource Limitations:** Edge devices often have limited processing power, storage, and memory, which can hinder scalability.
- Interoperability:
  - **Diverse Ecosystem:** Fog computing environments involve a diverse set of devices and platforms, making interoperability a challenge.
  - **Standards:** Lack of standardized protocols and interfaces can impede seamless communication between edge devices and fog nodes.
- Reliability and Redundancy:
  - **Device Failures:** Edge devices can fail or become unreliable due to various factors, necessitating redundancy and fault-tolerant mechanisms.
  - **High Availability:** Ensuring high availability of fog nodes and edge devices is crucial for mission-critical applications.
- Data Management:
  - **Data Governance:** Managing data generated at the edge, including storage, retrieval, and data lifecycle management, can be complex.
  - **Data Quality:** Ensuring data quality and consistency across distributed edge devices can be challenging.
- Latency:
  - **Real-time Requirements:** Meeting stringent real-time requirements in applications like autonomous vehicles and industrial automation can be difficult due to network latency and delays.
- Resource Allocation:
  - **Resource Allocation Policies:** Developing efficient resource allocation policies for edge devices to balance workloads and optimize resource utilization can be complex.
  - **Dynamic Environments:** Fog nodes must adapt to dynamic network conditions and device availability.
- Energy Efficiency:
  - **Energy Consumption:** Some edge devices operate on battery power, and energy-efficient processing is crucial for prolonging battery life.
  - **Resource-Intensive Processing:** Resource-intensive tasks at the edge can lead to excessive energy consumption.

- **Data Transfer Costs:**
  - **Bandwidth Costs:** Transmitting data from edge devices to centralized data centers may incur high bandwidth costs, especially in remote or constrained network environments.
- **Regulatory Compliance:**
  - **Data Sovereignty:** Compliance with data sovereignty laws and regulations can be challenging when data is processed and stored at the edge, potentially across different regions.
- **Cost Management:**
  - **Infrastructure Costs:** Building and maintaining a distributed fog computing infrastructure can be cost-intensive, and cost management strategies need to be in place.
- **Human Resources and Training:**
  - **Skilled Workforce:** Fog computing requires skilled personnel for deployment, management, and troubleshooting, which may be in short supply.
- **Environmental Impact:**
  - **Energy Consumption:** Fog computing can increase energy consumption, and its environmental impact needs to be considered, especially in green computing initiatives.

Addressing these challenges requires collaborative efforts from industry stakeholders, standardization bodies, and research communities to develop solutions, best practices, and frameworks that make fog computing more secure, scalable, efficient, and reliable. As fog computing continues to evolve, it is expected to become an integral part of the emerging edge computing ecosystem, contributing to the growth of IoT, smart cities, and various other applications.

The SDGs are a set of 17 global goals established by the United Nations in 2015 as part of the 2030 Agenda for Sustainable Development. These goals are designed to address a wide range of global challenges and to guide international efforts toward a more sustainable, equitable, and prosperous future for all. The SDGs cover various aspects of social, economic, and environmental development and are aimed at improving the well-being of people and the planet.

Here are the 17 SDGs [3]:

SDGs encompass a comprehensive set of objectives aimed at addressing global challenges and promoting sustainable development worldwide. These goals cover a wide range of issues, including poverty eradication, hunger elimination, ensuring access to quality education, achieving gender equality, and ensuring access to clean water and sanitation. Other goals focus on promoting affordable and clean energy, fostering economic growth and decent work, building resilient infrastructure, reducing inequalities, and creating sustainable cities and communities. Additionally, the SDGs address responsible consumption and production patterns, climate action, conservation of marine and terrestrial ecosystems, and promoting peace, justice, and strong institutions. Finally, the SDGs emphasize the importance of partnerships for achieving these goals, highlighting the need for collaboration and cooperation at local, national, and global levels to drive sustainable development forward. Each SDG is associated with specific targets and indicators to track progress toward its achievement. The goals are interconnected, and addressing one goal often has positive effects on others. The SDGs provide a framework for governments, organizations, businesses, and individuals to work together to create a more sustainable and equitable world. The target year for achieving the SDGs is 2030.

While fog computing has the potential to support many of the SDGs, it can also have certain negative impacts or challenges if not implemented carefully. Here are ways in which fog computing could potentially have adverse effects on the SDGs [4]:

- **Privacy Concerns:** Fog computing, particularly when processing data at the edge, may raise privacy concerns if not properly secured. Inadequate data protection measures can violate individuals' privacy rights, which is essential for achieving Goal 16.
- **Increased Energy Use:** Deploying a large number of edge devices and fog nodes could lead to increased energy consumption, potentially contributing to higher carbon emissions. This could counteract efforts to promote clean and affordable energy (Goal 7).
- **Resource Intensive:** Managing a complex fog computing infrastructure with numerous edge devices and nodes may require significant resources, including materials, energy, and manpower, which could strain resources instead of promoting sustainable infrastructure (Goal 9).
- **Unequal Access:** If fog computing infrastructure is not deployed equitably, it could exacerbate the digital divide, making advanced services accessible to some communities but not to others, thereby undermining efforts to reduce inequalities (Goal 10).
- **Cross-Border Data Handling:** Fog computing may involve processing data across different jurisdictions, raising issues related to data sovereignty, legal compliance, and international law, potentially impacting Goal 16.
- **Cybersecurity:** The distributed nature of fog computing introduces new cybersecurity risks. Inadequate security measures could lead to data breaches, financial losses, and even security threats to society, which are counterproductive to Goal 16.
- **Environmental Consequences:** The energy consumption and manufacturing of edge devices and fog nodes may have negative environmental impacts if not managed responsibly, potentially undermining efforts to combat climate change (Goal 13).
- **Data Reliability:** If fog computing systems are not properly maintained and monitored, data integrity issues could arise, leading to incorrect

decisions and potentially affecting Goal 16 (Peace, Justice, and Strong Institutions).

- **The Monopoly of Services:** In some cases, large corporations could dominate the fog computing ecosystem, potentially leading to the monopolization of services and resources, which could undermine efforts to reduce inequalities (Goal 10).
- **Job Displacement:** As automation and artificial intelligence (AI) are integrated into fog computing systems, there could be concerns about job displacement in traditional industries, potentially impacting efforts to promote decent work and economic growth (Goal 8).
- **Resource Scarcity:** Depending on the scale of fog computing deployment, there could be resource scarcity issues related to materials used in edge devices and fog nodes, impacting multiple SDGs, including those related to responsible consumption and production (Goal 12).

Innovation plays a crucial role in advancing the SDGs by addressing various challenges and creating sustainable solutions. Here are innovative solutions across different sectors that contribute to the achievement of the SDGs [5]:

- **Solar Microgrids:** Implementing decentralized solar microgrids in off-grid and remote areas to provide clean and reliable electricity, supporting Goal 7 on affordable and clean energy.
- **Precision Agriculture:** Leveraging IoT, drones, and data analytics for precision agriculture to optimize resource use, reduce waste, and improve crop yields, contributing to Goal 2 on zero hunger.
- **Telemedicine and Telehealth:** Expanding telemedicine and telehealth services to provide remote medical consultations, diagnosis, and treatment, enhancing access to healthcare as per Goal 3.
- **Online Learning Platforms:** Developing innovative online learning platforms and digital educational resources to promote quality education, aligning with Goal 4.
- **Carbon Capture Technologies:** Developing innovative carbon capture and storage technologies to reduce greenhouse gas emissions and combat climate change.
- **Marine Conservation Technologies:** Using advanced monitoring and surveillance technologies, such as underwater drones, to protect and conserve marine ecosystems and biodiversity.
- **Precision Forestry:** Applying precision forestry techniques to manage forests sustainably, combat deforestation, and protect terrestrial ecosystems as per Goal 15.
- **Gender-Inclusive Tech:** Promoting gender-inclusive technology solutions and initiatives to empower women and girls, aligning with Goal 5.
- **Off-Grid Solutions:** Developing innovative off-grid energy solutions like wind turbines, small-scale hydro, and energy storage systems to bring clean energy access to remote and underserved areas.

- **Waste-to-Energy:** Implementing waste-to-energy technologies to convert organic waste into clean energy and reduce landfill waste, contributing to a circular economy as outlined in Goal 12.
- **Digital Financial Services:** Expanding access to digital financial services and mobile banking to promote financial inclusion, reduce poverty (Goal 1), and support decent work and economic growth (Goal 8).
- **Inclusive Tech Incubators:** Establishing technology incubators and accelerators that focus on inclusive innovation to ensure that technological advancements benefit all segments of society.
- **Sensor Networks:** Developing low-cost water quality sensors and networks to monitor and improve water quality in rivers, lakes, and reservoirs.
- **Product Traceability:** Implementing blockchain and supply chain technologies to reduce waste.
- **Early Warning Systems:** Developing early warning systems that leverage IoT and data analytics to predict and respond to natural disasters, supporting multiple SDGs, including disaster resilience and climate action.

These innovative solutions demonstrate how technology, data, and creativity can be harnessed to address global challenges and accelerate progress toward the SDGs. Collaboration between governments, businesses, civil society, and innovators is essential to scale these solutions and achieve a more sustainable and equitable future for all.

The novelty of the work lies in its comprehensive exploration of how fog computing, as a distributed computing paradigm, contributes to the achievement of SDGs across various sectors. By highlighting specific examples and applications, the chapter demonstrates the innovative solutions that fog computing offers to address key challenges in areas such as medicine and power. Notably, the chapter emphasizes the role of fog computing in bringing computing resources closer to the point of action, thereby enabling real-time decision-making, improving efficiency, and promoting sustainability. Additionally, the work underscores the importance of fog computing in enhancing data privacy and security, particularly in comparison to centralized cloud computing models. Overall, by showcasing the diverse applications and benefits of fog computing in advancing the SDGs, the chapter contributes to a deeper understanding of how emerging technologies can drive sustainable development and foster inclusive and resilient communities.

The objectives of the chapter are as follows:

- Investigate the role of fog computing in addressing key challenges across various sectors to advance the SDGs.
- Explore the impact of fog computing on agriculture (SDG 2) by empowering precision farming through real-time data analytics, optimizing resource utilization, and promoting sustainable practices.

- Assess the contribution of fog computing to transportation (SDG 11) by enabling intelligent traffic management, real-time monitoring, and supporting autonomous vehicles to reduce congestion and improve road safety.
- Investigate how fog computing promotes affordable and clean energy (SDG 7) through smart grid management, energy optimization, and renewable energy integration to ensure efficient energy monitoring and grid stability.
- Evaluate the role of fog computing in addressing environmental sustainability (SDG 13 and SDG 14) by facilitating efficient waste management, pollution monitoring, and conservation efforts through real-time data analysis.
- Examine how fog computing enhances data privacy and security (SDG 16) by mitigating risks associated with centralized cloud computing and safeguarding sensitive data.

14.1.1 Research methodology

This research will employ a mixed-method approach, combining qualitative and quantitative techniques to comprehensively explore the role of fog computing in advancing the SDGs. The qualitative aspect will involve a thorough literature review and case studies to understand the current state of fog computing applications across different sectors and their impact on achieving the SDGs. The quantitative aspect will involve data analysis and modeling to quantify the benefits of fog computing in specific domains such as healthcare, agriculture, transportation, energy, and environmental conservation. Additionally, interviews and surveys will be conducted with experts and stakeholders to gather insights into the practical implementation challenges and opportunities associated with fog computing adoption. Overall, this research methodology aims to provide a holistic understanding of fog computing's contribution to sustainable development and identify strategies for leveraging its potential to accelerate progress toward achieving the SDGs.

14.2 HEALTHCARE SERVICES ASSOCIATED WITH SDGs

Healthcare services are closely connected to the SDGs as they play a critical role in improving medical, transportation, and overall quality of life. Here are some of the healthcare services that are directly related to specific SDGs [6]:

- **Access to Healthcare (SDG 3 – Good Health and Well-Being):**
  Ensuring equitable access to healthcare services, including primary care, specialty care, and preventive measures, contributes to achieving SDG 3 by improving health and well-being for all.

- **Immunization and Vaccination Programs (SDG 3):**
  Vaccination programs aim to prevent diseases, reduce child mortality (SDG 3.2), and contribute to ensuring healthy lives and well-being (SDG 3.4).
- **Maternal and Child Health Services (SDG 3):**
  Prenatal care, safe childbirth, and child health services are essential for reducing maternal and child mortality rates (SDG 3.1 and 3.2).
- **Mental Health Services (SDG 3):**
  Access to mental health services helps promote mental well-being, reduce mental health disorders, and prevent suicides (SDG 3.4).
- **Disease Prevention and Control (SDG 3):**
  Healthcare services related to disease prevention, early diagnosis, and treatment contribute to reducing the burden of communicable and non-communicable diseases (SDG 3.3 and 3.4).
- **Universal Health Coverage (SDG 3):**
  Achieving universal health coverage ensures that everyone can access the healthcare services they need without facing financial hardship (SDG 3.8).
- **Family Planning and Reproductive Health (SDG 3):**
  Access to family planning services contributes to reproductive health, gender equality (SDG 5), and reducing maternal mortality (SDG 3.1).
- **Health Education and Awareness (SDG 3):**
  Health education programs promote health literacy and empower individuals to make informed decisions about their health (SDG 3.7).
- **Telemedicine and Digital Health (SDG 3):**
  Telemedicine and digital health services improve access to healthcare, particularly in remote areas, supporting SDG 3's goal of ensuring healthy lives and well-being for all.
- **Emergency Medical Services (SDG 3):**
  Quick and efficient emergency medical services save lives and support SDG 3's targets related to reducing mortality rates.
- **Community Health Initiatives (SDG 3):**
  Community-based healthcare services, including health workers and clinics, play a vital role in delivering care to underserved populations (SDG 3.8).
- **Infectious Disease Control (SDG 3):**
  Healthcare services for disease surveillance, outbreak response, and vaccination campaigns are essential for achieving SDG 3's targets related to epidemics.
- **Health Research and Innovation (SDG 3):**
  Research and innovation in healthcare contribute to improved treatments, diagnostics, and healthcare delivery, supporting SDG 3's overall objectives.

- **Water, Sanitation, and Hygiene (WASH) in Healthcare Facilities (SDG 6):**
  Ensuring clean water and proper sanitation in healthcare facilities is essential for quality healthcare and contributes to SDG 6 on clean water and sanitation.
- **Partnerships for Healthcare (SDG 17):**
  Collaborative efforts among governments, organizations, and communities are crucial for strengthening healthcare systems and achieving SDG 17's objectives.

Effective healthcare services are integral to achieving not only SDG 3 but also other interconnected goals related to gender equality, poverty reduction, and sustainable development. They are essential for building healthier, more equitable, and sustainable societies.

Real-time data analysis plays a vital role in supporting climate change mitigation, biodiversity protection, and ecosystem health by providing timely insights and enabling informed decision-making. Additionally, fog computing enhances data privacy and security through its distributed architecture. Here's how these elements align with SDGs [7]:

- **Climate Change Mitigation (SDG 13 – Climate Action):**
  Monitoring Emissions: Real-time data analysis allows for continuous monitoring of greenhouse gas emissions from various sources, including industries, transportation, and agriculture. This data helps identify emission hotspots and assess progress in reducing emissions.
  Climate Modeling: Real-time analysis supports climate modeling and scenario simulations, enabling policymakers to evaluate the potential impacts of climate policies and interventions.
  Early Warning Systems: Real-time climate data analysis supports the development of early warning systems for extreme weather events, such as hurricanes, droughts, and wildfires, allowing communities to prepare and respond effectively.
  Fog Computing: Fog computing enhances climate data security by processing sensitive information locally at the edge, reducing the risk of data breaches and unauthorized access. This aligns with SDG 16 (Peace, Justice, and Strong Institutions) by promoting data privacy and security.
- **Biodiversity Protection (SDG 15 – Life on Land):**
  Wildlife Monitoring: Real-time data analysis helps monitor and track the movement and behaviors of endangered species. This information aids conservationists in protecting critical habitats and preventing poaching.
  Ecosystem Health: Continuous data analysis assesses the health of ecosystems by monitoring factors such as water quality, soil conditions, and vegetation cover. Early detection of ecosystem changes can prompt intervention to mitigate damage.

Illegal Logging Detection: Real-time data analysis can detect illegal logging activities in forests by analyzing patterns of tree removal. This supports SDG 15's goal of sustainable land management.

Fog Computing: Distributed fog computing ensures the security and privacy of sensitive biodiversity data collected in remote areas. It reduces the need to transmit data over long distances, minimizing the risk of data breaches.

- **Ecosystem Health (SDG 14 – Life Below Water, SDG 15 – Life on Land):**
  Water Quality Monitoring: Real-time analysis of water quality data helps assess the health of aquatic ecosystems. It aids in identifying pollution sources and supporting water conservation efforts.

  Coral Reef Protection: Real-time data analysis can monitor the condition of coral reefs, detecting signs of coral bleaching or disease outbreaks, and facilitating timely conservation measures.

  Distributed Monitoring: Fog computing's distributed architecture supports ecosystem health monitoring in remote and sensitive areas, maintaining data privacy and security while ensuring timely analysis.

In summary, real-time data analysis, coupled with fog computing's distributed nature, contributes to climate change mitigation, biodiversity protection, and ecosystem health. It enables proactive measures, supports informed decision-making, and enhances data privacy and security in alignment with SDG 16. These efforts collectively contribute to the achievement of multiple SDGs related to environmental sustainability, conservation, and peace and justice.

14.3 ROLE OF FOG COMPUTING IN SDGs HEALTHCARE SERVICES

Fog computing plays a significant role in supporting healthcare services, contributing to the achievement of several SDGs, particularly SDG 3 (Good Health and Well-Being). Here's how fog computing supports healthcare services and SDGs [8]:

- **Real-Time Patient Monitoring (SDG 3.4):**
  Remote Monitoring: Fog computing enables real-time monitoring of patient vitals, making it possible to detect health issues promptly and provide timely interventions, thus contributing to the goal of ensuring healthy lives and well-being.
- **Telemedicine and Remote Consultations (SDG 3.8):**
  High-Quality Communication: Low-latency fog computing facilitates high-quality video and audio communication for telemedicine consultations, improving access to healthcare services, especially in remote or underserved areas.

- **Data Privacy and Security (SDG 3.9):**
  Edge Processing: Fog computing allows sensitive patient data to be processed locally at the edge, enhancing data privacy and security, which is essential for building resilient healthcare systems.
- **Emergency Response (SDG 3.1):**
  Ambulance Services: Fog computing can equip ambulances with real-time diagnostic capabilities and video conferencing to consult with specialists during transit, improving emergency response and reducing mortality rates.
- **Resource Optimization (SDG 3.7):**
  Hospital Asset Management: Fog computing systems can track the location and condition of medical equipment in hospitals, reducing downtime and ensuring that healthcare resources are used effectively.
- **Medical Imaging (SDG 3.8):**
  Image Analysis: Fog computing enhances the speed and accuracy of medical image analysis, including radiology and pathology, aiding in faster diagnoses and treatment planning.
- **Health Data Management (SDG 3.8):**
  Electronic Health Records (EHRs): Fog computing helps manage and update electronic health records efficiently, ensuring that healthcare providers have access to accurate patient information when needed.
- **Disaster Response (SDG 3.9):**
  Mobile Clinics: Fog computing can support mobile medical clinics during disaster response efforts, offering critical healthcare services and data connectivity in areas with disrupted infrastructure.
- **Data Backup and Recovery (SDG 3.9):**
  Data Redundancy: Fog computing can implement data redundancy and backup solutions, ensuring the availability of critical healthcare data in the event of system failures or disasters.
- **Mental Health Services (SDG 3.4):**
  Teletherapy: Fog computing enables the delivery of teletherapy services, which is particularly valuable for mental health support, contributing to SDG 3's targets on well-being.
- **Health Education and Awareness (SDG 3.6):**
  Health Information Dissemination: Fog computing can facilitate the dissemination of health information and educational content, promoting health literacy and empowering individuals to make informed health choices.
- **Resource-Efficient Healthcare (SDG 12):**
  Energy Efficiency: Fog computing solutions in healthcare can optimize energy consumption, contributing to SDG 12's goals of responsible consumption and production.

Fog computing's ability to improve healthcare accessibility, quality, and efficiency aligns with SDG 3's targets and supports overall efforts to ensure good health and well-being for all.

Fog computing plays a crucial role in facilitating efficient waste management, pollution monitoring, and conservation efforts. This support extends to climate change mitigation, biodiversity protection, and ecosystem health. Here's how fog computing contributes to these environmental and conservation initiatives [9]:

- **Waste Management (SDG 12 – Responsible Consumption and Production):**
  Smart Waste Bins: Fog computing enables the deployment of smart waste bins equipped with sensors that monitor waste levels. When bins are nearing full capacity, real-time data is sent to waste management systems, optimizing collection routes and reducing unnecessary trips, thus promoting responsible waste management and resource efficiency.
- **Air Quality Monitoring (SDG 13 – Climate Action):**
  Air Quality Sensors: Fog computing networks support air quality sensors placed throughout urban areas. These sensors continuously monitor air pollution levels, including particulate matter and pollutants. Real-time data analysis helps authorities identify pollution sources, enforce regulations, and take corrective actions to improve air quality, contributing to climate action efforts.
- **Water Quality Monitoring (SDG 6 – Clean Water and Sanitation):**
  IoT Water Sensors: Fog computing integrates with IoT water quality sensors placed in rivers, lakes, and reservoirs. These sensors continuously measure parameters like pH, turbidity, and contaminant levels. Real-time data analysis ensures the timely detection of water pollution incidents, allowing for rapid response and protection of clean water sources.
- **Energy Efficiency (SDG 7 – Affordable and Clean Energy):**
  Energy Management Systems: Fog computing supports energy management systems in buildings and industrial facilities. Real-time data analysis identifies energy consumption patterns, optimizes energy usage, and promotes clean and affordable energy practices.
- **Wildlife Monitoring (SDG 15 – Life on Land):**
  Wildlife Sensors: Fog computing networks are utilized in conservation areas to monitor wildlife. Cameras and sensors capture data on animal movements, habitats, and behaviors in real-time. This data aids in biodiversity protection, habitat preservation, and species conservation.
- **Climate Change Mitigation (SDG 13 – Climate Action):**
  Environmental Sensors: Fog computing networks integrate various environmental sensors, including temperature, humidity, and CO2 sensors. Real-time data analysis provides valuable information for

climate change research, helping scientists and policymakers make informed decisions to mitigate the effects of climate change.

- **Ecosystem Health (SDG 15 – Life on Land, SDG 14 – Life Below Water):**
  Ecosystem Monitoring: Fog computing enables real-time monitoring of terrestrial and marine ecosystems. Sensors track factors such as soil moisture, water temperature, and ocean acidity, providing insights into ecosystem health. Timely data analysis helps protect and preserve biodiversity on land and in the oceans.
- **Natural Disaster Early Warning (SDG 11 – Sustainable Cities and Communities):**
  Disaster Sensors: Fog computing networks can be equipped with sensors that detect natural disaster indicators such as seismic activity, temperature fluctuations, or humidity levels. Real-time data analysis enables early warning systems, supporting disaster preparedness and resilient communities.

Fog computing enhances the efficiency of waste management, pollution monitoring, and conservation efforts. It contributes to the achievement of multiple SDGs by promoting responsible resource use, reducing environmental impacts, and supporting the well-being of ecosystems and communities.

While fog computing offers significant advantages for waste management, pollution monitoring, and conservation efforts, it also presents several challenges. Addressing these challenges is crucial to fully realizing the potential benefits of fog computing in these areas [10]:

- **Data Security and Privacy:**
  Challenge: Real-time data analysis involves the collection and transmission of sensitive environmental and waste-related data. Ensuring the security and privacy of this data is essential to prevent unauthorized access, data breaches, and misuse.
  Solution: Implement robust encryption, access controls, and authentication mechanisms to protect data. Compliance with data privacy regulations is vital.
- **Data Quality and Reliability:**
  Challenge: Environmental sensors may occasionally produce inaccurate or unreliable data, leading to incorrect decisions and actions. Fog computing systems must handle data quality issues effectively.
  Solution: Implement data validation and cleansing techniques to identify and filter out erroneous data. Implement redundancy and error-checking mechanisms.
- **Interoperability:**
  Challenge: Different environmental sensors and monitoring devices may use diverse data formats and communication protocols. Integrating these heterogeneous systems can be complex.

Solution: Develop standardized data formats and communication protocols. Use middleware and gateways to bridge between different devices and systems.

- **Scalability:**

  Challenge: As the number of sensors and devices in a fog computing network grows, managing scalability becomes challenging. Ensuring that the system can handle increasing data volumes and traffic is crucial.

  Solution: Design the fog computing architecture with scalability in mind. Use load balancing, distributed computing, and edge server clusters to handle increasing workloads.

- **Energy Efficiency:**

  Challenge: Many environmental sensors and fog nodes are deployed in remote or off-grid locations. Ensuring energy-efficient operation, especially for devices running on batteries or renewable energy sources, is essential.

  Solution: Optimize algorithms and data collection intervals to minimize energy consumption. Implement energy harvesting solutions where feasible.

- **Maintenance and Reliability:**

  Challenge: Environmental sensors and fog nodes may be exposed to harsh conditions, leading to equipment failures and the need for regular maintenance.

  Solution: Implement remote monitoring capabilities to detect equipment issues in real-time. Design ruggedized and reliable hardware components.

- **Data Overload:**

  Challenge: In some cases, the sheer volume of data generated by environmental sensors can overwhelm fog computing systems. Handling and processing this data efficiently can be challenging.

  Solution: Employ data reduction techniques, filtering, and aggregation at the edge to reduce the volume of data sent to the cloud. Prioritize critical data for immediate analysis.

- **Regulatory Compliance:**

  Challenge: Environmental monitoring and waste management activities may be subject to regulatory requirements and reporting obligations. Ensuring compliance with these regulations can be complex.

  Solution: Stay informed about relevant environmental regulations and integrate compliance measures into the fog computing system. Implement audit trails and reporting capabilities.

- **Costs and Budget Constraints:**

  Challenge: Deploying and maintaining fog computing infrastructure, including sensors and nodes, can be costly. Budget constraints may limit the extent of deployment.

Solution: Conduct cost-benefit analyses to justify investments in fog computing. Explore partnerships and funding opportunities to support environmental initiatives.

Overcoming these hurdles can lead to more effective and efficient waste management, pollution monitoring, and conservation efforts, ultimately contributing to the achievement of SDGs related to environmental sustainability and ecosystem health.

14.4 CASE STUDIES ON FOG COMPUTING CONNECTED SDGS HEALTHCARE

Here are a few case studies that illustrate how fog computing is connected to SDGs in the context of healthcare services [11, 12]:

1. **Project: Remote Patient Monitoring in Rural India**
   - **SDGs Addressed:** SDG 3 (Good Health and Well-Being), SDG 9 (Industry, Innovation, and Infrastructure)
   - **Description:** In a rural healthcare center in India, fog computing was implemented to enable remote patient monitoring. Patients with chronic conditions, such as diabetes and hypertension, were provided with wearable health devices. These devices collected real-time health data and transmitted it to fog nodes within the healthcare center. The fog nodes processed the data locally, and healthcare providers could access patient information via a secure portal. This solution improved access to healthcare services in remote areas, contributing to SDG 3's goal of ensuring healthy lives and well-being for all.
2. **Project: Telemedicine Network in Sub-Saharan Africa**
   - **SDGs Addressed:** SDG 3 (Good Health and Well-Being), SDG 10 (Reduced Inequalities)
   - **Description:** A telemedicine network was established in several sub-Saharan African countries to address the lack of access to specialized healthcare services in rural regions. Fog computing infrastructure was deployed to support real-time video consultations between local clinics and remote medical specialists. The low-latency fog network ensured high-quality communication, making it easier for healthcare professionals to diagnose and treat patients in underserved areas. This project contributed to reducing healthcare inequalities (SDG 10) and improving healthcare access (SDG 3).
3. **Project: Disaster-Ready Mobile Clinics**
   - **SDGs Addressed:** SDG 3 (Good Health and Well-Being), SDG 11 (Sustainable Cities and Communities), SDG 13 (Climate Action)

- **Description:** In regions prone to natural disasters, mobile clinics equipped with fog computing capabilities were deployed to provide emergency healthcare services. These mobile units were equipped with medical devices connected to fog nodes, allowing healthcare providers to deliver care and transmit patient data even in disaster-stricken areas with disrupted infrastructure. This initiative improved disaster response (SDG 11), reduced mortality rates (SDG 3), and contributed to climate resilience (SDG 13).

4. **Project: Telepsychiatry for Mental Health Support**
   - **SDGs Addressed:** SDG 3 (Good Health and Well-Being)
   - **Description:** In a densely populated urban area, a fog computing-based telepsychiatry service was established to address the growing demand for mental health support. Patients could access remote therapy sessions with mental health professionals through a secure and low-latency fog network. This initiative helped reduce the burden of mental health disorders (SDG 3.4) and improve mental well-being for individuals in the community.

5. **Project: IoT-Enabled Elderly Care Facilities**
   - **SDGs Addressed:** SDG 3 (Good Health and Well-Being), SDG 11 (Sustainable Cities and Communities)
   - **Description:** In an aging urban population, fog computing was employed to enhance care services in elderly care facilities. IoT devices and sensors were connected to fog nodes to monitor residents' health and safety. Alerts and notifications could be generated in real-time to address potential issues, ensuring the well-being of elderly residents and contributing to the goal of healthy lives and sustainable urban communities [13].

These case studies demonstrate how fog computing technologies are applied to improve healthcare services, address healthcare inequalities, and enhance disaster response, aligning with various SDGs, particularly SDG 3 (Good Health and Well-Being). They highlight the versatility and impact of fog computing in healthcare contexts across different regions and challenges.

14.5 CONCLUSIONS

The significance of fog computing in advancing the SDGs and its potential to drive efficiency, innovation, and sustainability across various sectors. Here's a breakdown of the key points in your conclusion:

- **Role of Fog Computing:** Fog computing is recognized as a crucial enabler for achieving the SDGs. Its ability to bring computing resources closer to where they are needed most enhances its impact on diverse sectors.

- **Sustainability**: Fog computing aligns with sustainability goals by supporting responsible consumption and production (SDG 12) through efficient resource management and reducing the environmental footprint of data processing.
- **Sectoral Impact**: Fog computing's positive influence extends across sectors, including healthcare, agriculture, transportation, energy, and environmental conservation, highlighting its versatility and adaptability.
- **Accelerating Progress**: Leveraging fog computing's potential accelerates progress toward sustainable development, emphasizing its role as a catalyst for achieving SDGs.
- **Inclusive and Resilient Communities**: Adopting fog computing technologies promotes inclusivity by ensuring that even remote and underserved communities have access to advanced services. It also contributes to community resilience in the face of challenges.
- **Vision for the Future**: The conclusion suggests that fog computing is instrumental in creating a better future, where inclusive and resilient communities thrive, and sustainable development is realized.

Finally, this chapter's conclusion effectively underscores the positive impact of fog computing on sustainable development and the broader societal benefits it offers. It highlights the importance of leveraging technological innovations to build a more equitable and sustainable world.

REFERENCES

1. Vilela, P.H., Rodrigues, J.J., Vilela, L.R., Mahmoud, M.M. and Solic, P., 2018, June. A critical analysis of healthcare applications over fog computing infrastructures. In *2018 3rd International Conference on Smart and Sustainable Technologies (SpliTech)* (pp. 1–5). IEEE.
2. Kowsigan, M., 2022, November. A study and research direction towards healthcare data management system in FoG and IoT networks. In *2022 International Conference on Augmented Intelligence and Sustainable Systems (ICAISS)* (pp. 1054–1060). IEEE.
3. Mishra, S.K., Puthal, D., Rodrigues, J.J., Sahoo, B. and Dutkiewicz, E., 2018. Sustainable service allocation using a metaheuristic technique in a fog server for industrial applications. *IEEE Transactions on Industrial Informatics*, 14(10), pp. 4497–4506.
4. Prajapati, C.C. and Singla, J., 2020, December. A comprehensive review on smart fog-based healthcare framework. In *2020 3rd International Conference on Intelligent Sustainable Systems (ICISS)* (pp. 622–627). IEEE.
5. Iftikhar, S., Golec, M., Chowdhury, D., Gill, S.S. and Uhlig, S., 2022, June. Fog computing-based router-distributor application for sustainable smart home. In *2022 IEEE 95th Vehicular Technology Conference:(VTC2022-Spring)* (pp. 1–5). IEEE.

6. Wadhwa, H. and Aron, R., 2018, December. Fog computing with the integration of internet of things: Architecture, applications and future directions. In *2018 IEEE Intl Conf on Parallel & Distributed Processing with Applications, Ubiquitous Computing & Communications, Big Data & Cloud Computing, Social Computing & Networking, Sustainable Computing & Communications* (pp. 987–994). IEEE.

7. Duggal, S. and Kaur, P., 2021, March. Fog computing based health care applications and frameworks: A review. In *2021 8th International Conference on Computing for Sustainable Global Development (INDIACom)* (pp. 238–243). IEEE.

8. Babar, K. and Shah, M.A., 2022, June. Scalable and sustainable mist computing-based architecture for Internet of Health Things. In *Competitive Advantage in the Digital Economy (CADE 2022)* (Vol. 2022, pp. 111–116). IET.

9. Kumar, A.P. and Reddy, C., 2022, November. Comprehensive analysis on low power health monitoring system using fog computing in passive optical networks. In *2022 Sixth International Conference on I-SMAC (IoT in Social, Mobile, Analytics and Cloud)(I-SMAC)* (pp. 476–484). IEEE.

10. Singh, G. and Singh, J., 2023, January. A fog computing based agriculture-IoT framework for detection of alert conditions and effective crop protection. In *2023 5th International Conference on Smart Systems and Inventive Technology (ICSSIT)* (pp. 537–543). IEEE.

11. Kumar, M., Dubey, K. and Pandey, R., 2021, January. Evolution of emerging computing paradigm cloud to fog: applications, limitations and research challenges. In *2021 11th international conference on cloud computing, data science & engineering (Confluence)* (pp. 257–261). IEEE.

12. El Idrissi, M., Elbeqqali, O. and Riffi, J., 2019, October. A review on relationship between IoT–cloud computing–fog computing (Applications and Challenges). In *2019 third international conference on intelligent computing in data sciences (ICDS)* (pp. 1–7). IEEE.

13. EL Idrissi, M., Elbeqqali, O. and Riffi, J., 2019, October. From cloud computing to fog computing: two technologies to serve IoT–a review. In *2019 IEEE international smart cities conference (ISC2)* (pp. 272–279). IEEE.

Chapter 15

Energy-efficient framework for cloud–fog environment for sustainable development

Gaurav Goel

Chandigarh Engineering College-CGC, Landran, India
UPES, Dehradun, India

Rajeev Tiwari

IILM University, Greater Noida, India

Mohit Kumar

Dr. B.R. Ambedkar National Institute of Technology, Jalandhar, India

Shilpi Harnal

Chandigarh University, Mohali, India

15.1 INTRODUCTION

The widespread use of cloud computing in recent years has completely changed how we handle, store, and retrieve data, providing computing environments with previously unheard-of levels of scalability and flexibility. But increasing demand for computing power combined with environmental issues demands a paradigm change to cloud–fog computing frameworks, as shown in Figure 15.1, that are sustainable and energy-efficient. Cloud computing has become a key component of contemporary information technology in a time of unparalleled data generation and digital change. Cloud systems provide unmatched scalability, accessibility, and computational power[1]. However, the need for energy to power these enormous and complex data infrastructures is growing along with the world's reliance on cloud services. Fog computing, which offers decentralized computing at the network's edge, emerges as a viable extension of the cloud paradigm in answer to this difficulty [2].

15.1.1 Background

Organizations can now take advantage of scalable and on-demand computing resources thanks to significant developments in data processing capabilities brought about by the cloud computing industry's explosive growth.

DOI: 10.1201/9781003494430-15

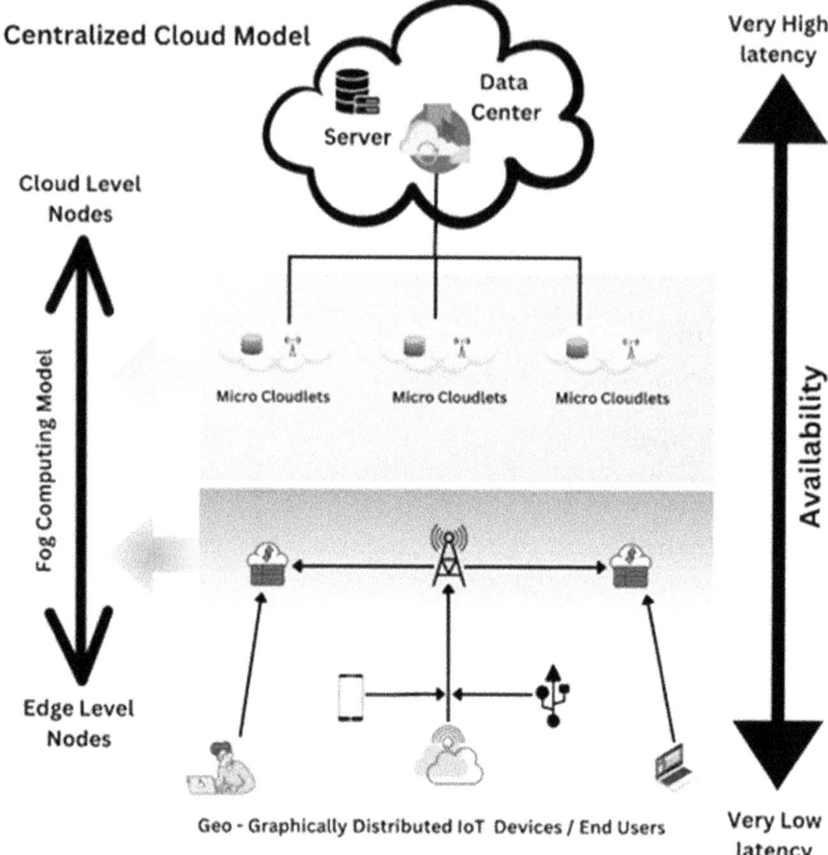

Figure 15.1 Traditional fog computing architecture.

Simultaneously, the development of fog computing [3], a cloud computing extension, moves computation closer to the network edge [4], providing advantages for real-time processing and low latency. Even though these developments have created previously unheard-of opportunities, data centers and computer infrastructure energy consumption are now a major worry.

Large-scale data centers, which define traditional cloud computing infrastructures, are infamous for their significant energy consumption and environmental effects. It is crucial to look at creative solutions that strike a balance between the requirement for processing capacity and environmental sustainability as the demand for cloud services keeps rising. Because fog computing is decentralized and distributed, it offers a chance to minimize the energy consumption and carbon impact of data processing [5].

15.1.2 Motivation

The pressing necessity to address the environmental effects of rising energy usage in cloud–fog computing settings is what drove this research. It is crucial to create frameworks that not only satisfy the rising computing demands but also follow sustainable principles as the digital world grows. Our goal is to minimize the environmental effect of cloud–fog computing while maintaining peak performance and effective use of resources by creating an energy-efficient framework.

15.1.3 Scope of the study

The complex interactions between fog and cloud computing environments and their combined effects on energy usage are the main topics of this study. The scope includes creating, putting into practice, and assessing an energy-efficient framework that balances the advantages of cloud and fog computing with environmental considerations. The suggested framework incorporates practical aspects, such as the incorporation of renewable energy sources.

15.2 RELATED WORK

The studies and associated work on the energy-efficient framework for cloud–fog sustainable development are described in this subsection. Analysis and discussion of relevant work have been done in this subsection.

15.2.1 Task scheduling and delay-based study

In order to address the challenge of effectively scheduling jobs in a Fog-cloud environment to enhance Quality of Service (QoS) for Internet of Things (IoT) applications in the IoT-Fog-cloud framework, Kumar et al. [6] have presented a hybrid electric earthworm optimization algorithm. Researchers employ the optimization technique known as "electric fish and earthworm" to lower system latency and energy usage. To improve the position-updating process and the overall efficiency of the proposed methodology, both active and passive electrolocation approaches are applied. This approach applies to real-world smart cities and gives considerable utility in vehicle network management. The authors discovered that their approach worked better than earlier processes after comparing cost, makespan, execution time, and energy consumption. The suggested method is evaluated using actual workloads from HPC2N and CEA-CURIE. There is no task offloading mechanism in this work.

For latency-sensitive IoT applications, Abu-Amssimir et al. [7] have proposed a greedy-edge-placement method that minimizes delay. The suggested method effectively reduces latency and maximizes throughput. The phase of choosing and placing applications for the proposed greedy-delay-minimizing technique is being worked on by researchers. Both the placement and selection stages—which selected fog nodes for module application deployment using the DFS algorithm—achieved reduced end-to-end latency. The recommended method successfully reduces network latency, bandwidth, and energy consumption while delivering high-quality services. The work's flaw is that in order to show network efficiency, reaction time might need to be computed.

A hybrid optimization technique based on the Firefly algorithm and Particle Swarm Optimization was described by Ogundoyin et al. [8]. Researchers talked about node capacity, energy consumption, and sojourn rate and trust. An individual objective function is consolidated using the liner-sum-weight method, and the weight vector for the aggregate function is determined using the best–worst method. The suggested strategy performs better than conventional approaches.

15.2.2 Energy minimization-based study

A unique method for vehicular fog computation was introduced by Hussain et al. [9]. The writers considered energy-sensitive automotive applications in addition to the delay problem. The focus of the research was on data dumping to RSUs and BSs (base stations) from automobile gadgets. In this paper, the authors propose a Swarm-Optimized Non-Dominated Sorting-Genetic Algorithm (SONG) as a multi-objective optimization problem. In comparison with the NSGA-2 and SMPSO techniques, the new method yields better quality than the earlier tactics. The network's reaction time is marginally faster than it was with earlier fixes.

Zeng et al. [10] addressed the issue of finding CPS in fog and investigated energy minimization using green energy. Different kinds of dispersed devices that generate distinct kinds of raw data sent from diverse sensors are included in a CPS system. The problem of launching the copy of CPS in fog has been examined in this study in relation to the variety of green energy while preserving the QoS; this is accomplished by linear programming while taking the bar balance rate and copy service location into consideration. Moreover, a heuristic approach and an enlarged test have been utilized to address the difficulty of the linear programming problem's computations. Through testing and trials, the researchers validated their heuristic algorithm, which they developed for energy optimization and efficiency. The stimulation findings validate the precision and effectiveness of their algorithm, meaning it performs in a manner that is nearly identical to the optimal option in terms of energy efficiency.

A practical and efficient method for dynamic resource provisioning in a fog computing environment was introduced by Faraji et al. [11]. This method took into account the environment's intrinsic dynamic nature as well as input workload unpredictability. According to this strategy, the resources needed for user and IoT device applications will be provided in accordance with the efficiency goals (cost minimization combined with maintenance of service quality). The outcomes of the stimulation tests demonstrate that, in comparison to comparable approaches, the suggested approaches lower overall costs and service delay while also lowering the rate of SLA violations. The effectiveness of the suggested solution has been assessed through a few tests using real workloads, and the simulation results show that, when compared to the current fundamental mechanisms, it has improved SLA violation and reduced latency and cost.

15.3 CHALLENGES IN ENERGY CONSUMPTION

In this section, we will discuss the challenges faced in energy consumption for cloud environments.

15.3.1 Data center energy consumption

Large-scale data centers, which are essential to traditional cloud computing, require a significant amount of energy for operation, maintenance, and cooling. Data centers are the backbone of cloud–fog computing, enabling the processing, management, and storage of enormous volumes of digital data. Nevertheless, these data centers' operation is expensive, especially when it comes to energy usage. Developing sustainable and energy-efficient methods requires an understanding of the size of data center energy use [12].

- **The Expansion of Digital Ecosystems:**
  The amount of data collected and processed globally has increased at an unprecedented rate due to the rising digitization of different parts of our lives. Due to the expansion of data centers in response to the increasing demand for computer resources [11], there has been a notable rise in energy usage.
- **Complexity and Scale:**
  These days, data centers are enormous structures made up of countless racks of servers, networking hardware, and storage devices. Due to their complexity and interdependence, these parts require a steady and significant electricity supply. This size and the intricate infrastructure of data centers highlight how difficult it is to efficiently manage and reduce energy use.
- **Overhead and Cooling Systems:**
  The requirement for effective cooling systems is one of the main variables influencing data center energy consumption. Servers and other hardware components produce heat during operation; therefore, continuous cooling is required to keep the environment at its best. The energy-intensive nature of these cooling systems contributes significantly to data centers' overall energy usage [13].

- **Constantly Running:**
  Data centers are constant energy consumers since they run around the clock to guarantee service availability. The energy requirements of data centers are further increased by this constant operation as well as the requirement for redundancy and backup systems.

- **Concentrations by Region:**
  Since they are located close to hubs for network connectivity, have stable power infrastructure, and have a pleasant environment, data centers are frequently concentrated in particular areas. The impact of data center energy usage is amplified regionally, posing particular problems to environmental sustainability and energy infrastructures.

- **Consequences for the Environment and Economy:**
  The amount of energy consumed by data centers has significant effects on the environment and the economy. Businesses that operate data centers face financial difficulties due to high running costs, which are mostly caused by power bills and cooling expenses. Concurrently, the carbon footprint linked to this energy usage exacerbates climate change and environmental issues [14].

- **Demand Prediction:**
  The need for data center resources is predicted to rise as digital services continue to spread and cutting-edge technologies like 5G, IoT, and artificial intelligence (AI) become more widely used. Recognizing and controlling this growing demand is essential to reducing the amount of energy used by data centers.

15.3.2 The carbon footprint and environmental effects of data center energy use

The swift expansion of digital infrastructure, marked by substantial requirements for data processing and storage, has led to noteworthy environmental apprehensions over data center energy use. It is essential to comprehend the data center operations' carbon footprint and environmental impact in order to create sustainable practices and lessen the wider ecological effects.

- **Energy Source and Emissions of Carbon:**
  The energy source that powers these facilities has a direct impact on the environmental effects of data center energy consumption. Conventional energy grids, which frequently get their power from fossil fuels, are used by many data centers. Fossil fuel burning produces greenhouse emissions that worsen climate change and increase the carbon footprint of data center operations.

- **Intensity of Carbon:**
  One important statistic for evaluating the environmental impact of data centers is carbon intensity, which is defined as the amount of carbon dioxide emitted per unit of energy consumed. A higher carbon

intensity indicates a greater contribution to global warming. To tackle this issue, data centers must improve their overall energy efficiency and switch to greener energy sources.

- **Integration of Renewable Energy:**
  One of the most important strategies for reducing the environmental impact is incorporating renewable energy sources into data center operations. The carbon footprint of data centers can be greatly reduced, and sustainability can be promoted by using renewable energy sources like solar, wind, hydro, and others in place of or in addition to conventional energy sources [15].

- **Efficiency of Power Usage:**
  An industry-standard metric for assessing a data center's energy efficiency is called Power Usage Effectiveness (PUE). A reduced PUE signifies increased energy efficiency. Energy-efficient technologies, including server virtualization, improved cooling systems, and intelligent workload management, must be implemented in order to maximize PUE.

- **Recovery of Waste Heat:**
  During operation, data centers produce large volumes of waste heat. Waste heat recovery systems can be used in place of releasing this heat into the atmosphere. These systems increase energy efficiency and sustainability by capturing and repurposing excess heat for the purpose of heating neighboring facilities or as an input for additional energy generation.

- **Life Cycle Evaluation:**
  Data center operations must undergo a life cycle assessment (LCA) in order to fully comprehend the environmental impact. The environmental implications of the full life cycle, including production, building, operation, and decommissioning, are taken into account in this study. LCAs offer insights for long-term changes by locating and addressing the most influential stages.

- **Compliance with Environmental Regulations:**
  Regulations are scrutinizing how data centers affect the environment, and governments and international organizations are putting more and more emphasis on sustainability requirements. To show their dedication to environmental responsibility and support larger environmental aims, data centers must abide by certain standards [16].

- **Taking Care of Business Socially:**
  Data center operators are continually looking for solutions to reduce their environmental impact as part of their corporate social responsibility programs. Data center operators are coordinating their operations with more general environmental aims through promoting sustainable practices, funding renewable energy initiatives, and participating in carbon offset schemes.

15.3.3 Scalability issues with data center energy use for sustainable development in cloud–fog

The scalability problems connected with data center operations are a significant component of the ongoing efforts to build an energy-efficient framework for cloud–fog sustainable development. Scalability's effects on energy consumption must be taken into consideration as digital ecosystems increase in order to guarantee that the development of cloud and fog computing is consistent with the objectives of economic and environmental sustainability [17].

- **Growing Requirements for Computation:**
  The rising computational demands on data centers are one of the main causes of scalability issues. The need for data center resources is increasing rapidly as more and more companies, organizations, and people rely on digital services and applications [18]. To effectively scale infrastructure, new ways are needed to address this difficulty.
- **Extension of Physical Space and Infrastructure:**
  Physical expansion is frequently necessary to increase data center capacity, necessitating more floor space, cooling infrastructure, and energy sources. This growth may put pressure on the region's infrastructure and resources, making it difficult to find acceptable sites for additional data centers and handle the environmental effects that come with them.
- **New Hardware's High Energy Intensity:**
  To fulfill computing demands, introducing more potent and energy-intensive hardware is a major challenge. While newer technologies frequently offer better performance, they may also require more energy, so choosing hardware carefully is necessary to strike a balance between performance and energy economy [19].
- **Ineffective Use of Resources at Low Loads:**
  Data centers frequently operate at different loads; thus, it might be difficult to maintain energy efficiency when demand is low. Wasted energy results from servers and infrastructure operating at suboptimal levels of utilization, which causes inefficiencies. Strategies for dynamic resource allocation are necessary to minimize energy waste and adjust to changing workloads.
- **Inadequate Utilization and Overstocking:**
  Scalability problems are made worse by overprovisioning, the practice of allocating more resources than are technically required to support anticipated future growth. Underuse of resources as a result reduces overall energy efficiency [20]. Finding a balance between the need for scalability and wise resource allocation strategies is necessary for optimizing energy usage.

- **Latency and Geographic Distribution:**
  To reduce latency and improve user experience, scalability usually requires the geographical dispersion of fog nodes and data centers [21]. However, this dispersed architecture may lead to problems with data synchronization, higher network traffic, and differences in energy efficiency between locations.
- **Financial Consequences:**
  Although scalability is necessary to handle increasing computational needs, there may be major financial ramifications. Increased operational expenditures, such as power bills, maintenance charges, and the requirement for qualified staff, may result from rapid expansion. A major problem is controlling these expenses while guaranteeing sustainable scalability.
- **Scalability and Modular Design:**
  Adopting a modular design strategy can address scalability concerns by allowing data centers to expand in smaller, more controllable increments. This strategy makes it possible to use resources more effectively, lessens the effects of overprovisioning, and makes it easier to integrate energy-efficient technologies as data centers grow.
- **Edge Computing and Adaptive Infrastructure:**
  By placing computer resources closer to end users, edge computing and adaptive infrastructure solutions can improve scalability. In particular for applications that are latency-sensitive, this strategy eases the burden on central data centers and allows for more effective resource usage [22].

An all-encompassing strategy that incorporates strategic planning, environmental practices, and technological innovation is needed to address the scalability issues with data center energy use. The cloud–fog ecosystem can attain sustainable growth while reducing its environmental impact by implementing energy-efficient technologies, maximizing resource consumption, and implementing scalable and adaptable infrastructure designs.

15.4 COMPONENTS OF AN ENERGY-EFFICIENT FRAMEWORK FOR CLOUD–FOG SUSTAINABLE DEVELOPMENT

Creating an energy-efficient cloud–fog computing architecture entails combining a number of different elements and tactics in order to maximize resource use, reduce energy usage, and improve sustainability all around. The following essential elements serve as the cornerstone of a successful framework for energy efficiency:

15.4.1 Optimal resource distribution in an energy-sparing cloud–fog architecture

One essential part of an energy-efficient cloud–fog computing paradigm is optimized resource allocation. It entails the intelligent allocation of workloads and computational activities among cloud and fog nodes in order to maximize resource usage and minimize energy consumption. In the cloud–fog ecosystem, this component is essential to reaching performance and sustainability goals.

- **Workload Characterization and Profiling:**
  Comprehending the attributes of diverse workloads is the initial phase toward optimizing resource distribution. Sorting tasks according to their computing complexity, importance, and data dependencies is known as workload profiling [23]. The framework learns more about workloads' resource needs by describing them, which helps with resource allocation decisions.
- **Algorithms for Intelligent Resource Allocation:**
  The key to increasing efficiency is putting clever algorithms for resource allocation into practice. These algorithms take into account variables including workload demand, node capabilities, and patterns of energy use. The system can adapt to changing situations by using machine learning techniques [1] to dynamically alter allocation tactics based on historical data.
- **Methods for Load Balancing:**
  An essential part of making sure workloads are split equally among cloud and fog nodes is dynamic load balancing. Task priority, network conditions, and current node utilization are all taken into account by load balancing algorithms [24]. This part reduces resource bottlenecks, improves responsiveness, and helps maintain energy efficiency.
- **Scalability and Elasticity:**
  Elasticity and scalability allow an energy-efficient framework to adjust to changing workloads. The system is able to automatically scale resources up or down in response to demand thanks to elastic resource allocation. In order to optimize resource allocation for efficiency, scalability guarantees that the framework can expand or contract in response to shifting computing requirements.
- **Energy-Conscious Planning:**
  When assigning jobs, energy-aware scheduling takes into account each node's energy efficiency. The framework may route jobs to nodes that optimize energy usage by taking into account the power efficiency of individual nodes as well as trends of energy consumption [25]. The carbon

footprint of computational jobs is reduced by this method, which enhances sustainability overall.

- **Allocating resources predictively**
  The framework uses historical data and trends to predict future resource demands by utilizing predictive analytics. By anticipating workload spikes, predictive resource allocation avoids resource shortages or overprovisioning. This proactive strategy lowers energy waste and improves overall efficiency.
- **Flexible Resource Administration:**
  Continuous monitoring of system performance, workload fluctuations, and environmental factors are all part of adaptive resource management. Resource allocations can be dynamically modified by the framework in reaction to changes, guaranteeing that resources are in line with the changing needs of cloud–fog applications. Real-time energy optimization is achieved by adaptive resource management [26].
- **Discreet Resource Distribution Procedures:**
  Exact control over CPU, memory, and storage resource allocation is possible through the use of fine-grained resource allocation policies. This level of detail guarantees that resources are allotted according to the particular requirements of each activity, avoiding overallocation and enhancing overall energy efficiency [27].
- **Loops of Feedback for Ongoing Improvement:**
  Continuous improvement is facilitated by including feedback loops into the resource allocation procedure. Through the analysis of energy usage data, performance measures, and user input, the framework is able to iteratively improve its approaches to resource allocation. This cyclical procedure helps with continuous optimization and adjusts to cloud–fog environments' dynamic nature.

An essential component of energy-efficient cloud–fog frameworks is optimized resource allocation, which fosters resilience, adaptation, and sustainability. In the quest for a greener and more sustainable digital future, companies may achieve a balance between performance and energy efficiency by strategically allocating resources based on workload characteristics and environmental factors.

15.4.2 Dynamic load balancing in an energy-efficient cloud–fog framework

An energy-efficient cloud–fog computing platform must include dynamic load balancing. The process entails constant monitoring and real-time distribution of computational jobs among cloud and fog nodes to guarantee the best possible use of resources, responsiveness, and energy efficiency. Important features of dynamic load balancing in this framework are as follows:

- **Monitoring Workloads in Real Time:**
  Real-time monitoring of the computational burden throughout the cloud–fog architecture is the first step in dynamic load balancing. To learn more about each node's current status, this entails regularly monitoring its performance indicators, task completion rates, and resource use [28].
- **Evaluation of Node Utilization:**
  Every cloud and fog node's utilization is evaluated by the framework, which takes into account variables including CPU utilization [29], memory availability, and storage capacity. By comprehending the present condition of every node, the system may detect any obstructions and unused assets.
- **Methods for Redistributing Tasks:**
  Dynamic load balancing optimizes task allocation by using different redistribution algorithms based on the evaluation of node utilization. This could entail transferring workloads from overloaded nodes to idle ones or allocating new jobs in a way that reduces competition for resources and optimizes throughput [30].
- **Considerations for Latency and Responsiveness:**
  The latency requirements of various services and applications are taken into consideration by dynamic load balancing solutions. Tasks with demanding real-time processing needs are prioritized, and load is adjusted to reduce latency, ensuring that the overall system remains responsive and fulfills user expectations.
- **Algorithms that Adapt:**
  By putting adaptive load balancing algorithms into practice, the framework may dynamically modify its tactics in response to shifting circumstances. Predictive analytics and machine learning can be used to foresee future workload [31] trends and proactively balance the workload ahead of impending demands.
- **Load Balancing Aware of the Network:**
  Dynamic load balancing takes into account network conditions in addition to individual nodes. For the best work allocation, cloud and fog nodes must communicate well. Network-aware load balancing [32] solutions optimize data transfer, reduce latency, and ensure that the communication infrastructure runs efficiently.
- **Prioritizing the Workload:**
  Different workloads could have different priorities. Task criticality is taken into consideration by dynamic load balancing, which guarantees that high-priority workloads are given preference when allocating resources. The user requirements and corporate goals are in line with this prioritizing.

- **Metrics for Energy Efficiency:**
  Dynamic load balancing not only takes performance optimization into account, but also energy efficiency indicators. The framework contributes to the sustainability objectives of the cloud–fog architecture by balancing the computational load in a way that lowers overall energy usage.
- **Resilience and Fault Tolerance:**
  Because dynamic load balancing redistributes jobs in reaction to node failures or declines in performance, it improves fault tolerance and system resilience. This flexible strategy guarantees that the framework will be able to function normally even in the face of unforeseen circumstances [33].
- **Constant Observation and Input:**
  A dynamic load balancing system functions continuously and is motivated by feedback. Feedback loops and ongoing monitoring enable the framework to change and get better over time. Environmental factors, user input, and performance indicators all play a part in the ongoing improvement of load balancing techniques.

To create a cloud–fog computing environment that is responsive, effective, and energy-conscious, dynamic load balancing is essential. Dynamic load balancing optimizes efficiency and sustainability by automatically shifting workloads depending on priorities and real-time conditions. This helps an energy-efficient framework work as a whole.

15.4.3 Integration of renewable energy in an energy-saving cloud–fog architecture

One essential element of a sustainable and energy-efficient cloud–fog computing infrastructure is the incorporation of renewable energy. The framework seeks to minimize the environmental impact, help create a greener digital ecosystem, and lessen dependency on conventional energy networks by utilizing clean and renewable energy sources, including hydroelectric, solar, and wind power.

- **Environmental Aspects and Site Selection:**
  An essential initial step is choosing locations for cloud and fog infrastructure that support the production of renewable energy. When choosing a location, elements including wind patterns, solar exposure [15], and accessibility to hydroelectric supplies are taken into account. This minimizes the impact on the environment while ensuring the best possible use of renewable energy sources.
- **Integration of Solar Power:**
  Using photovoltaic systems to harness solar power is a popular way to include renewable energy sources with cloud–fog infrastructure.

Sunlight is converted into electricity via solar panels that are mounted on data centers [21], fog nodes, or adjacent facilities. This decentralized strategy lessens reliance on the conventional power grid and promotes energy self-sufficiency.

- **Integration of Wind Power:**
Another renewable energy source that can be included in the framework is wind power. Wind energy is used to create electricity by the strategic placement of wind turbines in appropriate areas. Integration of wind power offers a dependable and sustainable energy source, especially in areas with regular wind patterns.

- **Integration of Hydroelectric Power:**
Hydroelectric power can be effectively integrated into the cloud–fog architecture in regions where water resources are available. A consistent and sustainable power source can be produced by using small-scale hydroelectric [34] generators to capture energy from moving water.

- **Options for Energy Storage:**
Renewable energy sources are generally intermittent, requiring energy storage technologies to provide a consistent power supply. Batteries and other sophisticated storage systems are examples of energy storage technologies. They store surplus energy produced [24] during peak hours and release it at times when the output of renewable energy is low.

- **Implementing Microgrids:**
Microgrids are small-scale energy systems that can function both separately and in tandem with the larger power grid. They provide an environmentally friendly option for cloud–fog infrastructure. Microgrid designs [35] that use renewable energy sources enhance the framework's energy resilience and enable autonomous operation when necessary.

- **Systems with Grid Interactivity:**
Grid-interactive technologies facilitate communication between cloud and fog infrastructure and the primary power grid. When renewable energy output is abundant, extra power can be returned to the grid to further the cause of sustainability and possibly earn money in the form of energy credits.

- **Energy-Saving Techniques:**
Energy efficiency measures within the cloud–fog architecture should be addressed prior to adding renewable energy sources. This involves putting in place cooling systems, hardware, and workload management techniques that are energy-efficient in order to minimize overall energy consumption and optimize the effects of renewable energy integration.

- **Evaluation of the Environmental Impact:**
Conducting an environmental impact assessment is necessary to analyze the overall advantages of integrating renewable energy sources. This evaluation takes into account things like lowering carbon emissions [14], using less non-renewable resources, and having a positive effect on the ecology.

Integrating renewable energy goes beyond environmental responsibility; it synchronizes cloud–fog computing with the objectives of global sustainability. By introducing renewable energy sources into the framework, enterprises contribute to a cleaner, more robust, and sustainable digital future.

15.4.4 Edge device optimization in an energy-efficient cloud–fog framework

A key element of an energy-efficient cloud–fog computing system is edge device optimization. Its main goal is to make computer equipment at the network's edge more sustainable and efficient. Organizations can decrease energy usage, enhance performance, and contribute to a more ecologically conscious cloud–fog ecosystem by deploying hardware and software enhancements.

- **Low-Power Electronic Parts:**
  Optimization starts with choosing low-power hardware components for edge devices. Using memory modules, storage units, and processors that use less energy is part of this. Retaining sufficient performance while reducing edge device power consumption overall is the aim [36].
- **Acceleration via Hardware:**
  Using hardware acceleration, including specialized processing units (GPUs, TPUs) [19], improves edge devices' computing performance. Compared to depending only on general-purpose processors, offloading specific activities to specialized hardware accelerators can result in faster processing speeds and lower energy consumption.
- **Effective Cooling Techniques:**
  Effective cooling solutions are beneficial for edge devices, particularly those that are located in remote or limited locations. Energy-efficient or passive active cooling techniques assist in preserving ideal operating temperatures [37] without using excessive amounts of energy. This is especially important in situations when installing conventional air conditioning might not be feasible.
- **Techniques for Power Management:**
  Strategies for dynamic power control are essential for optimizing energy use. This entails modifying component power states in accordance with workload. To optimize energy consumption [38], edge devices, for example, can scale up when demand grows and down when activity is low.
- **Software Development with Energy Awareness:**
  Software energy efficiency optimization is equally crucial. Edge apps should be made with the least amount of processing overhead possible, as little data transmission as possible, and the ability to sleep as needed. In edge computing, energy-conscious programming techniques help to improve overall energy efficiency [39].

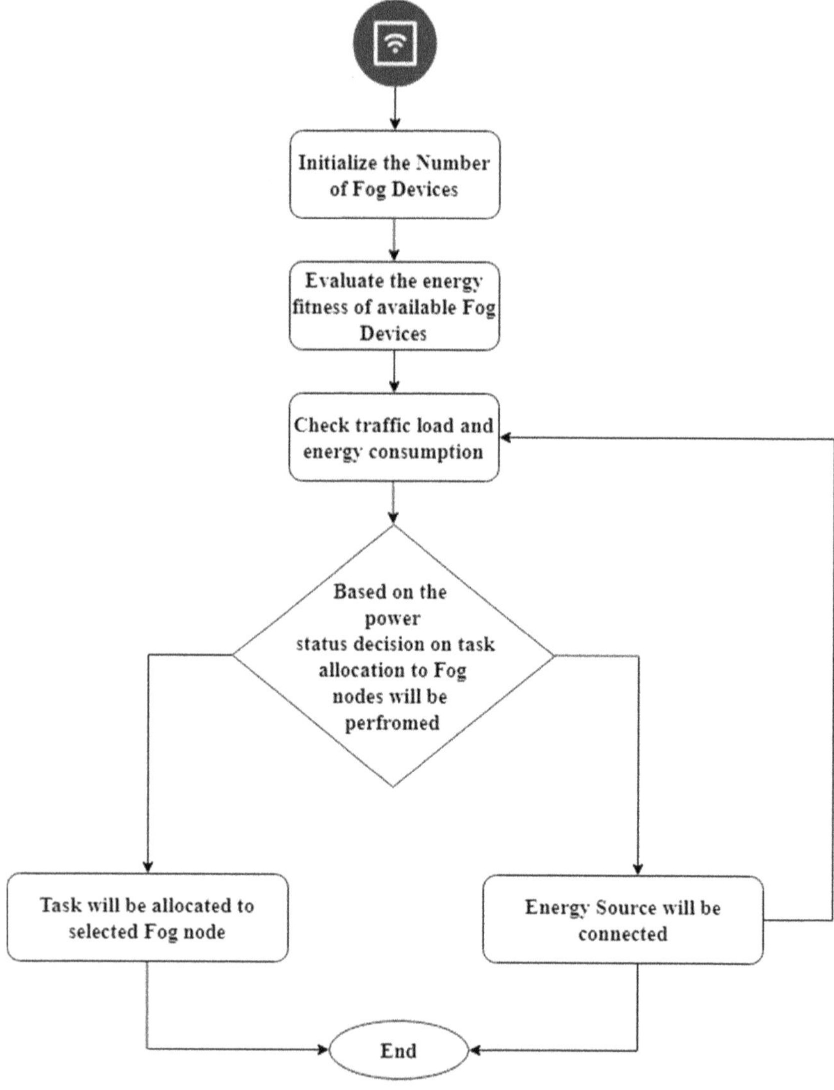

Figure 15.2 Methodology of energy-efficient framework.

- **Cutting-Edge Intelligence for Making Decisions:**
  Devices can make localized decisions without constantly depending on centralized cloud servers by implementing edge intelligence. In addition to cutting down on latency, this minimizes the requirement for continuous connection with the cloud, thereby saving energy [40]. Applications requiring real-time processing can benefit greatly from edge intelligence.

- **Edge-Based Load Balancing:**
  By using localized load balancing to distribute computing activities among edge devices in an efficient manner, the burden on individual devices is reduced. This guarantees a more equitable use of resources, avoids performance snags [41], and enhances the edge computing environment's overall energy efficiency.
- **Sleeping patterns and Wake-Up Techniques:**
  It is possible to program edge devices to go into sleep mode when they are idle. Devices with wake-up techniques can restart operations as necessary because they are activated by pre-programmed events or network requests. These cycles of sleep and wakefulness help save energy without sacrificing responsiveness.
- **Consolidation of Resources:**
  Consolidating resources refers to maximizing the utilization of computer power by consolidating several jobs onto a single device. By minimizing idle times and reducing the number of devices in use, this method improves energy efficiency [42]. To prevent resource contention, nevertheless, considerable thought must be given.
- **Ongoing analytics and Monitoring:**
  Evaluating performance and energy usage in real time is possible with the implementation of continuous monitoring and analytics on edge devices. Adaptive optimization is made possible by data-driven insights, which let the framework adapt dynamically to changing circumstances and steadily increase energy efficiency [4, 43, 44, 25].

Optimizing edge devices is a complex process that calls for a blend of smart management techniques, software optimizations, and hardware improvements. By emphasizing energy conservation at the edge, businesses can develop cloud–fog computing that is more resilient and adaptable.

15.5 METHODOLOGY OF ENERGY-EFFICIENT FRAMEWORK

A multidimensional methodology is required to establish an energy-efficient framework for a cloud–fog environment that promotes sustainable development, as shown in Figure 15.2. To identify current research, obstacles, and trends in energy efficiency within the cloud and fog computing sectors, a thorough analysis of the literature is first done. After that, specific objectives and goals are outlined, including targets like cutting energy use, making the best use of available resources, and lessening the impact on the environment.

Creating a thorough system architecture that smoothly combines cloud and fog computing resources is a crucial first step in guaranteeing peak performance with the least amount of energy consumption. Then, to maximize efficiency, resource management policies are developed to dynamically distribute workloads among nodes using strategies like task offloading and load balancing. To prioritize jobs according to energy requirements and deadlines, energy-aware scheduling algorithms are created, with an emphasis on adjusting system resources to workload variations [45]. To reduce data movement and transmission overhead, strong data management mechanisms are also put into place. These tactics include caching, replication, and compression techniques. To reduce the risks associated with node failures and ensure system uptime and dependability, fault tolerance methods are included. Thorough evaluation and validation of the framework's performance is carried out, and based on evaluation findings and input from stakeholders, ongoing optimization and improvement are carried out. Ultimately, thorough documentation and outreach initiatives guarantee that the framework's growth significantly advances sustainable development objectives in cloud–fog environments.

15.6 CASE STUDIES

In this section, discussion on case studies for Google's Commitment to Carbon Neutrality and Microsoft Datacenter sustainability is done.

15.6.1 Google's commitment to carbon neutrality

Google has committed to running its worldwide cloud network entirely on carbon-free energy by 2030, and in the interim, it has been taking major steps to lower its carbon footprint and encourage sustainability in its business practices.

15.6.1.1 Goals and techniques

Some goals and techniques for Google's Commitment to Carbon Neutrality are discussed below:

- **Increased Purchasing of Renewable Energy:**
  Google has persisted in purchasing and investing in renewable energy sources, including solar, wind, and other zero-carbon options. Its data centers and activities must be powered continuously by these efforts in order to avoid using conventional, carbon-intensive energy sources.
- **Creative Solutions for Energy Storage:**
  Google has probably investigated and put cutting-edge energy storage technologies into practice to combat the sporadic nature of renewable energy. This includes storing extra energy during high production periods for use during low production periods through the use of large-scale batteries and other energy storage technologies.

- **Global Renewable Energy Project Expansion:**
  Google is committed to using carbon-free energy around the clock for all of its activities, including new initiatives and growth. In order to guarantee a diverse and dependable supply of carbon-free energy, the corporation probably aims to create and promote renewable energy projects throughout the world.
- **Including Cutting-Edge Technologies:**
  When it comes to maximizing energy efficiency, Google has been at the forefront of implementing cutting-edge technology like AI and machine learning. Energy consumption can be reduced, and resource allocation can be dynamically adjusted with the help of predictive analytics and smart energy management systems.
- **Cooperation and Joint Ventures:**
  Given the magnitude of the problem, Google has probably partnered with governments, energy companies, and other institutions to build an ecosystem that facilitates the switch to carbon-free energy sources that run continuously.

15.6.2 Measuring and reporting progress

- **Openness and Responsibility:**
  Google consistently updates its progress toward carbon neutrality targets as part of its commitment to transparency. This comprises comprehensive data on the amount of energy used, the amount of renewable energy used, and the total carbon footprint of its operations.
- **Constant Observation and Improvement:**
  Google evaluates the functionality of its energy infrastructure on a regular basis using sophisticated monitoring tools. The implementation of continuous optimization guarantees that the organization recognizes areas for enhancement and adjusts its tactics in response to evolving energy environments.

15.6.3 Microsoft datacenter sustainability

In order to build a data center that is more resource-efficient and sustainable, Microsoft adopted the concepts of circular design. Reducing, reusing, and recycling materials are the main goals of circular design in order to reduce environmental effects.

15.6.3.1 Goals

- **Circular Economy Fundamentals:**
  implementing circular design principles to guarantee that equipment and materials used in the construction of data centers are obtained ethically and have the potential to be recycled or used for other purposes when their useful lives are over.

- **Effective Use of Water:**
  Reducing the water footprint of data center operations by putting water-saving technology and practices into practice.
- **Purchasing Renewable Energy:**
  Increasing the amount of energy generated by renewable sources in data centers, with an emphasis on long-term agreements that foster the industry's expansion.

As demonstrated by "Project Natick" and its circular data center design, Microsoft's dedication to data center sustainability highlights creative ways to lessen environmental effect. Microsoft leads the industry by embracing circular design principles, using renewable energy sources, and investigating alternate cooling techniques. These actions show how sustainable practices can be included in the foundation of data center operations.

15.6.3.2 Application

- **Ecological Sourcing:**
  Microsoft ensured that the materials used in the construction of data centers were procured responsibly and could be recycled by forming partnerships with suppliers who share their commitment to sustainable practices.
- **Measures of Water Efficiency:**
  Installing water-saving cooling methods and systems, including direct liquid cooling, will reduce the amount of water used in the data center.
- **Investing in Renewable Energy:**
  Microsoft encouraged the growth of clean energy production by investing in more renewable energy projects to power its data centers.

15.7 CONCLUSION

In conclusion, there is a lot of potential for the application of an energy-efficient framework for a cloud–fog environment to support sustainable development. An infrastructure that integrates renewable energy sources shows a dedication to sustainable energy practices. This lessens the impact on the environment and contributes to increased resilience against disruptions in conventional electricity infrastructures. Energy-efficient cloud–fog computing frameworks must be developed and put into place as the digital landscape develops further in order to guarantee sustainable growth. The suggested framework addresses the issues related to energy consumption and opens the door for a more sustainable and commercially feasible digital future by integrating intelligent resource allocation, dynamic load balancing, and renewable energy integration. By working together, researchers, legislators, and industry players can create a strong basis for an eco-friendly and energy-efficient cloud fog computing environment.

REFERENCES

[1] A. Shakarami, H. Shakarami, M. Ghobaei-Arani, E. Nikougoftar, and M. Faraji-Mehmandar, "Resource provisioning in edge/fog computing: A comprehensive and systematic review," *J. Syst. Archit.*, vol. 122, p. 102362, 2022, doi: 10.1016/j.sysarc.2021.102362

[2] S. Gupta et al., "Efficient prioritization and processor selection schemes for HEFT algorithm: A makespan optimizer for task scheduling in cloud environment," *Electron.*, vol. 11, no. 16, 2022, doi: 10.3390/electronics11162557

[3] M. Gorlatova, H. Inaltekin, and M. Chiang, "Characterizing task completion latencies in multi-point multi-quality fog computing systems," *Comput. Networks*, vol. 181, no. December 2019, p. 107526, 2020, doi: 10.1016/j.comnet.2020.107526

[4] J. Pereira, L. Ricardo, M. Luís, C. Senna, and S. Sargento, "Assessing the reliability of fog computing for smart mobility applications in VANETs," *Futur. Gener. Comput. Syst.*, vol. 94, pp. 317–332, 2019, doi: 10.1016/j.future.2018.11.043

[5] A. A. Butt, S. Khan, T. Ashfaq, S. Javaid, N. A. Sattar, and N. Javaid, "A cloud and fog based architecture for energy management of smart city by using meta-heuristic techniques," *2019 15th Int. Wirel. Commun. Mob. Comput. Conf. IWCMC 2019*, pp. 1588–1593, 2019, doi: 10.1109/IWCMC.2019.8766702

[6] C. Framework, "EEOA : Cost and energy efficient task scheduling in a cloud-fog framework," 2023.

[7] N. Abu-Amssimir and A. Al-Haj, "A QoS-aware resource management scheme over fog computing infrastructures in IoT systems," *Multimed. Tools Appl.*, 2023, doi: 10.1007/s11042-023-14856-6

[8] S. O. Ogundoyin and I. A. Kamil, "Optimal fog node selection based on hybrid particle swarm optimization and firefly algorithm in dynamic fog computing services," *Eng. Appl. Artif. Intell.*, vol. 121, p. 105998, 2023, doi: https://doi.org/10.1016/j.engappai.2023.105998

[9] M. M. Hussain et al., "SONG: A multi-objective evolutionary algorithm for delay and energy aware facility location in vehicular fog networks," *Sensors*, vol. 23, no. 2, 2023, doi: 10.3390/s23020667

[10] D. Zeng, L. Gu, and H. Yao, "Towards energy efficient service composition in green energy powered Cyber–Physical Fog Systems," *Futur. Gener. Comput. Syst.*, vol. 105, pp. 757–765, 2020, doi: 10.1016/j.future.2018.01.060

[11] M. Faraji Mehmandar, S. Jabbehdari, and H. Haj Seyyed Javadi, "A dynamic fog service provisioning approach for IoT applications," *Int. J. Commun. Syst.*, vol. 33, no. 14, pp. 1–30, 2020, doi: 10.1002/dac.4541

[12] M. S. Qureshi et al., "A comparative analysis of resource allocation schemes for real-time services in high-performance computing systems," *Int. J. Distrib. Sens. Networks*, vol. 16, no. 8, 2020, doi: 10.1177/1550147720932750

[13] R. Tiwari and N. Kumar, "Dynamic Web caching: For robustness, low latency & disconnection handling," *Proc. 2012 2nd IEEE Int. Conf. Parallel, Distrib. Grid Comput. PDGC 2012*, no. 1997, pp. 909–914, 2012, doi: 10.1109/PDGC.2012. 6449945

[14] M. Abdel-Basset, R. Mohamed, R. K. Chakrabortty, and M. J. Ryan, "IEGA: An improved elitism-based genetic algorithm for task scheduling problem in fog computing," *Int. J. Intell. Syst.*, vol. 36, no. 9, pp. 4592–4631, 2021, doi: 10.1002/ int.22470

[15] A. Toor, N. Sohail, A. Akhunzada, and J. Boudjadar, "Energy and performance aware fog computing : A case of DVFS and green renewable energy," *Futur. Gener. Comput. Syst.*, vol. 101, pp. 1112–1121, 2019, doi: 10.1016/j. future.2019.07.010

[16] N. Sarrafzade, R. Entezari-Maleki, and L. Sousa, "A genetic-based approach for service placement in fog computing," *J. Supercomput.*, vol. 78, no. 8, pp. 10854–10875, 2022, doi: 10.1007/s11227-021-04254-w

[17] D. Javaheri, S. Gorgin, J. A. Lee, and M. Masdari, "An improved discrete harris hawk optimization algorithm for efficient workflow scheduling in multi-fog computing," *Sustain. Comput. Informatics Syst.*, vol. 36, p. 100787, Dec. 2022, doi: 10.1016/J.SUSCOM.2022.100787

[18] X. Hu and C. Tang, "Secure outsourced computation of the characteristic polynomial and eigenvalues of matrix," pp. 4–9, 2015, doi: 10.1186/s13677-015-0033-9

[19] R. Tiwari, R. Sille, N. Salankar, & P. Singh "Utilization and energy consumption optimization for cloud computing environment," in *Cyber Security and Digital Forensics*, 2022, pp. 609–619.

[20] P. Chithaluru, R. Tiwari, and K. Kumar, "ARIOR: Adaptive ranking based improved opportunistic routing in wireless sensor networks," *Wirel. Pers. Commun.*, vol. 116, no. 1, pp. 153–176, 2021, doi: 10.1007/s11277-020-07709-0

[21] Z. A. Khan et al., "Energy management in smart sectors using fog based environment and meta-heuristic algorithms," *IEEE Access*, vol. 7, pp. 157254–157267, 2019, doi: 10.1109/ACCESS.2019.2949863

[22] A. Chhabra, G. Singh, and K. S. Kahlon, "Multi-criteria HPC task scheduling on IaaS cloud infrastructures using meta-heuristics," *Cluster Comput.*, vol. 24, pp. 885–918, 2021.

[23] N. Bulchandani, U. Chourasia, S. Agrawal, P. Dixit, and A. Pandey, "A survey on task scheduling algorithms in cloud computing," *Int. J. Sci. Technol. Res.*, vol. 9, no. 1, pp. 460–464, 2020.

[24] S. P. Singh, R. Kumar, A. Sharma, and A. Nayyar, "Leveraging energy-efficient load balancing algorithms in fog computing," *Concurr. Comput. Pract. Exp.*, vol. 34, no. 13, pp. 1–16, 2022, doi: 10.1002/cpe.5913

[25] P. G. V. Naranjo, Z. Pooranian, M. Shojafar, M. Conti, and R. Buyya, "FOCAN: A Fog-supported smart city network architecture for management of applications in the Internet of Everything environments," *J. Parallel Distrib. Comput.*, vol. 132, pp. 274–283, 2019, doi: 10.1016/j.jpdc.2018.07.003

[26] C. H. Hong and B. Varghese, "Resource management in fog/edge computing: A survey," *arXiv*, vol. 52, no. 5, 2018, doi: 10.1145/3326066

[27] H. Wadhwa and R. Aron, "TRAM: Technique for resource allocation and management in fog computing environment," *J. Supercomput.*, vol. 78, no. 1, pp. 667–690, 2022, doi: 10.1007/s11227-021-03885-3

[28] M. Kumar, A. Kishor, J. K. Samariya, and A. Y. Zomaya, "An autonomic workload prediction and resource allocation framework for fog enabled industrial IoT," *IEEE Internet Things J.*, p. 1, 2023, doi: 10.1109/JIOT.2023.3235107

[29] Sharma, G., Khurana, S., Harnal, S., & Lone, S. A. (2022). CSFPA: An intelligent hybrid workflow scheduling algorithm based upon global and local optimization approach in cloud. *Concurrency and Computation: Practice and Experience, 34*(23), e7176. https://doi.org/10.1002/cpe.7176

[30] I. Attiya, L. Abualigah, D. Elsadek, S. A. Chelloug, and M. Abd Elaziz, "An intelligent chimp optimizer for scheduling of IoT application tasks in fog computing," *Mathematics*, vol. 10, no. 7, pp. 1–18, 2022, doi: 10.3390/math1007 1100

[31] R. and A. A. and K. S. Goel Gaurav and Tiwari, "Workflow Scheduling Using Optimization Algorithm in Fog Computing," in *International Conference on Innovative Computing and Communications*, 2022, pp. 379–390.

[32] M. Mehmood, N. Javaid, J. Akram, S.H. Abbasi, A. Rahman, & F. Saeed, *Efficient Resource Distribution in Cloud*, vol. 1. Springer International Publishing, 2019. doi: 10.1007/978-3-319-98530-5

[33] Y. Ramzanpoor, M. Hosseini Shirvani, and M. Golsorkhtabaramiri, *Multiobjective fault-tolerant optimization algorithm for deployment of IoT applications on fog computing infrastructure*, vol. 8, no. 1. Springer International Publishing, 2022. doi: 10.1007/s40747-021-00368-z

[34] M. A. Sharkh and M. Kalil, "A dynamic algorithm for fog computing data processing decision optimization," *2020 IEEE Int. Conf. Commun. Work. ICC Work. 2020 - Proc.*, pp. 0–5, 2020, doi: 10.1109/ICCWorkshops49005.2020.9145296

[35] T. M. Maung, M. Mie, and S. Thwin, *Advances on Broadband and Wireless Computing, Communication and Applications*, vol. 25. Springer International Publishing, 2019. doi: 10.1007/978-3-030-02613-4

[36] H. Rafique, M. A. Shah, S. U. Islam, T. Maqsood, S. Khan, and C. Maple, "A Novel Bio-Inspired Hybrid Algorithm (NBIHA) for efficient resource management in fog computing," *IEEE Access*, vol. 7, pp. 115760–115773, 2019, doi: 10.1109/ACCESS.2019.2924958

[37] M. R. Alizadeh, V. Khajehvand, A. M. Rahmani, and E. Akbari, "Task scheduling approaches in fog computing: A systematic review," *Int. J. Commun. Syst.*, vol. 33, no. 16, pp. 1–36, 2020, doi: 10.1002/dac.4583

[38] Babbar, G., Bajaj, R., Mittal, N. et al. Hybrid model of alternating least squares and root polynomial technique for color correction. *Soft. Comput.* 27, 4321–4335 (2023). https://doi.org/10.1007/s00500-023-07831-8

[39] A. Chhabra, S. K. Sahana, N. S. Sani, A. Mohammadzadeh, and H. A. Omar, "Energy-aware bag-of-tasks scheduling in the cloud computing system using hybrid oppositional differential evolution-enabled whale optimization algorithm," *Energies*, vol. 15, no. 13, 2022, doi: 10.3390/en15134571

[40] L. Yin, J. Luo, and H. Luo, "Tasks scheduling and resource allocation in fog computing based on containers for smart manufacturing," *IEEE Trans. Ind. Informatics*, vol. 14, no. 10, pp. 4712–4721, 2018, doi: 10.1109/TII.2018.2851241

[41] M. K. Mishra, N. K. Ray, A. R. Swain, G. B. Mund, and B. S. P. Mishra, "An adaptive model for resource selection and allocation in fog computing environment," *Comput. Electr. Eng.*, vol. 77, pp. 217–229, 2019, doi: 10.1016/j.compeleceng.2019.05.010

[42] K. Kaur, S. Garg, G. Kaddoum, F. Gagnon, and D. N. K. Jayakody, "EnLoB: Energy and load balancing-driven container placement strategy for data centers," *2019 IEEE Globecom Work. GC Wkshps 2019 - Proc.*, pp. 1–6, 2019, doi: 10.1109/GCWkshps45667.2019.9024592

[43] T. Wang, Y. Liang, W. Jia, M. Arif, A. Liu, and M. Xie, "Coupling resource management based on fog computing in smart city systems," *J. Netw. Comput. Appl.*, vol. 135, no. February, pp. 11–19, 2019, doi: 10.1016/j.jnca.2019.02.021

[44] Singla, S., Sharma, P. & Sharma, P.K. Enhanced security using proxy signcryption technique for wireless mesh networks. *Int J Syst Assur Eng Manag* 14, 474–482, 2023. https://doi.org/10.1007/s13198-022-01820-0

[45] Sharma, G., Khurana, S., Harnal, S., & Lone, S. A. (2022). CSFPA: An intelligent hybrid workflow scheduling algorithm based upon global and local optimization approach in cloud. *Concurrency and Computation: Practice and Experience*, 34(23), e7176.

Index

For Product Safety Concerns and Information please contact our EU
representative GPSR@taylorandfrancis.com
Taylor & Francis Verlag GmbH, Kaufingerstraße 24, 80331 München, Germany